Modeling Data Irregularities and Structural Complexities in Data Envelopment Analysis

Modeling Data Irregularities and Structural Complexities in Data Envelopment Analysis

Edited by

Joe Zhu
Worcester Polytechnic Institute, U.S.A.

Wade D. Cook
York University, Canada

Joe Zhu
Worcester Polytechnic Institute
Worcester, MA, USA

Wade D. Cook
York University
Toronto, ON, Canada

Library of Congress Control Number: 2007925039

ISBN 978-0-387-71606-0   e-ISBN 978-0-387-71607-7

Printed on acid-free paper.

© 2007 by Springer Science+Business Media, LLC
All rights reserved. This work may not be translated or copied in whole or in part without the written permission of the publisher (Springer Science+Business Media, LLC, 233 Spring Street, New York, NY 10013, USA), except for brief excerpts in connection with reviews or scholarly analysis. Use in connection with any form of information storage and retrieval, electronic adaptation, computer software, or by similar or dissimilar methodology now know or hereafter developed is forbidden.
The use in this publication of trade names, trademarks, service marks and similar terms, even if the are not identified as such, is not to be taken as an expression of opinion as to whether or not they are subject to proprietary rights.

9 8 7 6 5 4 3 2 1

springer.com

To *Alec and Marsha Rose*

CONTENTS

1	**Data Irregularities and Structural Complexities in DEA** Wade D. Cook and Joe Zhu	1
2	**Rank Order Data in DEA** Wade D. Cook and Joe Zhu	13
3	**Interval and Ordinal Data** Yao Chen and Joe Zhu	35
4	**Variables with Negative Values in DEA** Jesús T. Pastor and José L. Ruiz	63
5	**Non-Discretionary Inputs** John Ruggiero	85
6	**DEA with Undesirable Factors** Zhongsheng Hua and Yiwen Bian	103
7	**European Nitrate Pollution Regulation and French Pig Farms' Performance** Isabelle Piot-Lepetit and Monique Le Moing	123
8	**PCA-DEA** Nicole Adler and Boaz Golany	139
9	**Mining Nonparametric Frontiers** José H. Dulá	155

10	**DEA Presented Graphically Using Multi-Dimensional Scaling** Nicole Adler, Adi Raveh and Ekaterina Yazhemsky	**171**
11	**DEA Models for Supply Chain or Multi-Stage Structure** Wade D. Cook, Liang Liang, Feng Yang, and Joe Zhu	**189**
12	**Network DEA** Rolf Färe, Shawna Grosskopf and Gerald Whittaker	**209**
13	**Context-Dependent Data Envelopment Analysis and its Use** Hiroshi Morita and Joe Zhu	**241**
14	**Flexible Measures—Classifying Inputs and Outputs** Wade D. Cook and Joe Zhu	**261**
15	**Integer DEA Models** Sebastián Lozano and Gabriel Villa	**271**
16	**Data Envelopment Analysis with Missing Data** Chiang Kao and Shiang-Tai Liu	**291**
17	**Preparing Your Data for DEA** Joe Sarkis	**305**
	About the Authors	**321**
	Index	**331**

Chapter 1

DATA IRREGULARITIES AND STRUCTURAL COMPLEXITIES IN DEA

Wade D. Cook[1] and Joe Zhu[2]
[1]Schulich School of Business, York University, Toronto, Ontario, Canada, M3J 1P3, wcook@shulich.yorku.ca

[2]Department of Management, Worcester Polytechnic Institute, Worcester, MA 01609, jzhu@wpi.edu

Abstract: Over the recent years, we have seen a notable increase in interest in data envelopment analysis (DEA) techniques and applications. Basic and advanced DEA models and techniques have been well documented in the DEA literature. This edited volume addresses how to deal with DEA implementation difficulties involving data irregularities and DMU structural complexities. Chapters in this volumes address issues including the treatment of ordinal data, interval data, negative data and undesirable data, data mining and dimensionality reduction, network and supply chain structures, modeling non-discretionary variables and flexible measures, context-dependent performance, and graphical representation of DEA.

Key words: Data Envelopment Analysis (DEA), Ordinal Data, Interval Data, Data Mining, Efficiency, Flexible, Supply Chain, Network, Undesirable

1. INTRODUCTION

Data envelopment analysis (DEA) was introduced by Charnes, Cooper and Rhodes (CCR) in 1978. DEA measures the relative efficiency of peer decision making units (DMUs) that have multiple inputs and outputs, and has been applied in a wide range of applications over the past 25 years, in settings that include hospitals, banks, maintenance crews, etc.; see Cooper, Seiford and Zhu (2004).

As DEA attracts ever-growing attention from practitioners, its application and use become a very important issues. It is, therefore, important to deal with computation/data issues in DEA. These include, for example, how to deal with inaccurate data, qualitative data, outliers, undesirable factors, and many others. It is as well critical, from a managerial perspective, to be able to visualize DEA results, when the data are more than 3-dimensional.

The current volume presents a collection of articles that address data issues in the application of DEA, and special problem structures with respect to the nature of DMUs.

2. DEA MODELS

In this section, we present some basic DEA models that will be used in later chapters. For a more detailed discussion on these and other DEA models, the reader is referred to Cooper, Seiford and Zhu (2004), and other DEA textbooks.

Suppose we have a set of n peer DMUs, $\{DMU_j : j = 1, 2, ..., n\}$, which produce multiple outputs y_{rj}, ($r = 1, 2, ..., s$), by utilizing multiple inputs x_{ij}, ($i = 1, 2, ..., m$). When a DMU_o is under evaluation by the CCR ratio model, we have (Charnes, Cooper and Rhodes, 1978)

$$\max \frac{\sum_{r=1}^{s} \mu_r y_{ro}}{\sum_{i=1}^{m} v_i x_{io}}$$

$$\text{s.t.} \quad \frac{\sum_{r=1}^{s} \mu_r y_{rj}}{\sum_{i=1}^{m} v_i x_{ij}} \leq 1, \quad j = 1, 2, ..., n \quad (1)$$

$$\mu_r, v_i \geq 0, \quad \forall r, i$$

In this model, inputs x_{ij} and outputs y_{rj} are observed non-negative data[1], and μ_r and v_i are the unknown weights, or decision variables.

A fully rigorous development would replace $u_r, v_i \geq 0$ with

---

[1] For the treatment of negative input/output data, please see Chapter 4.

$$\frac{\mu_r}{\sum_{i=1}^{m} v_i x_{io}}, \frac{\mu_r}{\sum_{i=1}^{m} v_i x_{io}} \geq \varepsilon > 0$$

where ε is a non-Archimedean element smaller than any positive real number.

Model (1) can be converted into a linear programming problem

$$\max \sum_{r=1}^{s} \mu_r y_{ro}$$

subject to

$$\sum_{r=1}^{s} \mu_r y_{rj} - \sum_{i=1}^{m} v_i x_{ij} \leq 0, \text{ all } j \quad (2)$$

$$\sum_{i=1}^{m} v_i x_{io} = 1$$

$$\mu_r, v_i \geq 0$$

In this model, the weights are usually referred to as multipliers. Therefore, model (2) is also called a multiplier DEA model. The dual program to (2) can be expressed as

$$\theta^* = \min \theta$$

subject to

$$\sum_{j=1}^{n} x_{ij} \lambda_j \leq \theta x_{io} \quad i = 1, 2, ..., m; \quad (3)$$

$$\sum_{j=1}^{n} y_{rj} \lambda_j \geq y_{ro} \quad r = 1, 2, ..., s;$$

$$\lambda_j \geq 0 \quad j = 1, 2, ..., n.$$

Model (3) is referred to as the envelopment model. To illustrate the concept of envelopment, we consider a simple numerical example used in Zhu (2003) as shown in Table 1-1 where we have five DMUs representing five supply chain operations. Within a week, each DMU generates the same profit of $2,000 with a different combination of supply chain cost and response time.

Table 1-1. Supply Chain Operations Within a Week

DMU	Inputs		Output
	Cost ($100)	Response time (days)	Profit ($1,000)
1	1	5	2
2	2	2	2
3	4	1	2
4	6	1	2
5	4	4	2

Source: Zhu (2003).

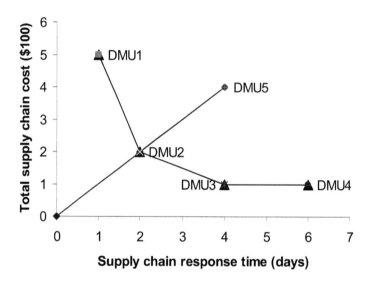

Figure 1-1. Five Supply Chain Operations

Figure 1-1 presents the five DMUs and the piecewise linear DEA frontier. DMUs 1, 2, 3, and 4 are on the frontier--or the *envelopment* frontier. If we apply model (3) to DMU5, we have,

Min θ
Subject to
$1 \lambda_1 + 2\lambda_2 + 4\lambda_3 + 6\lambda_4 + 4\lambda_5 \leq 4\theta$
$5 \lambda_1 + 2\lambda_2 + 1\lambda_3 + 1\lambda_4 + 4\lambda_5 \leq 4\theta$
$2 \lambda_1 + 2\lambda_2 + 2\lambda_3 + 2\lambda_4 + 2\lambda_5 \geq 2$
$\lambda_1, \lambda_2, \lambda_3, \lambda_4, \lambda_5 \geq 0$

This model has the unique optimal solution of $\theta^* = 0.5$, $\lambda_2^* = 1$, and $\lambda_j^* = 0$ ($j \neq 2$), indicating that DMU5 needs to reduce its cost and response time to the amounts used by DMU2 if it is to be efficient This example indicates that technical efficiency for DMU5 is achieved at DMU2.

Now, if we apply model (3) to DMU4, we obtain $\theta^* = 1$, $\lambda_4^* = 1$, and $\lambda_j^* = 0$ ($j \neq 4$), indicating that DMU4 is on the frontier. However, Figure 1-1 indicates that DMU4 can still reduce its response time by 2 days to achieve coincidence with DMU3. This input reduction is usually called input slack.

The nonzero slack can be found by using the following model

$$\max \sum_{i=1}^{m} s_i^- + \sum_{r=1}^{s} s_r^+$$

subject to

$$\sum_{j=1}^{n} x_{ij} \lambda_j + s_i^- = \theta^* x_{io} \quad i = 1, 2, \ldots, m; \tag{4}$$

$$\sum_{j=1}^{n} y_{rj} \lambda_j - s_r^+ = y_{ro} \quad r = 1, 2, \ldots, s;$$

$$\lambda_j, s_i^-, s_r^+ \geq 0 \ \forall i, j, r$$

where θ^* is determined by model (3) and is fixed in model (4).

For DMU4 with $\theta^* = 1$, model (4) yields the following model,

Max $s_1^- + s_2^- + s_1^+$
Subject to
$1 \lambda_1 + 2\lambda_2 + 4\lambda_3 + 6\lambda_4 + 4\lambda_5 + s_1^- = 6\theta^* = 6$
$5 \lambda_1 + 2\lambda_2 + 1\lambda_3 + 1\lambda_4 + 4\lambda_5 + s_2^- = 1\theta^* = 1$
$2 \lambda_1 + 2\lambda_2 + 2\lambda_3 + 2\lambda_4 + 2\lambda_5 - s_1^+ = 2$
$\lambda_1, \lambda_2, \lambda_3, \lambda_4, \lambda_5, s_1^-, s_2^-, s_1^+ \geq 0$

The optimal slacks are $s_1^{-*} = 2$, $s_2^{-*} = s_1^{+*} = 0$, with $\lambda_3^* = 1$ and all other $\lambda_j^* = 0$.

We now have

Definition 1 (DEA Efficiency): The performance of DMU_o is fully (100%) efficient if and only if both (i) $\theta^* = 1$ and (ii) all slacks $s_i^{-*} = s_r^{+*} = 0$.

Definition 2 (Weakly DEA Efficient): The performance of DMU_o is weakly efficient if and only if both (i) $\theta^* = 1$ and (ii) $s_i^{-*} \neq 0$ and/or $s_r^{+*} \neq 0$ for some i and r in some alternate optima.

Model (4) is usually called the second stage calculation of an envelopment model. In fact, the envelopment model can be written as:

$$\min \theta - \varepsilon \left(\sum_{i=1}^{m} s_i^- + \sum_{r=1}^{s} s_r^+ \right)$$

subject to

$$\sum_{j=1}^{n} x_{ij} \lambda_j + s_i^- = \theta x_{io} \quad i = 1, 2, \ldots, m; \quad (5)$$

$$\sum_{j=1}^{n} y_{rj} \lambda_j - s_r^+ = y_{ro} \quad r = 1, 2, \ldots, s;$$

$$\lambda_j, s_i^-, s_r^+ \geq 0 \ \forall i, j, r$$

where the s_i^- and s_r^+ are slack variables used to convert the inequalities in (3) to equivalent equations. This is equivalent to solving (5) in two stages by first minimizing θ, then fixing $\theta = \theta^*$ as in (4), where the slacks are to be maximized without altering the previously determined value of $\theta = \theta^*$. Formally, this is equivalent to granting "preemptive priority" to the determination of θ^* in (3). In this manner, the fact that the non-Archimedean element ε is defined to be smaller than any positive real number is accommodated without having to specify the value of ε (Cooper, Seiford and Zhu, 2004).

The above models are called input-oriented DEA models, as possible input reductions are of interest while the outputs are kept at their current levels. Similarly, output-oriented models can be developed. These models focus on possible output increases while the inputs are kept at their current levels. The interested reader should refer to Cooper, Seiford and Zhu (2004).

The models in Table 1-2 are also known as CRS (constant returns to scale) models. If the constraint $\sum_{j=1}^{n} \lambda_j = 1$ is adjoined, they are referred to as variable returns to scale (VRS) models (Banker, Charnes, Cooper, 1984). This is due to the fact that $\sum_{j=1}^{n} \lambda_j = 1$ changes the shape of DEA frontier, and is related to the concept of returns to scale.

Table 1-2. CCR DEA Model

Input-oriented	
Envelopment model	Multiplier model
$\min \theta - \varepsilon(\sum_{i=1}^{m} s_i^- + \sum_{r=1}^{s} s_r^+)$ subject to $\sum_{j=1}^{n} x_{ij}\lambda_j + s_i^- = \theta x_{io}$ $i=1,2,\ldots,m;$ $\sum_{j=1}^{n} y_{rj}\lambda_j - s_r^+ = y_{ro}$ $r=1,2,\ldots,s;$ $\lambda_j \geq 0$ $\quad j=1,2,\ldots,n.$	$\max z = \sum_{r=1}^{s} \mu_r y_{ro}$ subject to $\sum_{r=1}^{s} \mu_r y_{rj} - \sum_{i=1}^{m} v_i x_{ij} \leq 0$ $\sum_{i=1}^{m} v_i x_{io} = 1$ $\mu_r, v_i \geq \varepsilon > 0$
Output-oriented	
Envelopment model	Multiplier model
$\max \phi + \varepsilon(\sum_{i=1}^{m} s_i^- + \sum_{r=1}^{s} s_r^+)$ subject to $\sum_{j=1}^{n} x_{ij}\lambda_j + s_i^- = x_{io}$ $i=1,2,\ldots,m;$ $\sum_{j=1}^{n} y_{rj}\lambda_j - s_r^+ = \phi y_{ro}$ $r=1,2,\ldots,s;$ $\lambda_j \geq 0$ $\quad j=1,2,\ldots,n.$	$\min q = \sum_{i=1}^{m} v_i x_{io}$ subject to $\sum_{i=1}^{m} v_i x_{ij} - \sum_{r=1}^{s} \mu_r y_{rj} \geq 0$ $\sum_{r=1}^{s} \mu_r y_{ro} = 1$ $\mu_r, v_i \geq \varepsilon > 0$

3. DATA AND STRUCTURE ISSUES

The current volume deals with data irregularities and structural complexities in applications of DEA.

Chapter 2 (by Cook and Zhu) develops a general framework for modeling and treating qualitative data in DEA and provides a unified structure for embedding rank order data into the DEA framework. It is shown that the existing approaches for dealing with qualitative data are equivalent.

Chapter 3 (by Chen and Zhu) discusses how to use the standard DEA models to deal with imprecise data in DEA (IDEA), concentrating on interval and ordinal data. There are two approaches in dealing with such imprecise inputs and outputs. One uses scale transformations and variable alternations to convert the non-linear IDEA model into a linear program, while the other identifies a set of exact data from the imprecise inputs and outputs and then uses the standard linear DEA model. The chapter focuses on the latter IDEA approach that uses the standard DEA model. It is shown that different results are obtained depending on whether the imprecise data are introduced directly into the multiplier or envelopment DEA model; the presence of imprecise data invalidates the linear duality between the multiplier and envelopment DEA models. The approaches are illustrated with both numerical and real world data sets.

Chapter 4 (by Pastor and Ruiz) presents an overview of the different existing approaches dealing with the treatment of negative data in DEA. Both the classical approaches and the most recent contributions to this problem are presented. The focus is mainly on issues such as translation invariance and units invariance of the variables, classification invariance of the units, as well as efficiency measurement and target setting.

Chapter 5 (by Ruggiero) discusses existing approaches to measuring performance when non-discretionary inputs affect the transformation of discretionary inputs into outputs. The suitability of the approaches depends on underlying assumptions about the relationship between non-discretionary inputs and outputs. One model treats non-discretionary inputs like discretionary inputs but uses a non-radial approach to project inefficient decision making units (DMUs) to the frontier holding non-discretionary inputs fixed. Other approaches use multiple stage models with regression to control for the effect that non-discretionary inputs have on production.

Chapter 6 (by Hua and Bian) discusses the existing methods of treating undesirable factors in DEA. Under strongly disposable technology and weakly disposable technology, there are several approaches for treating undesirable outputs in the DEA literature. One such approach is the hyperbolic output efficiency measure that increases desirable outputs and decreases undesirable outputs simultaneously. Based on the classification invariance property, a linear monotone decreasing transformation is used to treat the undesirable outputs. A directional distance function is used to estimate the efficiency scores based on weak disposability of undesirable outputs. This chapter also presents an extended DEA model in which undesirable outputs and non-discretionary inputs are considered simultaneously.

Chapter 7 (by Piot-Lepetit and Le Moing) highlights the usefulness of the directional distance function in measuring the impact of the EU Nitrate directive, which prevents the free disposal of organic manure and nitrogen surplus. Efficiency indices for the production and environmental performance of farms at an individual level are proposed, together with an evaluation of the impact caused by the said EU regulation. This chapter extends the previous approach to good and bad outputs within the framework of the directional distance function, by introducing a by-product (organic manure), which becomes a pollutant once a certain level of disposability is exceeded.

Chapter 8 (by Adler and Golany) presents the combined use of principal component analysis (PCA) and DEA with the stated aim of reducing the curse of dimensionality that occurs in DEA when there is an excessive number of inputs and outputs in relation to the number of decision-making units. Various PCA-DEA formulations are developed in the chapter utilizing the results of principal component analyses to develop objective assurance region type constraints on the DEA weights. The first set of models applies PCA to grouped data representing similar themes, such as quality or environmental measures. The second set of models, if needed, applies PCA to all inputs and separately to all outputs, thus further strengthening the discrimination power of DEA. A case study of municipal solid waste managements in the Oulu district of Finland, which has been frequently analyzed in the literature, will illustrate the different models and the power of the PCA-DEA formulation.

Chapter 9 (by Dulá) deals with the extension of data envelopment analysis to the general problem of mining oriented outliers. DEA is firmly anchored in efficiency and productivity paradigms. This research claims new application domains for DEA by releasing it from these moorings. The same reasons why efficient entities are of interest in DEA apply to the geometric equivalent in general point sets since they are based on the data's magnitude limits relative to the other data points. A framework for non-parametric frontier analysis is derived from a new set of first principles.

Chapter 10 (by Adler, Raveh and Yazhemsky) presents the results of DEA in a two-dimensional plot. Presenting DEA graphically, due to its multiple variable nature, has proven difficult and some have argued that this has left decision-makers at a loss in interpreting the results. Co-Plot, a variant of multi-dimensional scaling, locates each decision-making unit in a two-dimensional space in which the location of each observation is determined by all variables simultaneously. The graphical display technique exhibits observations as points and variables (ratios) as arrows, relative to the same center-of-gravity. Observations are mapped such that similar decision-making units are closely located on the plot, signifying that they belong to a group possessing comparable characteristics and behavior.

Chapter 11 (by Cook, Liang, Yang and Zhu) presents several DEA-based approaches for characterizing and measuring supply chain efficiency. The models are illustrated in a seller-buyer supply chain context, when the relationship between the seller and buyer is treated leader-follower and cooperative, respectively. In the leader-follower structure, the leader is first evaluated, and then the follower is evaluated using information related to the leader's efficiency. In the cooperative structure, the joint efficiency which is modeled as the average of the seller's and buyer's efficiency scores is maximized, and both supply chain members are evaluated simultaneously.

Chapter 12 (by Färe, Grosskopf and Whittaker) describes network DEA models, where a network consists of sub-technologies. A DEA model typically describes a technology to a level of abstraction necessary for the analyst's purpose, but leaves out a description of the sub-technologies that make up the internal functions of the technology. These sub-technologies are usually treated as a "black box", i.e., there is no information about what happens inside them. The specification of the sub-technologies enables the explicit examination of input allocation and intermediate products that together form the production process. The combination of sub-technologies into networks provides a method of analyzing problems that the traditional DEA models cannot address.

Chapter 13 (Morita and Zhu) presents a context-dependent DEA methodology, which refers to a DEA approach where a set of DMUs is evaluated against a particular evaluation context. Each evaluation context represents an efficient frontier composed of DMUs in a specific performance level. The context-dependent DEA measures the attractiveness and the progress when DMUs exhibiting poorer and better performance are chosen as the evaluation context, respectively. This chapter also presents a slack-based context-dependent DEA approach. In DEA, nonzero input and output slacks are very likely to be present, after the radial efficiency score improvement. Slack-based context-dependent DEA allows us to fully evaluate the inefficiency in a DMU's performance.

Chapter 14 (by Cook and Zhu) presents DEA models to accommodate flexible measures. In standard DEA, it is assumed that the input versus output status of each of the chosen analysis measures is known. In some situations, however, certain measures can play either input or output roles. Consider using the number of nurse trainees on staff in a study of hospital efficiency. Such a factor clearly constitutes an output measure for a hospital, but at the same time is an important component of the hospital's total staff complement, hence is an input. Both an individual DMU model and an aggregate model are suggested as methodologies for deriving the most appropriate designations for flexible measures.

Chapter 15 (by Lozano and Villa) presents DEA models under situations where one or more inputs and/or outputs are integer quantities. Commonly, in these situations, the non-integer targets are rounded off. However, rounding off may easily lead to an infeasible target (i.e. out of the Production Possibility Set) or to a dominated operation point. In this chapter, a general framework to handle integer inputs and outputs is presented and a number of integer DEA models are reviewed.

Chapter 16 (by Sarkis) looks at some data requirements and characteristics that may ease the execution of the DEA models and the interpretation of DEA results.

REFERENCES

1. Banker, R., A. Charnes and W.W. Cooper, 1984, Some models for estimating technical and scale inefficiencies in data envelopment analysis, *Management Science* 30, 1078-1092.
2. Charnes A., W.W. Cooper and E. Rhodes, 1978, Measuring the efficiency of decision making units, *European Journal of Operational Research*, 2(6), 428-444.
3. Cooper, W.W., L.M. Seiford and J. Zhu, 2004, *Handbook of Data Envelopment Analysis*, Kluwer Academic Publishers, Boston.
4. Zhu, J. 2003, *Quantitative Models for Performance Evaluation and Benchmarking: Data Envelopment Analysis with Spreadsheets and DEA Excel Solver*, Kluwer Academic Publishers, Boston.

Chapter 2

RANK ORDER DATA IN DEA

Wade D. Cook[1] and Joe Zhu[2]
[1]*Schulich School of Business, York University, Toronto, Ontario, Canada, M3J 1P3, wcook@shulich.yorku.ca*

[2]*Department of Management, Worcester Polytechnic Institute, Worcester, MA 01609, jzhu@wpi.edu*

Abstract: In data envelopment analysis (DEA), performance evaluation is generally assumed to be based upon a set of quantitative data. In many real world settings, however, it is essential to take into account the presence of qualitative factors when evaluating the performance of decision making units (DMUs). Very often rankings are provided from best to worst relative to particular attributes. Such rank positions might better be presented in an ordinal, rather than numerical sense. The chapter develops a general framework for modeling and treating qualitative data in DEA, and provides a unified structure for embedding rank order data into the DEA framework. We show that the approach developed earlier in Cook et al (1993, 1996) is equivalent to the IDEA methodology given in Chapter 3. It is shown that, like IDEA, the approach given her for dealing with qualitative data lends itself to treatment by conventional DEA methodology.

Key words: Data envelopment analysis (DEA), efficiency, qualitative, rank order data

1. INTRODUCTION

In the data envelopment analysis (DEA) model of Charnes, Cooper and Rhodes (1978), each member of a set of n decision making units (DMUs) is to be evaluated relative to its peers. This evaluation is generally assumed to be based on a set of *quantitative* output and input factors. In many real world settings, however, it is essential to take into account the presence of *qualitative* factors when rendering a decision on the performance of a DMU. Very often it is the case that for a factor such as management competence, one can, at most, provide a *ranking* of the DMUs from best to worst relative to this attribute. The capability of providing a more precise, quantitative measure reflecting such a factor is often not feasible. In some situations such factors can be legitimately "quantified," but very often such quantification may be superficially forced as a modeling convenience.

In situations such as that described, the "data" for certain influence factors (inputs and outputs) might better be represented as rank positions in an ordinal, rather than numerical sense. Refer again to the management competence example. In certain circumstances, the information available may only permit one to put each DMU into one of L categories or groups (e.g. 'high', 'medium' and 'low' competence). In other cases, one may be able to provide a complete rank ordering of the DMUs on such a factor.

Cook, Kress and Seiford (1993), (1996) first presented a modified DEA structure, incorporating rank order data. The 1996 article applied this structure to the problem of prioritizing a set of research and development projects, where there were both inputs and outputs defined on a Likert scale. An alternative to the Cook et al approach was provided in Cooper, Park and Yu (1999) in the form of the imprecise DEA (IDEA) model. While various forms of imprecise data were examined, one major component of that research focused on ordinal (rank order) data. See Chapter 3 for a treatment of the specifics of IDEA. These two approaches to the treatment of ordinal data in DEA are further discussed and compared in Cook and Zhu(2006).

In the current chapter, we present a unified structure for embedding rank order or Likert scale data into the DEA framework. This development is very much related to the presentation in Cook and Zhu (2006).To provide a practical setting for the methodology to be developed herein, Section 2 briefly discusses the R&D project selection problem as presented in more detail in Cook et al (1996). Section 3 presents a continuous projection model, based on the conventional radial model of Charnes et al (1978). In Section 4 this approach is compared to the IDEA methodology of Cooper et al (1999). We demonstrate that IDEA for Likert Scale data is in fact equivalent to the earlier approach of Cook, Kress and Seiford (1996). Section 5 develops a discrete projection methodology that guarantees projection to points on the Likert Scale. Conclusions and further directions are addressed in Section 6.

2. ORDINAL DATA IN R&D PROJECT SELECTION

Consider the problem of selecting R&D projects in a major public utility corporation with a large research and development branch. Research activities are housed within several different divisions, for example, thermal, nuclear, electrical, and so on. In a budget constrained environment in which such an organization finds itself, it becomes necessary to make choices among a set of potential research initiatives or projects that are in competition for the limited resources. To evaluate the impact of funding (or not funding) any given research initiative, two major considerations generally must be made. First, the initiative must be viewed in terms of more than one factor or criterion. Second, some or all of the criteria that enter the evaluation may be qualitative in nature. Even when pure quantitative factors are involved, such as long term saving to the organization, it may be difficult to obtain even a crude estimate of the value of that factor. The most that one can do in many such situations is to classify the project (according to this factor) on some scale (high/medium/low or say a 5-point scale).

Let us assume that for each qualitative criterion each initiative is rated on a 5-point scale, where the particular point on the scale is chosen through a consensus on the part of executives within the organization. Table 2-1 presents an illustration of how the data might appear for 10 projects, 3 qualitative output criteria (benefits), and 3 qualitative input criteria (cost or resources). In the actual setting examined, a number of potential benefit and cost criteria were considered as discussed in Cook et al (1996).

We use the convention that for both outputs and inputs, a rating of 1 is "best," and 5 "worst." For outputs, this means that a DMU ranked at position 1 generates *more* output than is true of a DMU in position 2, and so on. For inputs, a DMU in position 1 consumes *less* input than one in position 2.

Table 2-1. Ratings by criteria

Project No.	Outputs			Inputs		
	1	2	3	4	5	6
1	2	4	1	5	2	1
2	1	1	4	3	5	2
3	1	1	1	1	2	1
4	3	3	3	4	3	2
5	4	3	5	5	1	4
6	2	5	1	1	2	2
7	1	4	1	5	4	3
8	1	5	3	3	3	3
9	5	2	4	4	2	5
10	5	4	4	5	5	5

Regardless of the manner in which such a scale rating is arrived at, the existing DEA model is capable only of treating the information as if it has cardinal meaning (e.g. something which receives a score of 4 is evaluated as being twice as important as something that scores 2). There are a number of problems with this approach. First and foremost, the projects' original data in the case of some criteria may take the form of an ordinal ranking of the projects. Specifically, the most that can be said about two projects i and j is that i is preferred to j. In other cases it may only be possible to classify projects as say 'high', 'medium' or 'low' in importance on certain criteria. When projects are rated on, say, a 5-point scale, it is generally understood that this scale merely provides a relative positioning of the projects. In a number of agencies investigated (for example, hydro electric and telecommunications companies), 5-point scales are common for evaluating alternatives in terms of qualitative data, and are often accompanied by interpretations such as

1 = Extremely important
2 = Very important
3 = Important
4 = Low in importance
5 = Not important,

that are easily understood by management. While it is true that market researchers often treat such scales in a numerical (i.e. cardinal) sense, it is not practical that in rating a project, the classification 'extremely important' should be interpreted literally as meaning that this project rates three times better than one which is only classified as 'important.' The key message here is that many, if not all criteria used to evaluate R&D projects are qualitative in nature, and should be treated as such. The model presented in the following sections extends the DEA idea to an ordinal setting, hence accommodating this very practical consideration.

3. MODELING LIKERT SCALE DATA: CONTINUOUS PROJECTION

The above problem typifies situations in which pure ordinal data or a mix of ordinal and numerical data, are involved in the performance measurement exercise. To cast this problem in a general format, consider the situation in which a set of N decision making units (DMUs), k=1,...N are to be evaluated in terms of R_1 numerical outputs, R_2 ordinal outputs, I_1 numerical inputs, and I_2 ordinal inputs. Let $Y_k^1 = (y_{rk}^1)$, $Y_k^2 = (y_{rk}^2)$ denote the R_1-

dimensional and R_2-dimensional vectors of outputs, respectively. Similarly, let $X_k^1 = (x_{ik}^1)$ and $X_k^2 = (x_{ik}^2)$ be the I_1 and I_2-dimensional vectors of inputs, respectively.

In the situation where all factors are quantitative, the conventional radial projection model for measuring the efficiency of a DMU is expressed by the ratio of weighted outputs to weighted inputs. Adopting the general variable returns to scale (VRS) model of Banker, Charnes and Cooper (1984), the efficiency of DMU "0" follows from the solution of the optimization model:

$$e_o = \max \left(\mu_o + \sum_{r \in R_1} \mu_r^1 y_{ro}^1 + \sum_{r \in R_2} \mu_r^2 y_{ro}^2 \right) / \left(\sum_{i \in I_1} \upsilon_i^1 x_{io}^1 + \sum_{i \in I_2} \upsilon_i^2 x_{io}^2 \right)$$

s.t.

$$\left(\mu_o + \sum_{r \in R_1} \mu_r^1 y_{rk}^1 + \sum_{r \in R_2} \mu_r^2 y_{rk}^2 \right) / \left(\sum_{i \in I_1} \upsilon_i^1 x_{ik}^1 + \sum_{i \in I_2} \upsilon_i^2 x_{ik}^2 \right) \le 1, \text{ all } k \quad (3.1)$$

$$\mu_r^1, \mu_r^2, \upsilon_i^1, \upsilon_i^2 \ge \varepsilon, \text{ all } r, i$$

Problem (3.1) is convertible to the linear programming format:

$$e_o = \max \mu_o + \sum_{r \in R_1} \mu_r^1 y_{ro}^1 + \sum_{r \in R_2} \mu_r^2 y_{ro}^2$$

$$\text{s.t.} \sum_{i \in I_1} \upsilon_i^1 x_{io}^1 + \sum_{i \in I_2} \upsilon_i^2 x_{io}^2 = 1 \quad (3.2)$$

$$\mu_o + \sum_{r \in R_1} \mu_r^1 y_{rk}^1 + \sum_{r \in R_2} \mu_r^2 y_{rk}^2 - \sum_{i \in I_1} \upsilon_i^1 x_{ik}^1 - \sum_{i \in I_2} \upsilon_i^2 x_{ik}^2 \le 0, \text{ all } k$$

$$\mu_r^1, \mu_r^2, \upsilon_i^1, \upsilon_i^2 \ge \varepsilon, \text{ all } r, i,$$

whose dual is given by

$$\min \theta - \varepsilon \sum_{r \in R_1 \cup R_2} s_r^+ - \varepsilon \sum_{i \in I_1 \cup I_2} s_i^-$$

s.t.

$$\sum_{k=1}^{N} \lambda_k y_{rk}^1 - s_r^+ = y_{ro}^1, \ r \in R_1$$

$$\sum_{n=1}^{N} \lambda_k y_{rk}^2 - s_r^+ = y_{ro}^2, \ r \in R_2$$

$$\theta x_{io}^1 - \sum_{k=1}^{N} \lambda_k x_{ik}^1 - s_i^- = 0, \ i \in I_1 \quad (3.2')$$

$$\theta x_{io}^2 - \sum_{k=1}^{N} \lambda_k x_{ik}^2 - s_i^- = 0, \ i \in I_2$$

$$\sum_{k=1}^{N} \lambda_k = 1$$

$$\lambda_k, s_r^+, s_i^- \ge 0, \text{ all } k, r, i, \theta \text{ unrestricted}$$

To place the problem in a general framework, assume that for each ordinal factor ($r \in R_2$, $i \in I_2$), a DMU k can be assigned to one of L rank positions, where $L \leq N$. As discussed earlier, L=5 is an example of an appropriate number of rank positions in many practical situations. We point out that in certain application settings, different ordinal factors may have different L-values associated with them. For exposition purposes, we assume a common L-value throughout. We demonstrate later that this represents no loss of generality.

One can view the allocation of a DMU to a rank position ℓ on an output r, for example, as having assigned that DMU an output *value* or *worth* $y_r^2(\ell)$. The implementation of the DEA model (3.1) (and (3.2)) thus involves determining two things:

(1) multiplier values μ_r^2, υ_i^2 for outputs $r \in R_2$ and inputs $i \in I_2$;
(2) rank position values $y_r^2(\ell)$, $r \in R_2$, and $x_i^2(\ell)$, $i \in I_2$, all ℓ.

In this section we show that the problem can be reduced to the standard VRS model by considering (1) and (2) simultaneously.

To facilitate development herein, define the L-dimensional unit vectors $\gamma_{rk} = (\gamma_{rk}(\ell))$, and $\delta_{ik} = (\delta_{ik}(\ell))$ where

$$\gamma_{rk}(\ell) = \begin{cases} 1 & \text{if DMU k is ranked in } \ell \text{ th position on output r} \\ 0, & \text{otherwise} \end{cases}$$

$$\delta_{ik}(\ell) = \begin{cases} 1 & \text{if DMU k is ranked in } \ell \text{ th position on input i} \\ 0, & \text{otherwise} \end{cases}$$

For example, if a 5-point scale is used, and if DMU #1 is ranked in $\ell = 3^{rd}$ place on ordinal output r=5, then $\gamma_{51}(3) = 1$, $\gamma_{51}(\ell) = 0$, for all other rank positions ℓ. Thus, y_{51}^2 is assigned the value $y_5^2(3)$, the *worth* to be credited to the 3rd rank position on output factor 5. It is noted that y_{rk}^2 can be represented in the form

$$y_{rk}^2 = y_r^2(\ell_{rk}) = \sum_{\ell=1}^{L} y_r^2(\ell) \gamma_{rk}(\ell),$$

where ℓ_{rk} is the rank position occupied by DMU k on output r. Hence, model (3.2) can be rewritten in the more representative format.

$$e_o = \max \mu_o + \sum_{r \in R_1} \mu_r^1 y_{ro}^1 + \sum_{r \in R_2} \sum_{\ell=1}^{L} \mu_r^2 y_r^2(\ell) \gamma_{ro}(\ell)$$

s.t. $\sum_{i \in I_1} \upsilon_i^1 x_{io}^1 + \sum_{i \in I_2} \sum_{\ell=1}^{L} \upsilon_i^2 x_i^2(\ell) \delta_{io}(\ell) = 1$ (3.3)

$\mu_o + \sum_{r \in R_1} \mu_r^1 y_{rk}^1 + \sum_{r \in R_2} \sum_{\ell=1}^{L} \mu_r^2 y_r^2(\ell) \gamma_{rk}(\ell) - \sum_{i \in I_1} \upsilon_i^1 x_{ik}^1 - \sum_{i \in I_2} \sum_{\ell=1}^{L} \upsilon_i^2 x_i^2(\ell) \delta_{ik}(\ell)$
≤ 0, all k

$\{Y_r^2 = (y_r^2(\ell)), X_i^2 = (x_i^2(\ell))\} \in \Psi$

$\mu_r^1, \upsilon_i^1 \geq \varepsilon$

In (3.3) we use the notation Ψ to denote the set of *permissible worth vectors*. We discuss this set below.

It must be noted that the same infinitesimal ε is applied here for the various input and output multipliers, which may, in fact, be measured on scales that are very different from another. If two inputs are, for example, x_{i1k}^1 representing 'labor hours', and x_{i2k}^1 representing 'available computer technology', the scales would clearly be incompatible. Hence, the likely sizes of the corresponding multipliers υ_{i1}^1, υ_{i2}^1 may be correspondingly different. Thrall (1996) has suggested a mechanism for correcting for such scale incompatibility, by applying a *penalty vector* G to augment ε, thereby creating differential lower bounds on the various υ_i, μ_r. Proper choice of G can effectively bring all factors to some form of common scale or unit. For simplicity of presentation we will assume the cardinal scales for all $r \in R_1$, $i \in I_1$ are similar in dimension, and that G is the unit vector. The more general case would proceed in an analogous fashion.

Permissible Worth Vectors

The values or worths $\{y_r^2(\ell)\}$, $\{x_i^2(\ell)\}$, attached to the ordinal rank positions for outputs r and inputs i, respectively, must satisfy the minimal requirement that it is *more* important to be ranked in ℓ^{th} position than in the $(\ell+1)^{st}$ position on any such ordinal factor. Specifically, $y_r^2(\ell) > y_r^2(\ell+1)$ and $x_i^2(\ell) < x_i^2(\ell+1)$. That is, for outputs, one places a higher weight on being ranked in ℓ^{th} place than in $(\ell+1)^{st}$ place. For inputs, the opposite is true. A set of linear conditions that produce this realization is defined by the set Ψ, where

$\Psi = \{(Y_r^2, X_r^2) | \ y_r^2(\ell) - y_r^2(\ell+1) \geq \varepsilon, \ \ell=1, \ldots L-1, \ y_r^2(L) \geq \varepsilon,$
$x_i^2(\ell+1) - x_i^2(\ell) \geq \varepsilon, \ \ell=1, \ldots L-1, \ x_i^2(1) \geq \varepsilon \}.$

Arguably, ε could be made dependent upon ℓ (i.e. replace ε by ε_ℓ). It can be shown, however, that all results discussed below would still follow. For convenience, we, therefore, assume a common value for ε. We now demonstrate that the nonlinear problem (3.3) can be written as a linear programming problem.

Theorem 3.1
Problem (3.3), in the presence of the permissible worth space Ψ, can be expressed as a linear programming problem.

<u>Proof</u>: In (3.3), make the change of variables $w_{r\ell}^1 = \mu_r^2 y_r^2(\ell)$, $w_{i\ell}^2 = \upsilon_i^2 x_i^2(\ell)$.

It is noted that in Ψ, the expressions $y_r^2(\ell) - y_r^2(\ell+1) \geq \varepsilon$, $y_r^2(L) \geq \varepsilon$ can be replaced by

$\mu_r^2 y_r^2(\ell) - \mu_r^2 y_r^2(\ell+1) \geq \mu_r^2 \varepsilon$, $\mu_r^2 y_r^2(L) \geq \mu_r^2 \varepsilon$, which becomes $w_{r\ell}^1 - w_{r\ell+1}^1 \geq \mu_r^2 \varepsilon$, $w_{rL}^2 \geq \mu_r^2 \varepsilon$.

A similar conversion holds for the $x_i^2(\ell)$. Problem (3.3) now becomes

$e_0 = \max \mu_0 + \sum_{r \in R1} \mu_r^1 y_{ro}^1 + \sum_{r \in R2} \sum_{\ell=1}^{L} w_{r\ell}^1 \gamma_{ro}(\ell)$

s.t. $\sum_{i \in I1} \upsilon_i^1 x_{io}^1 + \sum_{i \in I2} \sum_{\ell=1}^{L} w_{i\ell}^2 \delta_{io}(\ell) = 1$

$\mu_0 + \sum_{r \in R1} \mu_r^1 y_{rk}^1 + \sum_{r \in R2} \sum_{\ell=1}^{L} w_{r\ell}^1 \gamma_{rk}(\ell)$

$- \sum_{i \in I1} \upsilon_i^1 x_{ik}^1 - \sum_{i \in I2} \sum_{\ell=1}^{L} w_{i\ell}^2 \delta_{ik}(\ell) \leq 0$, all k (3.4)

$w_{r\ell}^1 - w_{r\ell+1}^1 \geq \mu_r^2 \varepsilon$, $\ell = 1, \ldots L-1$, all $r \in R_2$

$w_{rL}^1 \geq \mu_r^2 \varepsilon$, all $r \in R_2$

$w_{i\ell+1}^2 - w_{i\ell}^2 \geq \upsilon_i^2 \varepsilon$, $\ell = 1, \ldots L-1$, all $i \in I_2$

$w_{i1}^2 \geq \upsilon_i^2 \varepsilon$, all $i \in I_2$

$\mu_r^1, \upsilon_i^1 \geq \varepsilon$, all $r \in R_1, i \in I_1$

$\mu_r^2, \upsilon_i^2 \geq \varepsilon$, all $r \in R_2, i \in I_2$

Problem (3.4) is clearly in linear programming problem format.
QED

We state without proof the following theorem.

Theorem 3.2
At the optimal solution to (3.4), $\mu_r^2 = \upsilon_i^2 = \varepsilon$ for all $r \in R_2, i \in I_2$.

Problem (3.4) can then be expressed in the form:

$e_0 = \max \mu_0 + \sum_{r \in R_1} \mu_r^1 y_{ro}^1 + \sum_{r \in R2} \sum_{\ell=1}^{L} w_{r\ell}^1 \gamma_{ro}(\ell)$

s.t. $\sum_{i \in I1} \upsilon_i^1 x_{io}^1 + \sum_{i \in I2} \sum_{\ell=1}^{L} w_{i\ell}^2 \delta_{io}(\ell) = 1$

$$\mu_o + \sum_{r\in R1} \mu_r^1 y_{rk}^1 + \sum_{r\in R2} \sum_{\ell=1}^{L} w_{r\ell}^1 \gamma_{rk}(\ell) - \sum_{i\in I1} v_i^1 x_{ik}^1 - \sum_{i\in I2} \sum_{\ell=1}^{L} w_{i\ell}^2 \delta_{ik}$$
$(\ell) \leq 0$, all k
$-w_{r\ell}^1 + w_{r\ell+1}^1 \leq -\varepsilon^2$, $\ell=1,\ldots L-1$, all $r \in R_2$ (3.5)
$-w_{rL}^1 \leq -\varepsilon^2$, all $r \in R_2$
$-w_{i\ell+1}^2 + w_{i\ell}^2 \leq -\varepsilon^2$, $\ell=1, \ldots, L-1$, all $i \in I_2$
$-w_{i1}^2 \leq -\varepsilon^2$, all $i \in I_2$
$\mu_r^1, v_i^1 \geq \varepsilon$, $r \in R_1$, $i \in I_1$

It can be shown that (3.5) is equivalent to the *standard* VRS model. First we form the dual of (3.5).

$$\min \theta - \varepsilon \sum_{r\in R1} s_r^+ - \varepsilon \sum_{i\in I1} s_i^- - \varepsilon^2 \sum_{r\in R2} \sum_{\ell=1}^{L} \alpha_{r\ell}^1 - \varepsilon^2 \sum_{i\in I2} \sum_{\ell=1}^{L} \alpha_{i\ell}^2$$

s.t. $\sum_{k=1}^{N} \lambda_k y_{rk}^1 - s_r^+ = y_{ro}^1$, $r \in R_1$

$\theta x_{io}^1 - \sum_{k=1}^{N} \lambda_k x_{ik}^1 - s_i^- = 0$, $i \in I_1$

(3.5')

$$\left.\begin{array}{l} \sum_{k=1}^{N} \lambda_k \gamma_{rk}(1) - \alpha_{r1}^1 = \gamma_{ro}(1) \\ \sum_{k=1}^{N} \lambda_k \gamma_{rk}(2) + \alpha_{r1}^1 - \alpha_{r2}^1 = \gamma_{ro}(2) \\ \vdots \\ \sum_{k=1}^{N} \lambda_k \gamma_{rk}(L) + \alpha_{rL-1}^1 - \alpha_{rL}^1 = \gamma_{ro}(L) \end{array}\right\} r \in R_2$$

$$\left.\begin{array}{l} \delta_{io}(L)\theta - \sum_{k=1}^{N} \lambda_k \delta_{ik}(L) - \alpha_{iL}^2 = 0 \\ \delta_{io}(L-1)\theta - \sum_{k=1}^{N} \lambda_k \delta_{ik}(L-1) + \alpha_{iL}^2 - \alpha_{iL-1}^2 = 0 \\ \vdots \\ \delta_{io}(1)\theta - \sum_{k=1}^{N} \lambda_k \delta_{ik}(1) + \alpha_{i2}^2 - \alpha_{i1}^2 = 0 \end{array}\right\} i \in I_2$$

$\sum_{k=1}^{N} \lambda_k = 1$

$\lambda_k, s_r^+, s_i^-, \alpha_{r\ell}^1, \alpha_{i\ell}^2 \geq 0, \theta$ unrestricted.

Here, we use $\{\lambda_k\}$ as the standard dual variables associated with the N ratio constraints, and the variables $\{\alpha_{i\ell}^2, \alpha_{r\ell}^1\}$ are the dual variables associated with the rank order constraints defined by Ψ. The slack variables s_r^+, s_i^- correspond to the lower bound restrictions on μ_r^1, υ_i^1.

Now, perform simple row operations on (3.5') by replacing the ℓth constraint by the sum of the first ℓ constraints. That is, the second constraint (for those $r \in R_2$ and $i \in I_2$) is replaced by the sum of the first two constraints, constraint 3 by the sum of the first three, and so on. Letting

$$\overline{\gamma}_{rk}(\ell) = \sum_{n=1}^{\ell} \gamma_{rk}(n) = \gamma_{rk}(1) + \gamma_{rk}(2) + \ldots + \gamma_{rk}(\ell),$$

and

$$\overline{\delta}_{ik}(\ell) = \sum_{n=\ell}^{L} \delta_{ik}(n) = \delta_{ik}(L) + \delta_{ik}(L-1) + \ldots + \delta_{ik}(\ell),$$

problem (3.5') can be rewritten as:

$$\min \theta - \varepsilon \sum_{r \in R1} s_r^+ - \varepsilon \sum_{i \in I1} s_i^- - \varepsilon^2 \sum_{r \in R2} \sum_{l=1}^{L} \alpha_{rl}^1 - \varepsilon^2 \sum_{i \in I2} \sum_{l=1}^{L} \alpha_{il}^2$$

s.t. $\sum_{k=1}^{N} \lambda_k y_{rk}^1 - s_r^+ = y_{ro}^1, r \in R_1$

$\theta x_{io}^1 - \sum_{k=1}^{N} \lambda_k x_{ik}^1 - s_i^- = 0, i \in I_1$ (3.6')

$\sum_{k=1}^{N} \lambda_k \overline{\gamma}_{rk}(\ell) - \alpha_{rl}^1 = \overline{\gamma}_{ro}(\ell), r \in R_2, \ell = 1, \ldots, L$

$\theta \overline{\delta}_{io}(\ell) - \sum_{k=1}^{N} \lambda_k \overline{\delta}_{ik}(\ell) - \alpha_{il}^2 = 0, i \in I_2, \ell = 1, \ldots L$

$\sum_{k=1}^{N} \lambda_k = 1$

$\lambda_k, s_r^+, s_i^-, \alpha_{rl}^1, \alpha_{il}^2 \geq 0$, all i, r, ℓ, k, θ unrestricted in sign

The dual of (3.6') has the format:

$$e_o = \max \mu_o + \sum_{r \in R1} \mu_r^1 y_{ro}^1 + \sum_{r \in R2} \sum_{l=1}^{L} \overline{w}_{rl}^1 \overline{\gamma}_{ro}(\ell)$$

s.t. $\sum_{i \in I1} \upsilon_i^1 x_{io}^1 + \sum_{i \in I2} \sum_{l=1}^{L} \overline{w}_{il}^2 \overline{\delta}_{io}(\ell) = 1$ (3.6)

$\mu_o + \sum_{r \in R1} \mu_r^1 y_{rk}^1 + \sum_{r \in R2} \sum_{l=1}^{L} \overline{w}_{rl}^1 \overline{\gamma}_{rk}(\ell) - \sum_{i \in I1} \upsilon_i^1 x_{ik}^1 - \sum_{i \in I2} \sum_{\ell=1}^{L} \overline{w}_{il}^2 \overline{\delta}_{ik}(\ell) \leq 0$, all k

$\mu_r^1, \upsilon_i^1 \geq \varepsilon$, $\overline{w}_{rl}^1, \overline{w}_{il}^2 \geq \varepsilon^2$,

which is a form of the VRS model. The slight difference between (3.6) and the conventional VRS model of Banker et al. (1984), is the presence of a different ε (i.e., ε^2) relating to the multipliers $\overline{w}_{rl}^1$, $\overline{w}_{il}^2$, than is true for the multipliers μ_r^1, υ_i^1. It is observed that in (3.6') the common L-value can easily be replaced by criteria specific values (e.g. L_r for output criterion r).

The model structure remains the same, as does that of model (3.6). Of course, since the intention is to have an infinitesimal lower bound on multipliers (i.e., $\varepsilon \approx 0$), one can, from the start, restrict

$$\mu_r^1, \upsilon_i^1 \geq \varepsilon^2 \text{ and } \mu_r^2, \upsilon_i^2 \geq \varepsilon.$$

This leads to a form of (3.6) where all multipliers have the same infinitesimal lower bounds, making (3.6) precisely a VRS model in the spirit of Banker et a. (1984).

Criteria Importance

The presence of ordinal data factors results in the need to *impute* values $y_r^2(\ell)$, $x_i^2(\ell)$ to outputs and inputs, respectively, for DMUs that are ranked at positions on an L-point Likert or ordinal scale. Specifically, all DMUs ranked at that position will be credited with the same "amount" $y_r^2(\ell)$ of output r ($r \in R_2$) and $x_i^2(\ell)$ of input i ($i \in I_2$).

A consequence of the change of variables undertaken above, to bring about linearization of the otherwise nonlinear terms, e.g., $w_{r\ell}^1 = \mu_r^2 y_r^2(\ell)$, is that at the optimum, all $\mu_r^2 = \varepsilon^2$, $\upsilon_i^2 = \varepsilon^2$. Thus, all of the ordinal criteria are relegated to the status of being of *equal importance*. Arguably, in many situations, one may wish to view the relative importance of these ordinal criteria (as captured by the μ_r^2, υ_i^2) in the same spirit as we have viewed the data values $\{y_{rk}^2\}$. That is, there may be sufficient information to be able to *rank* these criteria. Specifically, suppose that the R_2 output criteria can be grouped into L_1 categories and the I_2 input criteria into L_2 categories. Now, replace the variables μ_r^2 by $\mu^2(m)$, and υ_i^2 by $\upsilon^2(n)$, and restrict:

$$\mu^2(m) - \mu^2(m+1) \geq \varepsilon, \; m=1,\ldots L_1-1$$
$$\mu^2(L_1) \geq \varepsilon$$
and
$$\upsilon^2(n) - \upsilon^2(n+1) \geq \varepsilon, \; n=1,\ldots,L_2-1$$
$$\upsilon^2(L_2) \geq \varepsilon.$$

Letting m_r denote the rank position occupied by output $r \in R_2$, and n_i the rank position occupied by input $i \in I_2$, we perform the change of variables

$$w_{r\ell}^1 = \mu^2(m_r) y_r^2(\ell)$$
$$w_{i\ell}^2 = \upsilon^2(n_i) x_i^2(\ell)$$

The corresponding version of model (3.4) would see the lower bound restrictions $\mu_r^2, \upsilon_i^2 \geq \varepsilon$ replaced by the above constraints on $\mu^2(m)$ and υ^2

(n). Again, arguing that at the optimum in (3.4), these variables will be forced to their lowest levels, the resulting values of the $\mu^2(m)$, $\upsilon^2(n)$ will be
$\mu^2(m) = (L_1+1-m)\varepsilon$, $\upsilon^2(n) = (L_2+1-n)\varepsilon$.
This implies that the lower bound restrictions on $w_{r\ell}^1$, $w_{i\ell}^2$ become

$$w_{r\ell}^1 \geq (L_1+1-m_r)\,\varepsilon^2,\; w_{i\ell}^2 \geq (L_2+1-n_i)\,\varepsilon^2.$$

Example

When model (3.6') is applied to the data of Table 2-1, the efficiency scores obtained are as shown in Table 2-2.

Table 2-2. Efficiency Scores (Non-ranked Criteria)

Project	1	2	3	4	5	6	7	8	9	10
Score	0.76	0.73	1.00	0.67	1.00	0.82	0.67	0.67	0.55	0.37

Here, projects 3 and 5 turn out to be 'efficient', while all other projects are rated well below 100%. In this particular analysis, ε was chosen as 0.03. In another run (not shown here) where $\varepsilon = 0.01$ was used, projects 3, 5 and 6 received ratings of 1.00, while all others obtained somewhat higher scores than those shown in Table 2-2. When a very small value of ε ($\varepsilon = 0.001$) was used, all except one of the projects was rated as efficient. Clearly this example demonstrates the same degree of dependence on the choice of ε as is true in the standard DEA model. See Ali and Seiford (1993).

From the data in Table 2-1 it might appear that only project 3 should be efficient since 3 dominates project 5 in all factors except for the second input where project 3 rates second while project 5 rates first. As is characteristic of the standard ratio DEA model, a single factor can produce such an outcome. In the present case this situation occurs because $w_{21}^2 = 0.03$ while $w_{22}^2 = 0.51$. Consequently, project 5 is accorded an 'efficient' status by permitting the gap between w_1^2 and w_2^2 to be (perhaps unfairly) very large. Actually, the set of multipliers which render project 5 efficient also constitute an optimal solution for project 3.

If we further constrain the model by implementing criteria importance conditions as defined in the previous section, the relative positioning of the projects changes as shown in Table 2-3.

Table 2-3. Efficiency Scores (Ranked Criteria)

Project	1	2	3	4	5	6	7	8	9	10
Score	0.71	0.72	1.00	0.60	1.00	0.80	0.62	0.63	0.50	0.35

Hence, criteria importance restrictions can have an impact on the efficiency status of the projects.

Two interesting phenomena characterize DEA problems containing ordinal data. If one examines in detail the outputs from the analysis of the example data, two observations can be made. First, it is the case that $\theta = 1$ for each project (whether efficient or inefficient). This means that each project is either on the frontier proper or an extension of the frontier. Second, if one were to use the CRS rather than VRS model, it would be observed that $\sum \lambda_k = 1$ for each project. The implication would seem to be that the two models (CRS and VRS) are equivalent in the presence of ordinal data. Moreover, since $\theta = 1$ in all cases, these models are as well equivalent to the additive model of Charnes et al. (1985). The following two theorems prove these results for the general case.

Theorem 3.3
In problem (3.6'), if I_2 is non empty, $\theta = 1$ at the optimum.

Proof:

By definition, $\bar{\delta}_{ik}(\ell) = \sum_{n=\ell}^{L} \delta_{ik}(n)$. Thus, $\bar{\delta}_{ik}(1) = 1$ for all k, and for any ordinal input i. From the constraint set of (3.6'), if $\sum_{k=1}^{N} \lambda_k = 1$, then since

$$\theta \bar{\delta}_{io}(1) \geq \sum_{k=1}^{N} \lambda_k \bar{\delta}_{ik}(1), \text{ and since } \sum_{k=1}^{N} \lambda_k \bar{\delta}_{ik}(1) = 1 \text{ (given that all}$$

members of $\{\bar{\delta}_{ik}(1)\}_{k=1}^{N}$ equal 1), it follows that $\theta \bar{\delta}_{io}(1) \geq 1$. But since $\bar{\delta}_{io}(1) = 1$, then $\theta \geq 1$, meaning that at the optimum $\theta = 1$.

QED

This rather unusual property of the DEA model in the presence of ordinal data is generally explainable by observing the dual form (3.5). It is noted that ε^2 plays the role of discriminating between the levels of relative importance of consecutive rank positions. If in the extreme case $\varepsilon = 0$, then any one rank position becomes as important as any other. This means that regardless of the rank position occupied by a DMU "o", that position can be credited with at least as high a weight as those assumed by the peers of that DMU. Hence, every DMU will be deemed technically efficient. It is only the presence of positive gaps (defined by ε^2) between rank positions that renders a DMU inefficient via the slacks.

Theorem 3.4

If R_2 and I_2 are both non empty, then in the CRS version of problem (3.6'), $\sum_{k=1}^{N} \lambda_k = 1$ at the optimum.

Proof: Reconsider problem (3.6'), but with the constraint $\sum_{k=1}^{N} \lambda_k = 1$ removed. As in Theorem 3.3, the constraint

$$\theta \bar{\delta}_{io}(1) = \bar{\delta}_{io}(1) \geq \sum_{k=1}^{N} \lambda_k \bar{\delta}_{ik}(1),$$

holds. But, since $\bar{\delta}_{ik}(1) = 1$ for all k, then it is the case that $\sum_{k=1}^{N} \lambda_k \leq 1$. On the output side, however, $\bar{\gamma}_{rk}(L) = 1$ for all k, and any ordinal output r. But, since

$$\sum_{k=1}^{N} \lambda_k \bar{\gamma}_{rk}(L) \geq \bar{\gamma}_{ro}(L),$$

it follows that $\sum_{k=1}^{N} \lambda_k \geq 1$. Thus, $\sum_{k=1}^{N} \lambda_k = 1$.

QED

From Theorem 3.4, it follows that the VRS and CRS models are equivalent. Moreover, from Theorem 3.3, one may view these two models as equivalent to the additive model in that the objective function of (3.6') is equivalent to maximizing the sum of all slacks.

It should be pointed out that the projection to the efficient frontier in model (3.6) treats the Likert scale [1,L] as if it were a continuum, rather than as consisting of a set of discrete rank positions. Specifically, at the optimum in (3.6), any given projected value, e.g. $\theta \bar{\delta}_{io}(\ell) - \alpha_{i\ell}^2$, $i \in I_2$, is not guaranteed to be one of the discrete points on the [1,L] scale. For this reason, we refer to (3.6) as a *continuous projection* model. In this respect, the model can be viewed as providing a form of *upper bound* on the extent of reduction that can be anticipated in the ordinal inputs. Suppose, for example, that at the optimum $\theta \bar{\delta}_{io}(\ell) - \alpha_{i\ell}^2 = 2.7$, a *rank position* between the legitimate positions 2 and 3 on an L-point scale. Then, arguably, it is possible for $\bar{\delta}_{io}(\ell)$ to be projected only to point 3, not further.

One can, of course, argue that the choosing of a specific number of rank positions L is generally motivated by an inability to be more discriminating (a larger L-value was not practical). At the proposed "efficient" rank position of 2.7, we are claiming that the DMU "o" will be using more input i than a DMU ranked in 2nd place, but less input than one ranked in 3rd place. Thus, to some extent, this projection automatically has created an L+1- point Likert scale, where previously the scale had contained only L points. The projection has permitted us to increase our degrees of discrimination.

In the special case treated by Cooper et al (1999) where L=N, this issue never arises, as every DMU is entitled to occupy its own rank position on an N-point scale.

In Section 5 we will re-examine the discrete nature of the L-point scale and propose a model structure accordingly. In the following Section 4 we evaluate the IDEA concept of Cooper et al (1999) in relation to the model developed above.

4. THE CONTINUOUS PROJECTION MODEL AND IDEA

Cooper, Park and Yu (1999) examine the DEA structure in the presence of *imprecise* data (IDEA) for certain factors. Zhu (2003) and others have extended Cooper, Park and Yu's (1999) earlier model. This is further elaborated in Chapter 3. One particular form of imprecise data is a full ranking of the DMUs in an ordinal sense. Representation of rank data via a Likert scale, with $L \leq N$ rank positions, can be viewed as a generalization of the Cooper, Park and Yu (1999) structure wherein $L = N$.

To demonstrate this, we consider a full ranking of the DMUs in an ordinal sense and, to simplify the presentation, we suppose weak ordinal data can be expressed as (see Cooper, Park and Yu (1999) and Zhu (2003)):

$$y_{r1} \leq y_{r2} \leq \ldots \leq y_{rk} \leq \ldots \leq y_{rn} \ (\ r \in R_2) \quad (4.1)$$
$$x_{i1} \leq x_{i2} \leq \ldots \leq x_{ik} \leq \ldots \leq x_{in} \ (i \in I_2) \quad (4.2)$$

When the set Ψ is expressed as (4.1) and (4.2), model (3.5) can be expressed as

$$e_o = \max \mu_o + \sum_{r \in R_1} \mu_r^1 y_{ro}^1 + \sum_{r \in R_2} w_{ro}^1$$
$$\text{s.t.} \sum_{i \in I_1} \upsilon_i^1 x_{io}^1 + \sum_{i \in I_2} w_{io}^2 = 1$$
$$\mu_o + \sum_{r \in R_1} \mu_r^1 y_{rk}^1 + \sum_{r \in R_2} w_{rk}^1 - \sum_{i \in I_1} \upsilon_i^1 x_{ik}^1 - \sum_{i \in I_2} w_{ik}^2 \leq 0, \text{ all k} \quad (4.3)$$
$$- w_{r,k+1}^1 + w_{rk}^1 \leq 0, \ k=1,\ldots N-1, \text{ all } r \in R_2$$
$$- w_{i,k+1}^2 + w_{ik}^2 \leq 0, \ k=1, \ldots, N-1, \text{ all } i \in I_2$$
$$\mu_r^1, \upsilon_i^1 \geq \varepsilon, r \in R_1, i \in I_1$$

We next define $t_{rk}^1 = v_i^1 x_{ik}^1$ ($r \in R_1$) and $t_{ik}^2 = \mu_r^1 y_{rk}^1$ ($i \in I_1$). Then model (4.3) becomes the IDEA model of Cooper, Park and Yu (1999)[1]:

$$e_o = \max \mu_o + \sum_{r \in R_1} t_{io}^2 + \sum_{r \in R_2} w_{ro}^1$$

s.t. $\sum_{i \in I1} t_{ro}^1 + \sum_{i \in I2} w_{io}^2 = 1$

$\mu_o + \sum_{r \in R1} t_{ik}^2 + \sum_{r \in R2} w_{rk}^1 - \sum_{i \in I1} t_{rk}^1 - \sum_{i \in I2} w_{ik}^2 \leq 0$, all k (4.4)

$-w_{r,k+1}^1 + w_{rk}^1 \leq 0$, k=1,...N-1, all $r \in R_2$

$-w_{i,k+1}^2 + w_{ik}^2 \leq 0$, k=1, ..., N-1, all $i \in I_2$

$t_{ik}^2, t_{rk}^1 \geq \varepsilon$, $r \in R_1, i \in I_1$

It is pointed out that in the original IDEA model of Cooper, Park and Yu (1999), scale transformations on the input and output data, i.e., $\hat{x}_{ij} = x_{ij}/\max_j \{x_{ij}\}$, $\hat{y}_{rj} = y_{rj}/\max_j \{y_{rj}\}$ are done before new variables are introduced to convert the non-linear DEA model with ordinal data into a linear program. However, as demonstrated in Zhu (2003), such scale transformations are unnecessary and redundant. As a result, the same variable alternation technique is used in Cook, Kress and Seiford (1993, 1996) and Cooper, Park and Yu (1999) in converting the non-linear IDEA model into linear programs. The difference lies in the fact that Cook, Kress and Seiford (1993, 1996) aim at converting the non-linear IDEA model into a conventional DEA model. To use the conventional DEA model based upon Cooper, Park and Yu (1999), one has to obtain a set of exact data from the imprecise or ordinal data (see Zhu, 2003).

Based upon the above discussion, we know that the equivalence between the model of section 3 and the IDEA model of Cooper, Park and Yu (1999) holds for any L if rank data are under consideration.

We finally discuss the treatment of strong versus weak ordinal relations in the model of Section 3. Note that Ψ in section 3, actually represents strong ordinal relations[2]. Cook, Kress, and Seiford (1996) points out that efficiency scores can depend on ε in set Ψ and propose a model to determine a proper ε. Zhu (2003) shows an alternative approach in determining ε. Further, as shown in Zhu (2003), part of weak ordinal relations can be replaced by strong ones without affecting the efficiency ratings and without the need for selecting the ε.

Alternatively, we can impose strong ordinal relations as $y_{rk} \geq \eta y_{r,k-1}$ and $x_{ik} \geq \eta x_{i,k-1}$ ($\eta > 1$).

---

[1] The original IDEA model of Cooper, Park and Yu (1999) is discussed under the model of Charnes, Cooper and Rhodes (1978).

[2] As shown in Zhu (2003), the expression in Ψ itself does not distinguish strong from weak ordinal relations if the IDEA model of Cooper, Park and Yu (1999, 2002) is used. Zhu (2003) proposes a correct way to impose strong ordinal relations in the IDEA model.

5. DISCRETE PROJECTION FOR LIKERT SCALE DATA: AN ADDITIVE MODEL

The model of the previous sections can be considered as providing a *lower bound* on the efficiency rating of any DMU. As discussed above, projections may be infeasible in a strict ordinal ranking sense. The DEA structure explicitly implies that points on the frontier constitute permissible targets. Formalizing the example of the Section 3, suppose that at two efficient (frontier) points k_1, k_2, it is the case that for an ordinal input $i \varepsilon$ I_2, the respective rank positions are $\delta_{ik_1}(2) = 1$ and $\delta_{ik_2}(3) = 1$, that is, DMU k_1 is ranked in 2nd place on input i, while k_2 is ranked at 3rd place. Since all points on the line (facet) joining these two frontier units are to be considered as allowable projection points, then any "rank position" between a rank of 2 and a rank of 3 is allowed. The DEA structure thus treats the rank range [1,L] as continuous, not discrete. In a "full" rank order sense, one might interpret the projected point as giving DMU "o" a ranking just one position worse than that of DMU k_1, and thereby displacing k_2 and giving it (k_2) a rank that is one position worse than it had prior to the projection. This would mean that all DMUs ranked at or worse than is true for DMU k_2, would also be so displaced.

If DMUs are not rank ordered in the aforementioned sense, but rather are assigned to L (e.g. L=5) rank positions, the described displacement does not occur. Specifically, if the ith ordinal input for DMU "o" is ranked in position ℓ_{ik} prior to projection toward the frontier, the only permissible other positions to which it can move are the discrete points ℓ_{ik} -1, ℓ_{ik} -2,...1. The modeling of such discrete projections cannot, however, be directly accomplished within the radial framework, where each data value (e.g. $\bar{\delta}_{i_o}(\ell)$) is to be reduced by the same proportion $1-\theta$.

The requirement to select from among a discrete set of rank positions (e.g. ℓ_{i_k} -1, ℓ_{i_k} -2,...,1) can be achieved from a form of the additive model as originally presented by Charnes et al (1985). As discussed in Section 3, however, the VRS (and CRS) model is equivalent to the additive model. Thus, there is no loss of generality. An integer additive model version of (3.6') can be expressed as follows: (We adopt here an 'invariant' form of the model, by scaling the objective function coefficients by the original data values.)

$$\max \sum_{r \in R_1} (s_r^+/y_{ro}^1) + \sum_{i \in I_1} (s_i^-/x_{io}^1) + \sum_{r \in R_2} (\sum_\ell \alpha_{r\ell}^1 / (\sum_\ell \bar{\gamma}_{r_o}(\ell)))$$
$$+ \sum_{i \in I_2} (\sum_\ell \alpha_{i\ell}^2 / \sum_\ell \bar{\delta}_{io}(\ell)) \qquad (5.1a)$$

s.t. $\sum_{k=1}^{N} \lambda_k y_{rk}^1 - s_r^+ = y_{ro}^1$, $r \in R_1$ (5.1b)

$\sum_{k=1}^{N} \lambda_k x_{ik}^1 + s_i^- = x_{io}^1$, $i \in I_1$ (5.1c)

$\sum_{k=1}^{N} \lambda_k \bar{\gamma}_{rk}(\ell) - \alpha_{r\ell}^1 \geq \bar{\gamma}_{ro}(\ell)$, $r \in R_2$, $\ell = 1,\ldots L$ (5.1d)

$\sum_{k=1}^{N} \lambda_k \bar{\delta}_{ik}(\ell) + \alpha_{i\ell}^2 \leq \bar{\delta}_{io}(\ell)$, $i \in I_2$, $\ell = 1,\ldots L$ (5.1e)

$\sum_{k=1}^{N} \lambda_k = 1$ (5.1f)

$\lambda_k \geq 0$, all k, $s_i^-, s_r^+ \geq 0$, $i \in I_1, r \in R_1, \alpha_{r\ell}^1, \alpha_{r\ell}^2$ integer, $r \in R_2, i \in I_2$ (5.1g)

The imposition of the integer restrictions on the $\alpha_{r\ell}^1$, $\alpha_{r\ell}^2$ is intended to create projections for inputs in I_2 and outputs in R_2 to points on the Likert Scale.

Theorem 5.1: The projections resulting from model (5.1) correspond to points on the Likert scale [1,L] for inputs $i \in I_2$ and outputs $r \in R_2$.

Proof: If for any $i \in I_2$, a DMU k is ranked at position ℓ_{ik}, then by definition

$$\bar{\delta}_{ik}(\ell) = \begin{cases} 0, & \ell > \ell_{ik} \\ 1, & \ell \leq \ell_{ik} \end{cases}$$

Similarly, for $r \in R_2$, if k is ranked at position ℓ_{rk}, then

$$\bar{\gamma}_{rk}(\ell) = \begin{cases} 0, & \ell < \ell_{rk} \\ 1, & \ell \geq \ell_{rk} \end{cases}.$$

At the optimum of (5.1) (let $\{\hat{\lambda}_k\}$ denote the optimal λ_k), the $\{\sum_k \hat{\lambda}_k \bar{\delta}_{ik}(\ell)\}_{\ell=1}^{L}$ form a non increasing sequence, i.e.

$$\sum_{k=1}^{n} \hat{\lambda}_k \bar{\delta}_{ik}(\ell) \geq \sum_{k=1}^{n} \hat{\lambda}_k \bar{\delta}_{ik}(\ell+1), \quad \ell=1,\ldots L-1. \quad \text{Similarly, the}$$

$\{\sum_{k} \hat{\lambda}_k \bar{\gamma}_{rk}(\ell)\}_{\ell=1}^{L}$ for a non decreasing sequence. In constraint (5.1e), we let $\ell_{i\lambda(i)}$ denote the value of ℓ such that

$$\sum_{k} \hat{\lambda}_k \bar{\delta}_{ik}(\ell) = 0 \text{ for } \ell \geq \ell_{i\lambda(i)}, \text{ and } \sum_{k} \hat{\lambda}_k \bar{\delta}_{ik}(\ell) > 0 \text{ for } \ell < \ell_{i\lambda(i)}.$$

Clearly, $\ell_{i\lambda(i)} \leq \ell_{io}$.

As well, at the optimum

$$\alpha_{i\ell}^2 = \begin{cases} 0 \text{ for } \ell > \ell_{i_o} \\ 1 \text{ for } \ell_{i\lambda(i)} \\ 0 \text{ for } \ell < \ell_{i\lambda(i)} \end{cases}$$

Hence, if we define the "revised" $\bar{\gamma}$ and $\bar{\delta}$ - values, by $\hat{\gamma}_{ro}(\ell) = \bar{\gamma}_{ro}(\ell) + \alpha_{r\ell}^1$ and $\hat{\delta}_{io}(\ell) = \bar{\delta}_{io}(\ell) - \alpha_{i\ell}^2$, then these define a proper rank position for input i (output (r)) of DMU "o". That is,

$$\hat{\delta}_{io}(\ell) = \begin{cases} 0, & \ell > \ell_{i\lambda(i)} \\ 1, & \ell \leq \ell_{i\lambda(i)} \end{cases},$$

and

$$\hat{\gamma}_{ro}(\ell) = \begin{cases} 1, & \ell \geq \ell_{r\lambda(r)} \\ 0, & \ell < \ell_{r\lambda(r)} \end{cases},$$

meaning that the projected rank position of DMU "o" on e.g. input i is $\ell_{i\lambda(i)}$.

QED

Unlike the radial models of Charnes et al (1978) and Banker et al (1984), the additive model does not have an associated convenient (or at least universally accepted) measure of efficiency. The objective function of (5.1) is clearly a combination of input and output projections. While various "slacks based" measures have been presented in the literature (see e.g. Cooper, Seiford and Tone (2000)), we propose a variant of the "Russell Measure" as discussed in Fare and Lovell (1978). Specifically, define

$$\theta_i^1 = 1 - s_i^- / x_{io}^1, i \in I_1; \theta_i^2 = 1 - \sum_{\ell=1}^{L} \alpha_{i\ell}^2 / \sum_{\ell=1}^{L} \bar{\delta}_{io}(\ell), i \in I_2$$

$$\phi_r^1 = 1 + s_r^+/y_{ro}^1, \text{ r} \in R_1; \ \phi_r^2 = 1 + \sum_{\ell=1}^{L}\alpha_{r\ell}^1 / \sum_{\ell=1}^{L}\bar{\gamma}_{ro}(\ell), \text{ r} \in R_2.$$

We define the efficiency measure of DMU "o" to be

$$\beta_1 = [\sum_{i \in I_1}\theta_i^1 + \sum_{i \in I_2}\theta_i^2 + \sum_{r \in R_1}(1/\phi_r^1) + \sum_{r \in R_2}(1/\phi_r^2)]/[|I_1|+|I_2|+|R_1|+|R_2|], \quad (5.2)$$

where $|I_1|$ denotes the cardinality of $I_1,\ldots$ etc.

The following property follows immediately from the definitions of θ_i^1, θ_i^2, ϕ_r^1, ϕ_r^2.

Property 5.1: The efficiency measure β_1 satisfies the condition $0 \leq \beta_1 \leq 1$.

It is noted that $\beta_1 = 1$ only in the circumstance that the DMU "o" is actually *on* the frontier. This means, of course that if a DMU is not on the frontier, but at the optimum of (5.1) has all $\alpha_{i\ell}^1$, $\alpha_{i\ell}^2 = 0$, it will be declared inefficient even though it is impossible for it to improve its position on the ordinal (Likert) scale. An additional and useful measure of *ordinal* efficiency is

$$\beta_2 = [\sum_{i \in I_2}\theta_i^2 + \sum_{r \in R_2}(1/\phi_r^2)]/[|I_1|+|R_2|]. \quad (5.3)$$

In this case $\beta_2 = 1$ for those DMUs for which further movement (improvement) along ordinal dimensions is not possible.

Example

Continuing the example discussed in the previous Section 3, we apply model (5.1) with the requisite requirement that only Likert scale projections are permitted in regard to ordinal factors. Table 2-4 presents the results.

Table 2-4. Efficiency Scores β_1

Project No.	β_1	Peers
1	.73	3
2	.61	3
3	1.00	3
4	.53	3
5	1.00	5
6	.75	3
7	.57	3
8	.52	3
9	.48	3
10	.30	3

As in the analysis of Section 3, all projects except for #3, #5 are inefficient. Interestingly, their *relative* sizes (in a rank order sense) agree with the outcomes shown in Table 2-2 of Section 3.

6. CONCLUSIONS

This chapter has examined the use of ordinal data in DEA, and specifically makes the connection between the earlier work of Cook et al (1993,1996) and the IDEA concepts discussed in Chapter 3. Two general models are developed, namely, continuous and discrete projection models. The former, aims to generate the maximum reduction in inputs (input-oriented model), without attention to the feasibility of the resulting projections in a Likert scale sense. The latter model specifically addresses the need to project to discrete points for ordinal factors. We prove that in the presence of ordinal factors, CRS and VRS models are equivalent. As well, it is shown that in a pure technical efficiency sense, $\theta = 1$. Thus, it is only the slacks that render a DMU inefficient in regard to ordinal factors. The latter also implies that projections in the VRS (and CRS) sense are the same as those arising from the additive model. This provides a rationale for reverting to the additive model to facilitate projection in the discrete (versus continuous) model described in Section 5.

REFERENCES

1. Ali, A.I. and L.M. Seiford. 1993. Computational accuracy and infinitesimals in data envelopment analysis. *INFOR* 31, 290-297.
2. Banker, R., A. Charnes and W.W. Cooper. 1984. Some models for estimating technical and scale inefficiencies in Data Envelopment Analysis. *Management Science* 30(9), 1078-1092.
3. Charnes, A. and W.W. Cooper, B. Golany, L. Seiford and J. Stutz. 1985. Foundations of Data Envelopment Analysis for Pareto-Koopmans efficient empirical production functions. *Journal of Econometrics* 30(1-2), 91-107.
4. Charnes, A., W.W. Cooper and E. Rhodes. 1978. Measuring the efficiency of decision making units. *European Journal of Operational Research* 2, 429-444.
5. Cook, W.D., M. Kress and L.M. Seiford. 1993. On the use of ordinal data in Data Envelopment Analysis. *Journal of the Operational Research Society* 44, 133-140.

6. Cook, W. D., M. Kress and L.M. Seiford. 1996. Data Envelopment Analysis in the presence of both quantitative and qualitative factors. *Journal of the Operational Research Society* 47, 945-953.
7. Cook, W.D. and J. Zhu. 2006. Rank order data in DEA: A general framework. *European Journal of Operational Research* 174, 1021-1038.
8. Cooper, W.W., K.S. Park and G. Yu. 1999. IDEA and AR-IDEA: Models for dealing with imprecise data in DEA. *Mgmt. Sci.* 45, 597-607.
9. Cooper, W.W., K.S. Park, G. Yu. 2002. An illustrative application of IDEA (Imprecise Data Envelopment Analysis) to a Korean mobile telecommunication company. *Operations Research* 49(6), 807-820.
10. Cooper, W.W., L. Seiford and K. Tone. 2000. *Data Envelopment Analysis*, Kluwer Academic Publishers, Boston.
11. Fare, R. and C.A.K. Lovell. 1978. Measuring the technical efficiency of production. *Journal of Economic Theory* 19(1), 150-162.
12. Thrall, R.M. 1996. Duality, classification and slacks in DEA. *Annals of Operations Research* 66, 109-138.
13. Zhu, J. 2003. Imprecise Data Envelopment Analysis (IDEA): A review and improvement with an application. *European Journal of Operational Research* 144, 513-529.

The material in this chapter is based upon the earlier works of Cook, W.D., M. Kress and L.M. Seiford, 1993, "On the use of ordinal data in Data Envelopment Analysis", *Journal of the Operational Research Society* 44, 133-140, Cook, W. D., M. Kress and L.M. Seiford. 1996, "Data Envelopment Analysis in the presence of both quantitative and qualitative factors", *Journal of the Operational Research Society* 47, 945-953, and Cook, W.D. and J. Zhu, 2006, Rank order data in DEA: A general framework, *European Journal of Operational Research* 174, 1021-1038, with permission from Palgrave and Elsevier Science

Chapter 3

INTERVAL AND ORDINAL DATA
How Standard Linear DEA Model Treats Imprecise Data

Yao Chen[1] and Joe Zhu[2]

[1]*School of Management, University of Massachusetts, Lowell, MA 01854, Yao_Chen@uml.edu*

[2]*Department of Management, Worcester Polytechnic Institute, Worcester, MA 01609, jzhu@wpi.edu*

Abstract: The standard Data Envelopment Analysis (DEA) method requires that the values for all inputs and outputs are known exactly. When some inputs and output are imprecise data, such as interval or bounded data, ordinal data, and ratio bounded data, the resulting DEA model becomes a non-linear programming problem. Such a DEA model is called imprecise DEA (IDEA) in the literature. There are two approaches in dealing with such imprecise inputs and outputs. One approach uses scale transformations and variable alternations to convert the non-linear IDEA model into a linear program. The other identifies a set of exact data from the imprecise inputs and outputs and then uses the standard linear DEA model. This chapter focuses on the latter IDEA approach that uses the standard DEA model. This chapter shows that different results are obtained depending on whether the imprecise data are introduced directly into the multiplier or envelopment DEA model. Because the presence of imprecise data invalidates the linear duality between the multiplier and envelopment DEA models. The multiplier IDEA (MIDEA), developed based upon the multiplier DEA model, presents the best efficiency scenario whereas the envelopment IDEA (EIDEA), developed based upon the envelopment DEA model, presents the worst efficiency scenario. Weight restrictions are often redundant if they are added into MIDEA. Alternative optimal solutions on the imprecise data can be determined using the recent sensitivity analysis approach. The approaches are illustrated with both numerical and real world data sets.

Key words: Data Envelopment Analysis (DEA), Performance, Efficiency, Imprecise data, Ordinal data, Interval data, Bounded data, Multiplier, Envelopment

1. INTRODUCTION

Data Envelopment Analysis (DEA) developed by Charnes, Cooper and Rhodes (1978) (CCR) assumes that data on the inputs and outputs are known exactly. However, this assumption may not be true. For example, some outputs and inputs may be only known as in forms of bounded or interval data, ordinal data, and ratio bounded data. Cook, Kress and Seiford (1993), (1996) were the first who developed a modified DEA sturcture where the inputs and outputs are represented as rank positions in an ordinal, rather than numerical sense (see chapter 2).

If we incorporate such imprecise data information directly into the standard linear CCR model, the resulting DEA model is a non-linear and non-convex program. Such a DEA model is called imprecise DEA (IDEA) in Cooper, Park and Yu (1999) who discuss how to deal with bounded data and weak ordinal data and provide a unified IDEA model when weight restrictions are also present[1]. In a similar work, Kim, Park and Park (1999) discuss how to deal with bounded data, (strong and weak) ordinal data, and ratio bounded data.

As shown in Cook and Zhu (2006), the IDEA approach of Kim, Park and Park (1999) and Cooper, Park and Yu's (1999) approach is actually a direct result of Cook, Kress and Seiford (1993; 1996) with respect to the use of variable alternations.

Zhu (2003a; 2004) on the other hand shows that the non-linear IDEA can be solved in the standard linear CCR model via identifying a set of exact data from the imprecise input and output data. This approach allows us to use all existing DEA techniques to analyze the performance of DMUs and additional evaluation information (e.g., performance benchmarks, paths for efficiency improvement, and returns to scale (RTS) classification) can be obtained.

Chen, Seiford and Zhu (2000) and Chen (2006) calls the existing IDEA approaches multiplier IDEA (MIDEA) because these approaches are based upon the DEA multiplier models. These authors also show that IDEA models can be built on the envelopment DEA models. That is, the interval data and ordinal data can be introduced directly into the envelopment DEA model. We can the resulting DEA approach as envelopment IDEA (EIDEA). It is shown that EIDEA yields the worst scores whereas the MIDEA yields the best efficiency scores. Using the techniques developed in Zhu (2003a; 2004), the EIDEA can also be converted into linear DEA models.

---

[1] Zhu (2003a) shows that such weight restrictions are redudant when ordinal and ratio bounded data are present. This can substantially reduce the computation burden.

Despotis and Smirlis (2002) also develop a general structure to convert interval data in dealing with the imprecise data in DEA. Kao and Liu (2000) treat the interval data as fuzzy DEA approach.

The current chapter will only focus on the approach Zhu (2003a; 2004) and Chen (2006) where identification of a set of exact data allows us to use the existing standard DEA codes. For other approaches to interval data and ordinal data, the interested reader is referred to Cooper and Park (2006) and chapter 2.

The remainder of this chapter is organized as follows. The next section presents the multiplier and primal DEA models with some specific forms of imprecise data. We then presents the Multiplier IDEA (MIDEA) approach. We show how to convert the MIDEA model into linear programs. We then presents the Envelopment IDEA (EIDEA) approach described in Chen (2006)[2]. Conclusions are given in the last section.

2. IMPRECISE DATA

Suppose we have a set of n peer DMUs, $\{DMU_j : j = 1, 2, \ldots, n\}$, which produce multiple outputs y_{rj}, ($r = 1, 2, \ldots, s$), by utilizing multiple inputs x_{ij}, ($i = 1, 2, \ldots, m$). When a DMU_o is under evaluation by the CCR model, we have the multiplier DEA model

Maximize $\pi_o = \sum_{r=1}^{s} \mu_r y_{ro}$

subject to

$$\sum_{r=1}^{s} \mu_r y_{rj} - \sum_{i=1}^{m} \omega_i x_{ij} \leq 0 \quad \forall\, j$$

$$\sum_{i=1}^{m} \omega_i x_{io} = 1$$

$$\mu_r, \omega_i \geq 0 \quad \forall\, r, i$$

(1)

The dual program to (1) – the envelopment DEA model can be written as

---

[2] In Chen (2006), envelopment IDEA (EIDEA) is called primal IDEA (PIDEA).

$$\theta_o^* = \min \theta_o$$
subject to

$$\sum_{j=1}^{n} \lambda_j x_{ij} \leq \theta_o x_{io} \quad i = 1, 2, \ldots, m;$$

$$\sum_{j=1}^{n} \lambda_j y_{rj} \geq y_{ro} \quad r = 1, 2, \ldots, s; \quad (2)$$

$$\lambda_j \geq 0 \quad j = 1, 2, \ldots, n.$$

In the discussion to follow, we suppose the imprecise data take the forms of bounded data, ordinal data, and ratio bounded data as follows:

Interval or Bounded data

$$\underline{y}_{rj} \leq y_{rj} \leq \overline{y}_{rj} \text{ and } \underline{x}_{ij} \leq x_{ij} \leq \overline{x}_{ij} \text{ for } r \in \boldsymbol{BO}, i \in \boldsymbol{BI} \quad (3)$$

where $\underline{y}_{rj}$ and $\underline{x}_{ij}$ are the lower bounds and $\overline{y}_{rj}$ and $\overline{x}_{ij}$ are the upper bounds, and $\boldsymbol{BO}$ and $\boldsymbol{BI}$ represent the associated sets for bounded outputs and bounded inputs respectively.

Weak ordinal data

$$y_{rj} \leq y_{rk} \text{ and } x_{ij} \leq x_{ik} \text{ for } j \neq k, r \in \boldsymbol{DO}, i \in \boldsymbol{DI}$$

or to simplify the presentation,
$$y_{r1} \leq y_{r2} \leq \ldots \leq y_{rk} \leq \ldots \leq y_{rn} \ (r \in \boldsymbol{DO}) \quad (4)$$
$$x_{i1} \leq x_{i2} \leq \ldots \leq x_{ik} \leq \ldots \leq x_{in} \ (i \in \boldsymbol{DI}) \quad (5)$$

where $\boldsymbol{DO}$ and $\boldsymbol{DI}$ represent the associated sets for weak ordinal outputs and inputs respectively.

Strong ordinal data
$$y_{r1} < y_{r2} < \ldots < y_{rk} < \ldots < y_{rn} \ (r \in \boldsymbol{SO}) \quad (6)$$
$$x_{i1} < x_{i2} < \ldots < x_{ik} < \ldots < x_{in} \ (i \in \boldsymbol{SI}) \quad (7)$$

where $\boldsymbol{DO}$ and $\boldsymbol{DI}$ represent the associated sets for strong ordinal outputs and inputs respectively.

Ratio bounded data

$$L_{rj} \le \frac{y_{rj}}{y_{rj_o}} \le U_{rj} \ (j \ne j_o)(r \in \mathbf{RO}) \tag{8}$$

$$G_{ij} \le \frac{x_{ij}}{x_{ij_o}} \le H_{ij} \ (j \ne j_o) \ (i \in \mathbf{RI}) \tag{9}$$

where L_{rj} and G_{ij} represent the lower bounds, and U_{rj} and H_{ij} represent the upper bounds. **RO** and **RI** represent the associated sets for ratio bounded outputs and inputs respectively.

If we incorporate (3)-(9) into model (1), we have the multiplier IDEA (MIDEA) model

$$\max \pi_o = \sum_{r=1}^{s} \mu_r y_{ro}$$

$$s.t. \ \sum_{r=1}^{s} \mu_r y_{rj} - \sum_{i=1}^{m} \omega_i x_{ij} \le 0$$

$$\sum_{i=1}^{m} \omega_i x_{io} = 1 \tag{10}$$

$$(x_{ij}) \in \Theta_i^-$$

$$(y_{rj}) \in \Theta_r^+$$

$$\mu_r, \omega_i \ge 0$$

where $(x_{ij}) \in \Theta_i^-$ and $(y_{rj}) \in \Theta_r^+$ represent any of or all of (3)-(9).

If we incoporate (3)-(9) into model (2), we then have the envelopment IDEA (EIDEA) model. Obviously, model (10) is non-linear and non-convex, because some of the outputs and inputs become unknown decision variables. We will discuss how to solve these two non-linear IDEA models.

Cooper, Park and Yu (1999) and Kim, Park and Park (1999) show that model (10) can be converted into the following linear programming problem when scale transformations and variable alternations are applied:

$$\max \sum_r Y_{r0}$$

subject to

$$\sum_r Y_{rj} - \sum_i X_{ij} \le 0 \; \forall j$$

$$\sum_i X_{i0} = 1 \tag{11}$$

$$(X_{ij}) \in H_i^- \; \forall i$$

$$(Y_{rj}) \in H_r^+ \; \forall r$$

$$X_{ij}, Y_{rj} \ge 0 \; \forall i, r$$

where $X_{ij} = \hat{x}_{ij} \hat{\omega}_i$, $Y_{rj} = \hat{y}_{rj} \hat{\mu}_r$, $\hat{\omega}_i = \omega_i \cdot \max_j \{x_{ij}\}$, $\hat{\mu}_r = \mu_r \cdot \max_j \{y_{rj}\}$, $\hat{x}_{ij} = x_{ij} / \max_j \{x_{ij}\}$, $\hat{y}_{rj} = y_{rj} / \max_j \{y_{rj}\}$, $X_{ij}^o = \hat{x}_{ij}^o \hat{\omega}_i$, $Y_{rj}^o = \hat{y}_{rj}^o \hat{\mu}_r$, $\hat{x}_{ij}^o = \max_j \{\hat{x}_{ij}\}$, and $\hat{y}_{rj}^o = \max_j \{\hat{y}_{rj}\}$. Also, Θ_r^+ and Θ_i^- are transformed into H_r^+ and H_i^-.

Obviously, the standard (linear) CCR DEA model cannot be used and a set of special computation codes is needed for each evaluation, since a different objective function ($\sum Y_{ro}$) and a new constraint ($\sum X_{io}$) are present in model (11) for each DMU under evaluation. Note also that the number of new variables (Y_{rj} and X_{ij}) increases substantially as the number of DMUs increases.

Zhu (2003a) provides an improvement by only using variable alternations. That is, define $X_{ij} = x_{ij} \omega_i$, $Y_{rj} = y_{rj} \mu_r$ in model (1) when imprecise data are present, and the scale transformation is not needed. This simple approach is actually used in Cook, Kress and Seiford (1993). The interested reader is referred to Cook and Zhu (2006) and Chapter 2 for the detailed discussion and a general framework for dealing with ordinal data. The interested reader is also referred to Cooper and Park (2006) for a more detailed discussion on the above IDEA approach.

3. MULTIPLIER IDEA (MIDEA): STANDARD DEA MODEL APPROACH

The following theorem provides the theoretical foundation to the approach developed in Zhu (2003a; 2004) when the standard multiplier CCR model (1) is used to solve the IDEA model (10).

Theorem 1: Suppose Θ_r^+ and Θ_i^- are given by (3), then for DMU_o the optimal value to (10) can be achieved at $y_{ro} = \bar{y}_{ro}$ and $x_{io} = \underline{x}_{io}$ for DMU_o and $y_{rj} = \underline{y}_{rj}$ and $x_{ij} = \bar{x}_{ij}$ for DMU_j ($j \neq o$).

[Proof] Suppose we have a set of optimal solutions associated with π_o^* in (1), u_r^*, v_i^*, y_{rj}^* ($r \in BO$) and x_{ij}^* ($i \in BI$) such that

$$\sum_{r \in BO} \mu_r^* y_{rj}^* + \sum_{r \notin BO} \mu_r^* y_{rj} - \sum_{i \in BI} \omega_i^* x_{ij}^* - \sum_{i \notin BI} \omega_i^* x_{ij} \leq 0 \ (\forall \ j), \ \sum_{i \in BI} \omega_i^* x_{io}^* + \sum_{i \notin BI} \omega_i^* x_{io}$$

$= 1$, $\underline{y}_{rj} \leq y_{rj}^* \leq \bar{y}_{rj}$ ($r \in BO$) and $\underline{x}_{ij} \leq x_{ij}^* \leq \bar{x}_{ij}$ ($i \in BI$).

Now, on the basis of this set of optimal solutions, we define

$$\mu_r = \mu_r^* \frac{y_{ro}^*}{\bar{y}_{ro}} \ (r \in BO), \ \mu_r = \mu_r^* \ (r \notin BO), \ v_i = v_i^* \frac{x_{io}^*}{\underline{x}_{io}} \ (i \in BI) \text{ and } v_i$$

$= v_i^*$ ($i \notin BI$). Thus, for $j \neq o$, we have

$$\sum_{r \in BO} \mu_r \underline{y}_{rj} + \sum_{r \notin BO} \mu_r y_{rj} - \sum_{i \in BI} \omega_i \bar{x}_{ij} - \sum_{i \notin BI} \omega_i x_{ij} = \sum_{r \in BO} \mu_r^* y_{rj}^* \frac{y_{ro}^*}{\bar{y}_{ro}} \frac{\underline{y}_{rj}}{y_{rj}^*} +$$

$$\sum_{r \notin BO} \mu_r^* y_{rj} - \sum_{i \in BI} \omega_i^* x_{ij}^* \frac{x_{io}^*}{\underline{x}_{io}} \frac{\bar{x}_{ij}}{x_{ij}^*} - \sum_{i \notin BI} \omega_i^* x_{ij} \leq$$

$$\sum_{r \in BO} \mu_r^* y_{rj}^* + \sum_{r \notin BO} \mu_r^* y_{rj} - \sum_{i \in BI} \omega_i^* x_{ij}^* - \sum_{i \notin BI} \omega_i^* x_{ij}$$

and for $j = o$, we have

$$\sum_{r \in BO} \mu_r \bar{y}_{ro} + \sum_{r \notin BO} \mu_r y_{ro} - \sum_{i \in BI} \omega_i \underline{x}_{io} - \sum_{i \notin BI} \omega_i x_{io} =$$

$$\sum_{r \in BO} \mu_r^* y_{rj}^* + \sum_{r \notin BO} \mu_r^* y_{rj} - \sum_{i \in BI} \omega_i^* x_{ij}^* - \sum_{i \notin BI} \omega_i^* x_{ij} = \pi_o^* - 1.$$

Therefore, u_r, v_i, $\bar{y}_{ro}$ ($r \in BO$), $\underline{y}_{rj}$ ($j \neq o, r \in BO$), $\underline{x}_{io}$ ($i \in BI$), and $\bar{x}_{ij}$ ($j \neq o, i \in BI$) are also optimal. This completes the proof.

Theorem 1 is true due to that fact that increases on output values (decreases on input values) for DMU_o under evaluation or (and) decreases on output values (increases on input values) for other DMUs will not deteriorate the efficiency of DMU_o under evaluation by the multiplier DEA model.

3.1 Converting the bounded data into a set of exact data

Theorem 1 shows that when DMU_o is under evaluation, we can have a set of exact data via setting $y_{ro} = \bar{y}_{ro}$ and $x_{io} = \underline{x}_{io}$ for DMU_o and $y_{ij} = \underline{y}_{rj}$ and $x_{ij} = \bar{x}_{ij}$ for DMU_j ($j \neq o$) while model (10) maintains the efficiency rating for DMU_o. Note that in this case, model (10) is no longer a non-linear program, but a (linear) multiplier CCR model

$\pi_o^* =$ Maximize $\sum_{r \in BO} \mu_r \bar{y}_{ro} + \sum_{r \notin BO} \mu_r y_{ro}$
subject to

$$\sum_{r \in BO} \mu_r \underline{y}_{rj} + \sum_{r \notin BO} \mu_r y_{rj} - \sum_{i \in BI} \omega_i \bar{x}_{ij} - \sum_{i \notin BI} \omega_i x_{ij} \leq 0 \quad \forall \ j \neq o$$

$$\sum_{r \in BO} \mu_r \bar{y}_{ro} + \sum_{r \notin BO} \mu_r y_{ro} - \sum_{i \in BI} \omega_i \underline{x}_{io} - \sum_{i \notin BI} \omega_i x_{io} \leq 0 \quad (12)$$

$$\sum_{i \in BO} \omega_i \underline{x}_{io} + \sum_{i \notin BO} \omega_i x_{io} = 1$$

$$\mu_r, \omega_i \geq 0 \quad \forall \ r, i$$

where y_{rj} ($r \notin \mathbf{BO}$), and x_{ij} ($i \notin \mathbf{BI}$) are exact data.

We can also use the obtained exact data and apply them to the envelopment model (2), namely

$\theta_o^* = \min \theta_o$
subject to

$$\sum_{j \neq o} \lambda_j \bar{x}_{ij} + \lambda_o \underline{x}_{io} \leq \theta_o \underline{x}_{io} \quad i \in BI;$$

$$\sum_{j=1}^{n} \lambda_j x_{ij} \leq \theta_o x_{io} \quad i \notin BI; \quad (13)$$

$$\sum_{j \neq o} \lambda_j \underline{y}_{rj} + \lambda_o \bar{y}_{ro} \geq \bar{y}_{ro} \quad r \in BO;$$

$$\sum_{j=1}^{n} \lambda_j y_{rj} \geq y_{ro} \quad r \notin BO;$$

$$\lambda_j \geq 0 \quad j = 1, 2, \ldots, n.$$

3.2 Converting the weak ordinal data into a set of exact data

Consider DMU_k. Suppose we solve model (10) when Θ_i^- and Θ_r^+ are in forms of (4) and (5), and obtain a set of optimal solutions y_{rj}^* and x_{ij}^* with the optimal value π_k^*. We have

$$y_{r1}^* \le y_{r2}^* \le \ldots \le y_{r,k-1}^* \le y_{rk}^* \le y_{r,k+1}^* \le \ldots \le y_{rn}^* \quad (r \in DO) \tag{14}$$

$$x_{i1}^* \le x_{i2}^* \le \ldots \le x_{i,k-1}^* \le x_{ik}^* \le x_{i,k+1}^* \le \ldots \le x_{in}^* \quad (i \in DI) \tag{15}$$

Note that $\rho\, y_{rj}^*$ ($r \in DO$) and $\rho\, x_{ij}^*$ ($i \in DI$) are also optimal for DMU_k where ρ is a positive constant, because of the units invariant property. Therefore, we can always set $y_{rk}^* = x_{ik}^* = 1$. Then, we have a set of optimal solutions on weak ordinal outputs and inputs such that (14) and (15) can be expressed as[3]

$$0 \le y_{r1}^* \le y_{r2}^* \le \ldots \le y_{r,k-1}^* \le y_{rk}^* (=1) \le y_{r,k+1}^* \le \ldots \le y_{rn}^* \le M \quad (r \in DO) \tag{16}$$

$$0 \le x_{i1}^* \le x_{i2}^* \le \ldots \le x_{i,k-1}^* \le x_{ik}^* (=1) \le x_{i,k+1}^* \le \ldots \le x_{in}^* \le M \quad (i \in DI) \tag{17}$$

where M is very close to $+\infty$.[4]

Now, for the outputs and inputs in weak ordinal relations, we set up the following intervals,

$$y_{rj} \in [0, 1] \text{ and } x_{ij} \in [0, 1] \text{ for } DMU_j \ (j = 1, \ldots, k-1) \tag{18}$$

$$y_{rj} \in [1, M] \text{ and } x_{ij} \in [1, M] \text{ for } DMU_j \ (j = k+1, \ldots, n) \tag{19}$$

Based upon Theorem 1, we know that for $r \in DO$ and $i \in DI$, π_k^* remains the same and (18) and (19) are satisfied if $y_{rk} = x_{ik} = 1$ for DMU_k and $y_{rj} = 0$ (lower bound, $\underline{y}_{rj}$), $x_{ij} = 1$ (upper bound, $\bar{x}_{ij}$) for DMU_j ($j = 1$,

---

[3] This procedure appears to be unworkable when weight restrictions are present. However, we will see in Theorem 2, such weight restrictions are redundant and should be removed before the analysis. As a result, the current procedure is not affected.

[4] In computation, M does not have to be set equal to a very large number. In the application section in this chapter, M is set equal to 33.

..., k-1) and $y_{rj} = 1$ (lower bound, $\underline{y}_{rj}$), $x_{ij} = M$ (upper bound, $\bar{x}_{ij}$) for DMU_j $(j = k+1, ..., n)^5$.

3.3 Numerical Illustration

Table 3-1 presents the data set used by Cooper, Park and Yu (1999). Suppose we have cost and judgment as two inputs and revenue as the only output. Based on Theorem 1, we use the lower bound of judgment input as the exact input value for each DMU under evaluation and the upper bounds as the exact input values for other DMUs. For example, for DMU1, we use $x_{21} = 0.6$ (lower bound) and $x_{22} = 0.9$, $x_{24} = 0.8$ (upper bounds). In addition to the efficiency scores, Table 3-2 presents the slacks and referent DMUs based upon model (13).

Table 3-1. Exact and Imprecise Data

	Outputs		Inputs	
	Exact	Ordinal	Exact	Bound
	Revenue (y_{1j})	Satisfaction (y_{2j})	Cost (x_{1j})	Judgment (x_{2j})
DMU1	2000	4	100	[0.6, 0.7]
DMU2	1000	2	150	[0.8, 0.9]
DMU3	1200	5	150	1
DMU4	900	1	200	[0.7, 0.8]
DMU5	600	3	200	1

Source: Cooper, Park and Yu (1999).

Table 3-2. MIDEA Results When Bounded Data are Present[†]

DMU	Efficiency score θ_o^*	slack Cost	Referent DMU λ_j^*
1	1	0	$\lambda_1^* = 1$
2	0.4375	15.625	$\lambda_1^* = 0.5$
3	0.42	3	$\lambda_1^* = 0.6$
4	0.45	45	$\lambda_1^* = 0.45$
5	0.21	12	$\lambda_1^* = 0.3$

† Model (13) is used with two inputs of cost and judgment and one output of revenue. The ordinal output of satisfaction is not included in calculations.

Note that it is very difficult to retrieve the optimal values on the bounded input (output) if one uses the variable-alternation algorithm. However, based upon Theorem 2 and the recent development on sensitivity analysis by Zhu (2001), we can determine the range of multiple optimal solutions on

---

[5] See Chen (2006) for detailed discussion and alternative ways of setting the exact data when weak ordinal relations are present.

bounded data for DMU_o (and other DMUs). That is, we calculate the following linear program (Zhu, 2001),

$$\tilde{\theta}_o^* = \min \tilde{\theta}_o$$

subject to

$$\sum_{j \neq o} \lambda_j \bar{x}_{ij} + \lambda_o \underline{x}_{io} \leq \tilde{\theta}_o x_{io} \quad i \in BI;$$

$$\sum_{j \neq o} \lambda_j x_{ij} \leq x_{io} \quad i \notin BI \text{ (exact data)}; \quad (20)$$

$$\sum_{j \neq o} \lambda_j \underline{y}_{rj} + \lambda_o \bar{y}_{ro} \geq \bar{y}_{ro} \quad r \in BO;$$

$$\sum_{j \neq o} \lambda_j y_{rj} \geq y_{ro} \quad r \notin BO \text{ (exact data)};$$

$$\lambda_j \geq 0 \quad j \neq o$$

Consider DMU1 in Table 3-1, we have

$$\min \tilde{\theta}_1$$
subject to
$$150\lambda_2 + 150\lambda_3 + 200\lambda_4 + 200\lambda_5 \leq 100 \quad \text{(Cost)}$$
$$0.9\lambda_2 + 1\lambda_3 + 0.8\lambda_4 + 1\lambda_5 \leq 0.6\tilde{\theta}_1 \quad \text{(Judgement)}$$
$$1000\lambda_2 + 1200\lambda_3 + 900\lambda_4 + 600\lambda_5 \geq 2000 \quad \text{(Revenue)}$$
$$\lambda_j \geq 0, j = 2,3,4,5.$$

Now, suppose the lower bound of judgment input for DMU1 can be increased by σ, and the upper bounds of judgment input for other DMUs can be decreased by σ'. Based on Zhu (2001), if $\sigma \sigma' = \tilde{\theta}_1^* = 25/9$, then DMU1 remains efficient ($\pi_1^* = \theta_1^* = 1$). Thus, the efficiency stability region for judgment input is larger than the range of [0.6, 0.7] for DMU1. When DMU1 is under evaluation, any DMU1's judgment input value within the range of [0.6, 0.7] is an optimal solution such that $\pi_1^* = \theta_1^* = 1$ remains true. Table 3-3 reports the optimal values of judgment input by using model (20). As a result, the (multiple) optimal solutions on the bounded input can be retrieved for the variable-alternation algorithm.

We next convert the ordinal data into a set of exact data using three DMUs, namely, DMU4, DMU3 and DMU5, in Table 3-1. We have (i) DMU4 has the lowest rank. We use "1" as the satisfaction output value for all DMUs; (ii) DMU3 has the highest rank. We use "1" for the satisfaction output value for DMU3 and use 0 for other DMUs; and (iii) DMU5 is ranked third. We use "1" for the satisfaction output value for DMUs 2, 4 and 5 and use "0" for other DMUs.

Table 3-3. Alternative Optimal Solutions

	DMU Under Evaluation				
	DMU1	DMU2	DMU3	DMU4	DMU5
DMU1	[0.6, 0.7]	0.7	0.7	0.7	0.7
DMU2	[0.8, 0.9]	0.8	[0.8, 0.9]	[0.8, 0.9]	[0.8, 0.9]
DMU4	[0.7, 0.8]	[0.7, 0.8]	[0.7, 0.8]	0.7	[0.7, 0.8]

Table 3-4. Converting Ordinal Data Into Exact Data

	DMU under evaluation				
Satisfaction	DMU1	DMU2	DMU3	DMU4	DMU5
y_{21}	1	1	0	1	1
y_{22}	0	1	0	1	0
y_{23}	1	1	1	1	1
y_{24}	0	0	0	1	0
y_{25}	0	1	0	1	1

Table 3-5. MIDEA Results

	Efficiency score	slack		Referent DMU
DMU	θ_o^*	Revenue	Cost	λ_j^*
1	1	0	0	$\lambda_1^* = 1$
2	0.875	1000	31.25	$\lambda_1^* = 1$
3	1	0	0	$\lambda_3^* = 1$
4	1	1100	100	$\lambda_1^* = 1$
5	0.7	1400	40	$\lambda_1^* = 1$

Table 3-4 reports the set of exact data for the satisfaction output across all five DMUs when a specific DMU is under evaluation. Table 3-5 reports the results from model (13) when we use the exact data from Table 3-4. It can be seen that both MIDEA approaches yield the identical efficiency scores. The standard DEA approach indicates that DMUs 2, 4 and 5 have non-zero slack values.

3.4 Application

This section applies this standard DEA-based MIDEA approach to the 33 telephone offices in Kim, Park and Park (1999). Table 3-6 reports the data with man power (x_1), operating costs (x_2) and number of telephone lines (x_3) as the three inputs and local revenues (y_1), long distance revenues (y_2), international revenues (y_3), operation/maintenance level (y_4) and customer satisfaction (y_5) as the five outputs. Note that y_4 and y_5 are ordinal data. With respect to Kim, Park and Park (1999), the current paper reports y_4 differently with "1" for the worst and "33" for the best ("5" for the best in y_5), since larger output values are preferred in DEA.

Table 3-6. Data for Telephone Offices

DMU	x_1	x_2	x_3	y_1	y_2	y_3	y_4 [a]	y_5 [b]
1	239	7.03	158	47.1	16.67	34.04	28	2
2	261	3.94	163	37.47	14.11	19.97	26	3
3	170	2.1	90	20.7	6.8	12.64	19	3
4	290	4.54	201	41.82	11.07	6.27	23	4
5	200	3.99	140	33.44	9.81	6.49	30	2
6	283	4.65	214	42.43	11.34	5.16	21	4
7	286	6.54	197	47.03	14.62	13.04	9	2
8	375	6.22	314	55.48	16.39	7.31	14	1
9	301	4.82	257	49.2	16.15	6.33	8	3
10	333	6.87	235	47.12	13.86	6.51	6	2
11	346	6.46	244	49.43	15.88	8.87	18	2
12	175	2.06	112	20.43	4.95	1.67	32	5
13	217	4.11	131	29.41	11.39	4.38	33	2
14	441	7.71	214	61.2	25.59	33.01	16	3
15	204	3.64	163	32.27	9.57	3.65	15	4
16	216	3.24	154	32.81	11.46	9.02	25	2
17	347	5.65	301	59.01	17.82	8.19	29	1
18	288	4.66	212	42.27	14.52	7.33	24	4
19	185	3.37	178	32.95	9.46	2.91	7	2
20	242	5.12	270	65.06	24.57	20.72	17	1
21	234	2.52	126	31.55	8.55	7.27	27	2
22	204	4.24	174	32.47	11.15	2.95	22	3
23	356	7.95	299	66.04	22.25	14.91	13	2
24	292	4.52	236	49.97	14.77	6.35	12	3
25	141	5.21	63	21.48	9.76	16.26	11	2
26	220	6.09	179	47.94	17.25	22.09	31	2
27	298	3.44	225	42.35	11.14	4.25	4	2
28	261	4.3	213	41.7	11.13	4.68	20	5
29	216	3.86	156	31.57	11.89	10.48	3	3
30	171	2.45	150	24.09	9.08	2.6	10	5
31	123	1.72	61	11.97	4.78	2.95	5	1
32	89	0.88	42	6.4	3.18	1.48	2	5
33	109	1.35	57	10.57	3.43	2	1	4

Source: Kim et al. (1999). [a] Ordinal ranks (33 = the best; 1 = the worst). [b] Five ordinal scales (5 = the best; 1 = the worst).

Table 3-7. Exact Data When DMU29 is under Evaluation

DMU	Weak ordinal relations		Strong ordinal relations		Ratio bounded data	
	y_4	y_5	y_4	y_5	y_4	y_5
1	33	0	5.42743	0	5.42743	0.45662
2	33	33	4.74053	1	4.74053	1
3	33	33	2.95216	1	2.95216	1
4	33	33	3.86968	1.68000	3.86968	1.68000
5	33	0	6.21387	0	6.21387	0.45662
6	33	33	3.37993	1.68000	3.37993	1.68000
7	33	0	1.50073	0	1.50073	0.45662
8	33	0	2.10485	0	2.10485	0.20850
9	33	33	1.40255	1	1.40255	1
10	33	0	1.22504	0	1.22504	0.45662
11	33	0	2.75903	0	2.75903	0.45662
12	33	33	7.11426	2.82240	7.11426	2.82240
13	33	0	7.61226	0	7.61226	0.45662
14	33	33	2.40985	1	2.40985	1
15	33	33	2.25219	1.68000	2.25219	1.68000
16	33	0	4.43040	0	4.43040	0.45662
17	33	0	5.80735	0	5.80735	0.20850
18	33	33	4.14056	1.68000	4.14056	1.68000
19	33	0	1.31080	0	1.31080	0.45662
20	33	0	2.57853	0	2.57853	0.20850
21	33	0	5.07237	0	5.07237	0.45662
22	33	33	3.61653	1	3.61653	1
23	33	0	1.96715	0	1.96715	0.45662
24	33	33	1.83846	1	1.83846	1
25	33	0	1.71819	0	1.71819	0.45662
26	33	0	6.64884	0	6.64884	0.45662
27	33	0	1.07000	0	1.07000	0.45662
28	33	33	3.15882	2.82240	3.15882	2.82240
29	33	33	1	1	1	1
30	33	33	1.60578	2.82240	1.60578	2.82240
31	33	0	1.14490	0	1.14490	0.20850
32	0	33	0	2.82240	0.90909	2.82240
33	0	33	0	1.68000	0.82645	1.68000

Table 3-8. Efficiency Scores without Weight Restrictions

	Weak ordinal relations		Strong ordinal relations		Ratio bounded data	
DMU	efficiency	RTS	efficiency	RTS	efficiency	RTS
1	1	CRS	1	CRS	1	CRS
2	1	CRS	1	CRS	1	CRS
3	1	CRS	1	CRS	1	CRS
4	0.94322	DRS	0.88097	DRS	0.86641	DRS
5	1	DRS	0.99714	IRS	0.98523	IRS
6	0.91747	DRS	0.85080	DRS	0.83543	DRS
7	0.87524	DRS	0.87167	DRS	0.86893	DRS
8	0.72117	DRS	0.72117	DRS	0.72117	DRS
9	0.88812	DRS	0.83096	DRS	0.82752	DRS
10	0.76098	DRS	0.75611	DRS	0.75260	DRS
11	0.78351	DRS	0.77986	DRS	0.77699	DRS
12	1	CRS	1	CRS	1	CRS
13	1	CRS	1	CRS	1	CRS
14	1	CRS	1	CRS	1	CRS
15	1	CRS	0.85782	DRS	0.84854	DRS
16	1	CRS	0.96872	IRS	0.94574	IRS
17	0.91257	DRS	0.88730	DRS	0.86063	DRS
18	0.94468	DRS	0.85697	DRS	0.84235	DRS
19	1	CRS	0.79177	IRS	0.78686	IRS
20	1	CRS	1	CRS	1	CRS
21	1	CRS	1	CRS	1	CRS
22	1	CRS	0.80056	IRS	0.78029	IRS
23	0.86482	DRS	0.85685	DRS	0.85363	DRS
24	0.95990	DRS	0.90019	DRS	0.89546	DRS
25	1	CRS	1	CRS	1	CRS
26	1	CRS	1	CRS	1	CRS
27	0.97963	DRS	0.97963	DRS	0.97858	DRS
28	0.97363	DRS	0.90326	DRS	0.88385	DRS
29	0.95283	DRS	0.81847	IRS	0.81728	IRS
30	1	CRS	0.93254	DRS	0.92583	DRS
31	1	CRS	0.79833	IRS	0.79414	IRS
32	1	CRS	1	CRS	1	CRS
33	1	CRS	0.88997	IRS	0.88288	IRS

We set M equal to 33 for the two outputs in weak ordinal relations. Based upon (18) and (19) and Theorem 1, we can have a set of exact data when a telephone office is under evaluation. (The second and third columns of

Table 3-7 report a set of exact data on y_4 and y_5 when DMU29 is under evaluation.)

We use model (13) to analyze the efficiency, because in addition to the efficiency scores, it provides (i) the efficient projections; (ii) benchmarks with magnitudes; and (iii) information on RTS classification[6]. The second column of Table 3-8 records the efficiency scores under the assumption of weak rankings. It can be seen that our approach yields the same efficient DMUs and larger efficiency scores for inefficient DMUs compared to those in Kim, Park and Park (1999) where so called strict rankings (strong ordinal relations) are assumed[7]

3.5 Converting the strong ordinal data and ratio bounded data into a set of exact data

Recall that $y_{rk} - y_{r,k-1} \geq \chi$ and $x_{ik} - x_{i,k-1} \geq \chi$ are not valid forms to represent strict rankings under model (10). We propose the following correct and valid modifications when DMU_k is under evaluation

$$y_{r,k+1} \geq \chi_r y_{r,k} \ (\chi_r > 1) \text{ and } x_{ik} \geq \eta_i x_{i,k-1} \ (\eta_i > 1) \qquad (21)$$

Since (21) is units invariant, it allows scale transformations. $y_{r,k+1} \geq \chi_r y_{r,k}$ and $x_{ik} \geq \eta_i x_{i,k-1}$ in (10) are equivalent to $Y_{r,k+1} \geq \chi_r Y_{r,k}$ and $X_{ik} \geq \eta_i X_{i,k-1}$ in (11) respectively, if the scale-transformation and variable-alternation based approach is used. Note that (21) may allow all data equal to zero. However, the proposed method of finding exact data does not allow such cases to occur.

Based upon the discussion on converting weak ordinal data into a set of exact data, we can set $y_{rk} = 1$ and $x_{ik} = 1$, and further we have[8]

$$\begin{cases} y_{rj} = \chi_r^{j-k} \text{ for } DMU_j \ (j = k+1, ..., n) \\ x_{ij} = \eta_i^{j-k} \text{ for } DMU_j \ (j = 1, ..., k-1) \end{cases}$$

when χ_r and η_i are given.

Furthermore, parts of (8) and (9) actually represent strong ordinal relations when $L_{rj} = \chi_r$ or $1/H_{ij} = \eta_i$.

---

[6] Because of the possible multiple optimal solutions, method described in Zhu and Shen (1995) is used. (See Seiford and Zhu (1999) for a detailed discussion on RTS estimation.)

[7] This indicates that optimal solution is not achieved in Kim, Park and Park (1999). As noted in Zhu (2003a;b), this is due to the fact that incorrect forms of ordinal relations are used in Kim, Park and Park (1999).

[8] We can easily select a set of exact data for y_{rj} (j = 1, ..., k) and x_{ij} (j = k+1,..., n). For example, we can set these y_{rj} very close to zero and meanwhile (22) is satisfied.

In Kim, Park and Park (1999), $L_4 = 1.07$ for the forth output and $L_5 = 8.04$ for the fifth output. Thus, if only strong ordinal relations are assumed, we have

$$y_{r,k+1} \geq L_r\, y_{r,k} \quad \text{or} \quad \frac{y_{r,k+1}}{y_{r,k}} \geq L_r \quad (r = 4, 5) \tag{22}$$

When DMU_k is under evaluation, we let $y_{rk} = 1$ and we have a set of exact data consisting of (i) $y_{rj} = L_r^{j-k}$ ($r = 4, 5; j = k+1, \ldots, n$) and (ii) $y_{rj} = \varepsilon_j \approx 0$ such that $y_{r,j+1} \geq L_r\, y_{rj}$ ($j = 1, \ldots, k-1$). The fourth and fifth column of Table 3-2 present a set of exact data on y_4 and y_5 when DMU29 is under evaluation and strong ordinal relations in (22) are imposed.

The fourth and fifth columns in Table 3-8 report the efficiency scores and the RTS classification when (correct and valid) strong ordinal rankings are assumed. It can be seen that 8 efficient DMUs under weak ordinal relations become inefficient under strong ordinal relations.

Moreover, note that if we assume strong ordinal relations as in (22), too much flexibility may still be allowed in $y_{rj} = \varepsilon_j \approx 0$ ($j = 1, \ldots, k-1$). Therefore, we introduce

$$y_{r,k} \leq U_r\, y_{r,k-1} \quad \text{or} \quad \frac{y_{r,k}}{y_{r,k-1}} \leq U_r \quad (r = 4, 5) \tag{23}$$

to further restrict the values on y_{rj} ($j = 1, \ldots, k-1$).

Ratio bounded data (22) and (23) can also be converted into a set of exact data via the following two steps.

Step 1: Set $y_{rj_o} = 1$ and $x_{ij_o} = 1$

Step 2: We have bounded data for other DMUs: $L_{rj} \leq y_{rj} \leq U_{rj}$ and $G_{ij} \leq x_{ij} \leq H_{ij}$ ($j \neq j_o$) which can further be converted into exact data.

Step 1 is valid is because there are no other constraints on y_{rj_o} and x_{ij_o}. However, if y_{rj_o} and x_{ij_o} can take values within certain ranges as given in (3), we have two cases associated with step 1. (Case 1) If DMU_{j_o} is under evaluation, we set $y_{rj_o} = \overline{y}_{rj_o}$ and $x_{ij_o} = \underline{x}_{ij_o}$. (Case 2) If DMU_{j_o} is not under evaluation, we set $y_{rj_o} = \underline{y}_{rj_o}$ and $x_{ij_o} = \overline{x}_{ij_o}$.

Now, when DMU_k is under evaluation, if we set $y_{rk} = 1$, then we have $y_{rj} = 1/U_r^{k-j}$ ($j = 1, \ldots, k-1$) where $r = 4$ and 5 and $U_4 = 1.10$ and $U_5 = 23.04$. The last two columns of Table 3-7 report a set of exact data on y_4 and y_5 when DMU29 is under evaluation and when ratio scale bounds are introduced.

When ratio bounded data are imposed, the efficient DMUs stay the same and the efficiency scores for inefficient DMUs only slightly drop (see the sixth column in Table 3-8) compared to those under strong ordinal relations. Note that the objective of model (10) is maximization and better efficiency ratings are obtained in the current study compared to those in Kim, Park and Park (1999)[9].

Turning to the RTS classifications in Table 3-8, note that when weak ordinal relations are replaced by strong ones, the RTS classification of some DMUs changes. However, the RTS classification stays the same when strong ordinal relations are replaced by ratio bounded data. This may be due to the fact that ratio bounded can be viewed as strong ordinal relations.

4. TREATMENT OF WEIGHT RESTRICTIONS

The above discussion and the proposed method assume that weight restrictions related to imprecise data in forms (4)-(9) are not present. The following theorem shows that adding weight restrictions related to imprecise outputs and inputs in forms (4)-(9) does not change the efficiency ratings. Thus, these particular weight restrictions are redundant and can be removed before solving the model (10). As a result, the standard DEA model based approach can be used.

Theorem 2: Suppose the *f*th input and the *d*th output are imprecise data given by (4)-(9), and π_o^* is the optimal value to (10), then π_o^* remains unchanged if the following weight restrictions related to the *f*th input and *d*th output are added into model (10)

$$\alpha_f \leq \frac{f(\mu_r, r \notin DO \cap RO)}{\mu_f} \leq \beta_f \quad (f \in DO \text{ or } RO) \quad (24)$$

$$A_d \leq \frac{f(\omega_i, i \notin DI \cap RI)}{\omega_d} \leq B_d \quad (d \in DI \text{ or } RI) \quad (25)$$

where $f(\bullet)$ is a function on μ_r (ω_i) related to exact outputs (inputs).

[Proof]: See Zhu (2003a).[10]

Theorem 2 indicates that the optimal value to model (10) is equal to the optimal value to the following model (model (10) with weight restrictions (24) and (25))

---

[9] This implies that Kim, Park and Park (1999) did not provide the optimized results.
[10] This theorem is true because (4)-(9) are units invariant.

$$\max \pi_o = \sum_{r=1}^{s} \mu_r y_{ro}$$

$$\text{s.t.} \quad \sum_{r=1}^{s} \mu_r y_{rj} - \sum_{i=1}^{m} \omega_i x_{ij} \leq 0$$

$$\sum_{i=1}^{m} \omega_i x_{io} = 1$$

$$(x_{ij}) \in \Theta_i^-$$
$$(y_{rj}) \in \Theta_r^+ \tag{26}$$

$$\alpha_f \leq \frac{f(\mu_r, r \notin DO \cap RO)}{\mu_f} \leq \beta_f \quad f \in DO \text{ or } RO$$

$$A_d \leq \frac{f(\omega_i, i \notin DI \cap RI)}{\omega_d} \leq B_d \quad d \in DI \text{ or } RI$$

$$\mu_r, \omega_i \geq 0$$

In other words, the same efficiency ratings can be obtained by either solving model (10) or model (25). If one obtains the efficiency ratings under model (10), the same efficiency ratings are obtained for model (26). The method developed in the previous sections can be applied to solve model (10) without affecting the efficiency ratings. As a result, model (26) is solved indirectly.

Although we cannot set exact data for partial data in model (26), Theorem 2 provides an alternative where the efficiency ratings under (26) are obtained via setting exact data in model (10) and solving model (10). i.e., the proposed approach is not affected by the presence of weight restrictions, since solving model (10) with proposed approach is equivalent to solving model (26) directly[11]. To obtain the efficiency ratings under model (26), we only need to solve model (10) via setting exact data in model (10), because the weight restrictions represented by (24) and (25) are redundant.

Note that Kim, Park and Park (1999) incorporate the six sets of weight restrictions. Of which two are given as

---

[11] Note that the following two cases are different. Case I: Setting the variables by exact data in model (10) and then finding the efficiency ratings. (This provides the same efficiency ratings as those obtained from solving model (26) directly.) Case II: Setting the variables by a set of exact data in model (26) and finding the efficiency scores under model (26) (This leads to a different problem.) Case I represents the objective of the current study or the objective of solving model (26).

$$3.5 \leq \frac{\mu_1 + \mu_2 + \mu_3}{\mu_4} \leq 8 \text{ and } 7 \leq \frac{\mu_1 + \mu_2 + \mu_3}{\mu_5} \leq 18$$

Based upon Theorem 2, these two restrictions are redundant constraints in (1)[12]. Therefore, we only incorporate the remaining three sets of weight restrictions specified in Kim, Park and Park (1999)

$$0.8 \leq \frac{v_1}{v_2} \leq 3, \frac{v_3}{v_2} \leq 1, 1 \leq \frac{\mu_2}{\mu_1} \leq 2, 1 \leq \frac{\mu_3}{\mu_1} \leq 2$$

The second and third columns of Table 3-9 report the efficiency scores under ratio bounded data and weak ordinal relations, respectively, when the above weight restrictions are incorporated in model (10) or (12). In this case, model (12) is used. Only seven DMUs, namely DMUs 1, 12, 13, 20, 25, 26, and 32, are efficient with ratio bounded data, and 13 with weak ordinal relations.

We next use this data set to very Theorem 2. Table 3-10 provides two sets of optimal solutions from model (12) when DMU2 is under evaluation. The second and third columns give the exact data on y_4 and y_5 for all 33 DMUs, respectively. We have $\mu_1^* = \mu_3^* = 0.006116$, $\mu_2^* = 0.012233$, $\mu_4^* = 0.188621$ and $\mu_5^* = 0.069929$ with $\pi_2^* = 0.78247$. In this case, $(\mu_1^* + \mu_2^* + \mu_3^*)/\mu_4^* = 0.129705$ and $(\mu_1^* + \mu_2^* + \mu_3^*)/\mu_5^* = 0.349855$.

Now, we let $\gamma_4 = 4 \bullet \mu_4^*/(\mu_1^* + \mu_2^* + \mu_3^*) = 30.83921$ and $\gamma_5 = 8 \bullet \mu_5^*/(\mu_1^* + \mu_2^* + \mu_3^*) = 22.86662$. We next multiply the two previous outputs by 30.83921 and 22.86662, respectively and obtain a new set of exact data on y_4 and y_5 (see the fourth and fifth columns in Table 3-10). From model (11), we have $\tilde{\mu}_1^* = \tilde{\mu}_3^* = \tilde{\mu}_4^* = 0.006116$, $\tilde{\mu}_2^* = 0.012233$, and $\tilde{\mu}_5^* = 0.003058$ with $\pi_2^* = 0.78247$. In this case, $\tilde{\mu}_1^* + \tilde{\mu}_2^* + \tilde{\mu}_3^*/\tilde{\mu}_4^* = 4$ and $\tilde{\mu}_1^* + \tilde{\mu}_2^* + \tilde{\mu}_3^*/\tilde{\mu}_5^* = 8$. This numerical example indicates that the weight restrictions related to the ranking data are redundant in model (12) (or model (10)) and do not affect the proposed approach.

---

[12] In this case, $f(\bullet) = \sum_{r=1}^{3} \mu_r$.

Table 3-9. Efficiency Score with Weight Restrictions

DMU	Ratio bounded data	Weak ordinal relations	Kim et al. (1999)
1	1	1	1
2	0.78247	1	0.9875
3	0.78413	1	1
4	0.62340	0.81352	0.6721
5	0.90220	0.96949	0.8534
6	0.62553	0.83149	0.6415
7	0.65928	0.68760	0.6431
8	0.50440	0.56613	0.534
9	0.61945	0.77865	0.641
10	0.51727	0.55135	0.5063
11	0.54875	0.61023	0.5928
12	1	1	1
13	1	1	0.9251
14	0.79655	0.90892	0.9011
15	0.69698	1	0.6724
16	0.70725	0.91297	0.8679
17	0.61849	0.63957	0.7234
18	0.67527	0.86072	0.7511
19	0.60573	0.77961	0.5869
20	1	1	1
21	0.75114	0.89855	0.8761
22	0.67899	1	0.6817
23	0.69944	0.72766	0.6818
24	0.63477	0.79219	0.6627
25	1	1	0.8612
26	1	1	1
27	0.49219	0.55548	0.5554
28	0.76056	0.89112	0.7193
29	0.69469	0.92363	0.7278
30	0.84849	1	0.8360
31	0.50766	1	0.6307
32	1	1	1
33	0.69603	0.92018	0.7069

Table 3-10. Redudant Weight Restrictions

	Original exact data		New exact data		Efficiency for DMU2: $\pi_2^* = 0.78247$
DMU	y_4	y_5	y_4	y_5	Optimal weights
1	1.1449	0.456621	35.30781	10.44138	Original
2	1	1	30.83921	22.86662	$\omega_1^* = 0.002129$, $\omega_2^* = \omega_3^* = 0.002661$
3	0.513158	1	15.82539	22.86662	$\mu_1^* = \mu_3^* = 0.006116$,
4	0.751315	1.68	23.16996	38.41592	$\mu_2^* = 0.012233$, $\mu_4^* = 0.188621$
5	1.310796	0.456621	40.42392	10.44138	$\mu_5^* = 0.069929$
6	0.620921	1.68	19.14872	38.41592	
7	0.197845	0.456621	6.101374	10.44138	$\omega_1^*/\omega_2^* = 0.8$, $\omega_3^*/\omega_2^* = 1$
8	0.318631	0.208503	9.826323	4.767752	$\mu_2^*/\mu_1^* = 2$, $\mu_3^*/\mu_1^* = 1$
9	0.179859	1	5.546703	22.86662	
10	0.148644	0.456621	4.584052	10.44138	
11	0.466507	0.456621	14.38672	10.44138	
12	1.50073	2.8224	46.28134	64.53874	
13	1.605781	0.456621	49.52104	10.44138	
14	0.385543	1	11.88985	22.86662	
15	0.350494	1.68	10.80896	38.41592	
16	0.909091	0.456621	28.03565	10.44138	
17	1.225043	0.208503	37.77936	4.767752	New
18	0.826446	1.68	25.48695	38.41592	$\widetilde{\omega}_1^* = 0.002129$, $\widetilde{\omega}_2^* = \widetilde{\omega}_3^* = 0.002661$
19	0.163508	0.456621	5.042458	10.44138	$\widetilde{\mu}_1^* = \widetilde{\mu}_3^* = \widetilde{\mu}_4^* = 0.006116$,
20	0.424098	0.208503	13.07884	4.767752	$\widetilde{\mu}_2^* = 0.012233$, $\widetilde{\mu}_5^* = 0.003058$
21	1.07	0.456621	32.99796	10.44138	
22	0.683013	1	21.0636	22.86662	$\widetilde{\omega}_1^*/\widetilde{\omega}_2^* = 0.8$, $\widetilde{\omega}_3^*/\widetilde{\omega}_2^* = 1$
23	0.289664	0.456621	8.933021	10.44138	$\widetilde{\mu}_2^*/\widetilde{\mu}_1^* = 2$, $\widetilde{\mu}_3^*/\widetilde{\mu}_1^* = 1$
24	0.263331	1	8.120928	22.86662	$(\widetilde{\mu}_1^* + \widetilde{\mu}_2^* + \widetilde{\mu}_3^*)/\widetilde{\mu}_4^* = 4$
25	0.239392	0.456621	7.382662	10.44138	$(\widetilde{\mu}_1^* + \widetilde{\mu}_2^* + \widetilde{\mu}_3^*)/\widetilde{\mu}_5^* = 8$
26	1.402552	0.456621	43.25359	10.44138	
27	0.122846	0.456621	3.788473	10.44138	
28	0.564474	2.8224	17.40793	64.53874	
29	0.111678	1	3.444066	22.86662	
30	0.217629	2.8224	6.711511	64.53874	
31	0.135131	0.208503	4.16732	4.767752	
32	0.101526	2.8224	3.130969	64.53874	
33	0.092296	1.68	2.846336	38.41592	

[a]DMU2 is under evaluation by model (12).

5. ENVELOPMENT IDEA (EIDEA)

Chen, Seiford and Zhu (2000) and Chen (2006) point out that when the IDEA is developed based upon the envelopment DEA model, e.g., model (2), we get different efficiency results. When we assume that all output and input values are exact, models (1) and (2) yield the same efficiency score for a specific DMU under evaluation. However, the presence of imprecise data invalidates the linear duality between models (1) and (2) and consequently, model (1) is not equivalent to model (2). The EIDEA yields the worst efficiency scores. The invalidation of linear duality leads to an efficiency gap.

Consider four DMUs given in Chen (2006) as shown in Table 3-11 with two inputs and a single output of unity. Only DMU1, DMU2 and DMU4 have bounded data on the first input. The last column shows the efficiency scores based upon MIDEA.

Table 3-11. Four DMUs and Their MIDEA Scores

	Output (y)	Input 1 (x1)	Input 2 (x2)	MIDEA Score
DMU1	1	[1, 3]	5	1
DMU2	1	[3, 4]	3	1
DMU3	1	7	1	1
DMU4	1	[8, 9]	6	0.5

Figure 3-1 plots the four DMUs from Chen (2006). Because of the bounded data, DMU1 and DMU2 are represented by AA' and BB', respectively[13]. The (imprecise) efficient frontier is represented by the area bounded by AA', A'B', B'C, CB, and BA.

When DMU4 is under evaluation by the MIDEA, DMU4 is replaced by point D and A'B'C is used as the exact efficient frontier. For DMU4, B' is a referent DMU which can be achieved via (i) setting $x_{12} = 4$ (upper bound) for DMU2 or (ii) setting $x_{11} = 1$ (lower bound) for DMU1 and then using the convex combination of A and C with $\lambda_1 = \lambda_2 = 0.5$.

When DMU4 is under evaluation by the EIDEA, DMU4 will be evaluated against ABC rather than A'B'C. Consequently, the efficiency of DMU4 decreases. Upper bound on the first input of DMU4 and lower bounds on the first input of other DMUs would be used as the exact data when we evaluate DMU4 using EIDEA.

---

[13] DMU4 is represented by a line segment.

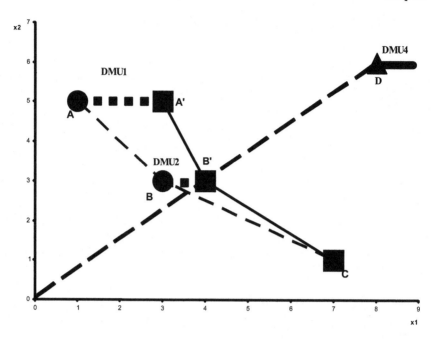

Figure 3-1. DEA Frontier with Imprecise Data

Theorem 3: Suppose for DMU_o, θ_o^* is the optimal value to (2) when (some) outputs and inputs are only known to be within specific bounds given by (3). This θ_o^* can be achieved with
(i) $x_{ij} = \underline{x}_{ij}$ for DMU_j $(j \neq o)$;
(ii) $y_{rj} = \overline{y}_{rj}$ for DMU_j $(j \neq o)$;
(iii) $y_{ro} = \underline{y}_{ro}$ for DMU_o;
(iv) $x_{io} = \overline{x}_{io}$ for DMU_o.

[Proof] See Chen (2006).

Theorem 3 is true due to the fact that input decreases/output increases in DMU_o or (and) input decreases/output increases in other DMUs will deteriorate the efficiency of DMU_o.

Theorem 4 indicating that EIDEA can be executed by setting the lower output bounds and upper input bounds as the exact output and input values for DMU_o and by setting upper output bounds and lower input bounds as the exact output and input values for the remaining DMUs.

Recall that in MIDEA, ordinal data and ratio bound data are converted into a set of exact data via the bounded data. By the same fashion, ordinal data and ratio bound data can be converted into a set of exact data under EIDEA.

Chen (2006) applies the EIDEA to the five DMUs Table 3-1. Table 3-12 reports the efficiency results when model (2) is used after a set of exact data is obtained based upon Theorem 3. It can be seen that only DMU1 is efficient under EIDEA.

Table 3-12. EIDEA Results

DMU	Efficiency score θ_o^*	slack		Referent DMU λ_j^*
		Revenue	Cost	
1	1	0	0	$\lambda_1^* = 1$
2	1/3	0	100	$\lambda_1^* = 0.5$
3	2/3	800	50	$\lambda_1^* = 1$
4	27/80	0	155	$\lambda_1^* = 0.45$
5	9/50	0	170	$\lambda_1^* = 0.3$

6. CONCLUSIONS

The current chapter presents how the standard linear DEA models are used to deal with interval or bounded data or ordinal data. Although the current chapter discusses specific forms of imprecise data[14], the results are true for any types of imprecise data (see Theorem 1 in Zhu (2004)).

The IDEA approach using the standard DEA model indicates that one has to decide whether the multiplier or envelopment DEA model will be used to deal with the imprecise data. The multiplier IDEA (MIDEA) yields the best efficiency scores whereas the envelopment IDEA (EIDEA) yields the worst efficiency scores. We should note that the MIDEA can also yield the same worst EIDEA efficiency scores if we set the exact data in a reversed direction.

The current chapter discusses IDEA procedure based upon the CCR model. Similar discussion can be developed based upon other DEA models.

We finally provide the following theorem to show that for inefficient DMUs, their multiplier DEA efficiency is always achieved at the bounds for interval data.

Theorem 4: For DMU_o, if $\pi_o^* < 1$ in (1), then the optimality must be achieved at $x_{io} = \underline{x}_{io}$ and $y_{ro} = \overline{y}_{ro}$ for DMU_o.

---

[14] We should note that the specific forms of imprecise data are probably the only imprecise data types that will occur in real application.

[Proof] Note that by defining $\omega_i = tv_i$, $\mu_r = tu_r$, and $t = (\sum v_i x_{io})^{-1}$, model (1) is equivalent to

$$\pi_o^* = \max \frac{\sum_{r=1}^{s} u_r y_{ro}}{\sum_{i=1}^{m} v_i x_{io}} \text{ subject to } \frac{\sum_{r=1}^{s} u_r y_{rj}}{\sum_{i=1}^{m} v_i x_{ij}} \leq 1 \quad \forall \ j \text{ and } u_r, v_i \geq \varepsilon.$$

Next, suppose the optimality is achieved at u_r^*, v_i^*, y_{rj}^* ($r \in \boldsymbol{BO}$) and x_{ij}^* ($i \in \boldsymbol{BI}$) with $\pi_o^* = \dfrac{\sum_{r\in BO} u_r^* y_{ro}^* + \sum_{r\notin BO} u_r^* y_{ro}}{\sum_{i\in BI} v_i^* x_{io}^* + \sum_{i\notin BI} v_i^* x_{io}} < 1$ and $\underline{y}_{ro} \leq y_{ro}^* < \bar{y}_{ro}$ ($r \in \boldsymbol{BO}$) and $\underline{x}_{io} < x_{io}^* \leq \bar{x}_{io}$ ($i \in \boldsymbol{BI}$).

Obviously, when $x_{io} = \underline{x}_{io}$ and $y_{ro} = \bar{y}_{ro}$, we have $\pi_o^* = \dfrac{\sum_{r\in BO} u_r^* y_{ro}^* + \sum_{r\notin BO} u_r^* y_{ro}}{\sum_{i\in BI} v_i^* x_{io}^* + \sum_{i\notin BI} v_i^* x_{io}} < \dfrac{\sum_{r\in BO} u_r^* \bar{y}_{ro} + \sum_{r\notin BO} u_r^* y_{ro}}{\sum_{i\in BI} v_i^* \underline{x}_{io} + \sum_{i\notin BI} v_i^* x_{io}}$. A contradiction[15]. This shows that the optimality must be achieved at $x_{io} = \underline{x}_{io}$ and $y_{ro} = \bar{y}_{ro}$ for DMU_o.

---

[15] If $\dfrac{\sum_{r\in BO} u_r^* \bar{y}_{ro} + \sum_{r\notin BO} u_r^* y_{ro}}{\sum_{i\in BI} v_i^* \underline{x}_{io} + \sum_{i\notin BI} v_i^* x_{io}} = \tilde{h}_o > 1$, we can always redefine the weights by dividing each u_r^* by $\tilde{h}_o$. As a result, the new optimal value is equal to one, and all $\dfrac{\sum_{r=1}^{s} u_r y_{rj}}{\sum_{i=1}^{m} v_i x_{ij}}$ decrease and are not greater than one. This will still give a contradiction. Thus, the theorem is true.

REFERENCES

1. Charnes, A, W.W. Cooper, E. Rhodes (1978), "Measuring the efficiency of decision making units", European Journal of Operational Research, 2, 429-444.
2. Chen, Y. (2006), "Imprecise DEA -- Envelopment and multiplier model approaches", Asian Pacific Journal of Operations Research, (in press).
3. Chen, Y., L.M. Seiford and Joe Zhu (2000), "Imprecise data envelopment analysis", Working paper.
4. Cook, W.D., M. Kress, L.M. Seiford (1993), "On the use of ordinal data in data envelopment analysis", Journal of Operational Research Society, 44, 133-140.
5. Cook, W.D., M. Kress, L.M. Seiford (1996), "Data envelopment analysis in the presence of both quantitative and qualitative factors", Journal of Operational Research Society, 47, 945-953.
6. Cook, W.D. and J. Zhu (2006), "Rank order data in DEA: A general framework", European Journal of Operational Research, (in press).
7. Cooper, W.W., Z.M. Huang, V. Lelas, S.X. Li, and O.B. Olesen (1998), "Chance Constrained Programming Formulations for Stochastic Characterizations of Efficiency and Dominance in DEA," Journal of Productivity Analysis, 9, No. 1, 53-79.
8. Cooper, W.W., and K.S. Park (2006), "Imprecise DEA–Telecommunications," in N. Avkrian, *Productivity Analysis in the Service Sector with Data Envelopment Analysis*, Chapter 21.
9. Cooper, W.W., K.S. Park, G. Yu (1999), "IDEA and AR-IDEA: Models for dealing with imprecise data in DEA," Management Science, 45, 597-607.
10. Cooper, W.W., K.S. Park, G. Yu (2001), "IDEA (imprecise data envelopment analysis) with CMDs (column maximum decision making units)", J. Oprl. Res. Soc. 52, 176-181.
11. Despotis, D. K. and Y.G. Smirlis (2002), "Data envelopment analysis with imprecise data," European Journal of Operational Research, 140, 24-36.
12. Kao, C., and S.T. Liu (1999), "Fuzzy efficiency measures in data envelopment analysis," Fuzzy Sets and Systems, 2000, 113:427-437.
13. Kim, S.H., C.G. Park, K.S. Park (1999), "An application of data envelopment analysis in telephone offices evaluation with partial data," *Computers & Operations Research*, 26;59-72.
14. Seiford, L.M. and J. Zhu (1999), "An investigation of returns to scale under data envelopment analysis", OMEGA, 27/1, 1-11.

15. Thompson, R.G., P.S. Dharmapala, E.J. Gatewood, S. Macy and R.M. Thrall (1996), "DEA/Assurance region SBDC efficiency and unique projections", Operations Research, 44, 533-542.
16. Zhu, J. (2001), "Super-efficiency and DEA sensitivity analysis," European Journal of Operational Research, 129, No. 2, 443-455.
17. Zhu, J. (2003a), "Imprecise data envelopment analysis (IDEA): A review and improvement with an application", European Journal of Operational Research, Vol. 144, Issue 3, 513-529.
18. Zhu, J. (2003b), Efficiency evaluation with strong ordinal input and output measures, European Journal of Operational Research, Vol. 146, Issue 3, 477-485.
19. Zhu, J. (2004), "Imprecise DEA via standard linear DEA models with a revisit to a Korean mobile telecommunication company", Operations Research, Vol. 52, No. 2, 323-329.
20. Zhu, J, and Z. Shen (1995), "A discussion of testing DMUs' returns to scale", European Journal of Operational Research, 81, 590-596.

Part of the material in this chapter is based upon Chen, Y., L.M. Seiford and Joe Zhu (2000), "Imprecise data envelopment analysis", working paper, and part of the material in this chapter is adapted from European Journal of Operational Research, Vol 144, Issue 3, Zhu, J., Imprecise data envelopment analysis (IDEA): A review and improvement with an application, 513-529, 2003, with permission from Elsevier Science

Chapter 4

VARIABLES WITH NEGATIVE VALUES IN DEA

Jesús T. Pastor and José L. Ruiz
Centro de Investigación Operativa, Universidad Miguel Hernández, Avd. de la Universidad, s/n, 03202-Elche, Alicante, Spain, jtpastor@umh.es, jlruiz@umh.es

Abstract: In this chapter we present an overview of the different existing approaches dealing with the treatment of negative data in DEA. We discuss both the classical approaches and the most recent contributions to this problem. The focus is mainly on issues such as translation invariance and units invariance of the variables, classification invariance of the units, as well as efficiency measurement and target setting.

Key words: Data Envelopment Analysis, Negative data.

1. INTRODUCTION

The variables -inputs and outputs- of any DEA model were initially assumed to be positive (Charnes et al., 1978). This strict initial condition was relaxed by Charnes et al. (1986). After considering the ratio form of the radial models, they end up requiring only non-negativity, provided each unit under evaluation has at least one positive input and one positive output. Going one step further, the aim of this chapter is to show how to deal with unrestricted in sign DEA variables. A considerable amount of DEA models have been proposed and used during the last twenty-seven years. For the sake of clarity, three of the initial DEA models will be called "basic" models: the CCR model (Charnes et al., 1978), the BCC model (Banker et al., 1984) and the additive model (Charnes et al., 1985). As it is well known, the first two are radial models and the last two yield variable returns to scale frontier.

Historically, the usual way of handling the aforementioned type of non-standard variables in a DEA framework was simply to either perform a frequently unjustified change of variables or to reduce the sample of units by eliminating those with negative values in some variable (see, for example, Charnes et al., 1989). Below, we list several applications dealing with negative data which were forerunners of the first theoretical papers on this issue:

(1) Pastor (1993) considered two unrestricted in sign outputs in order to evaluate a sample of 23 bank branches. These outputs were "increments of time deposits and of demand deposits over a one-year period".

(2) Zhu (1994) (also Seiford and Zhu (2002)) used the variable "profit and taxes" as an output in a DEA study of 35 Chinese textile firms, which may be negative in cases of financial loss. In fact, the values for five of the firms were negative.

(3) Thore et al. (1994) evaluated a sample of U.S. computer companies, where one of the outputs was "income before taxes".

(4) Lovell (1995) considered as output of an FDH model the "rate of growth of GDP per capita", a variable which ranges from -0.083 to 0.128.

It was in 1995 when the first theoretical paper devoted to negative data in DEA was published. Nevertheless, we should mention the previous paper by Ali and Seiford (1990) dealing with the translation invariance property, although it should be pointed out that these authors were only concerned with the presence of zero values and that negative values were not even mentioned in that paper. Pastor (1994) was the first to tackle this issue and to provide a translation invariance classification of the three basic DEA models. His findings were published in Lovell and Pastor (1995) and Pastor (1996). Zhu (1994) performed also an interesting application following the same path. Translation invariance guarantees that a DEA model using the original -negative- data and the same model using the translated -positive- data are equivalent, i.e., both have the same optimal solution(s). In other words, in the presence of a translation invariant DEA model we can always translate the data and solve it as if the data were positive. We will refer to this procedure as the "classical approach". We are going to compare the three basic DEA models according to their translation invariance capabilities, concluding that the winner is the additive model and the loser the CCR model. The BCC model stays in between.

Lastly, three new contributions to the treatment of variables with negative values in DEA have been proposed: An a priori approach for dealing with interval scale variables (Halme et al., 1998); an a posteriori approach for correcting the classification of units as efficient or inefficient in the presence of negative data (Seiford and Zhu, 2002); and, finally, a new DEA model due to Portela et al. (2004), based on directional distance functions, which has some of the good properties of both the radial and the additive models. Roughly speaking, the first contribution recommends changing the specification of the model by substituting each interval scale variable by its two components, whenever the interval scale variable is the result of a difference between two ratio scale variables. Halme et al. point out that the radial DEA models can only be used with this latter type of variables. We will comment on this issue later on, introducing some refinements. The second contribution judges as unacceptable that, in a radial output-oriented model, a unit with negative outputs is rated as efficient while another unit with positive outputs is classified as inefficient. In that case, the proposal is to re-evaluate the efficient units with negative outputs by projecting them onto one of the extended facets defined by the remaining efficient units with positive data. The third contribution introduces a new DEA model, called range directional model (RDM), which provides an efficiency score that results from the comparison of the unit under assessment with the so-called ideal point (or zenith, in the terminology of multi-objective programming) by using the corresponding directional distance function.

The approach in the latter paper has two main drawbacks, as acknowledged by the authors: it does not guarantee to provide efficient projections on the frontier and it does not assure yielding the closest targets on any criterion. The very last contribution in this field, due to Pastor and Ruiz (2005), shows how to deal with them by means of recent versions of the additive model, gathered under the name of weighted additive models. We further show the equivalence of one of our new models with the RDM.

This chapter is organized as follows: In section 2 we present the "classical" translation invariant approach. In section 3 we revise the a priori approach and in the next section we do it with the a posteriori approach. Section 5 is devoted to the RDM. Section 6 presents our weighted additive models. The performance of all the mentioned methods is illustrated by means of a numerical example in the last section.

2. THE CLASSICAL APPROACH: THE TRANSLATION INVARIANT DEA MODELS

It is well-known that the DEA models are linear programming (LP) problems, which is one of their most important features, since they can be easily solved. Therefore, each DEA model has associated both a primal and a dual formulation. In this context, these are called the primal "envelopment formulation" and the dual "multiplier formulation", respectively. The first radial DEA model was a fractional programming problem, known as the "ratio form" (Charnes et al., 1978). The ratio form can be linearized and converted into the multiplier form by means of the change of variables proposed by Charnes and Cooper (1962). Generally speaking, the envelopment formulation of a DEA model allows us to classify each DMU (decision making unit) as weak-efficient or inefficient and, consequently, identifies all the units that belong to the DEA weak efficient frontier. Moreover, the radial DEA are only able to detect radial inefficiencies. Hence, they need to solve another model in a second stage in order to identify which weak efficient units are efficient, i.e., which weak efficient units do not have non-radial inefficiencies. The additive-type models do classify the DMUs directly as efficient or inefficient, since they detect any kind of inefficiencies. (For a refined partition of the sample of DMUs, including extreme efficient points and strictly weak efficient points see Charnes et al. (1986) and Charnes et al. (1991)). The multiplier formulation of a DEA model identifies supporting hyperplanes to the frontier at one (or more) efficient point so that it provides the multipliers that represent a relative value system of the inputs and outputs considered in the analysis. These multipliers are exactly the coefficients of the variables in the hyperplane equation.

The translation invariance property allows the envelopment form of many DEA models to deal with negative data. The most favorable case corresponds to the additive model, which is fully translation invariant as stated below. The same happens with the family of weighted additive models, which have as objective function a weighted sum of the slacks instead of the sum of the slacks themselves, as in the case of the additive model. Mathematically, this means to maximize a weighted L_1-distance to the efficient frontier instead of maximizing the L_1-distance. The multiplier form of the DEA models does not have the same translation invariance properties as the envelopment form, in concordance with our geometrical intuition (see Thrall, 1996). In fact, the supporting hyperplanes containing facets of the efficiency frontier are parallel to the original ones. For this reason, in what follows, we will mainly focus on the envelopment formulation of the DEA models.

Proposition 1. The envelopment form of the (weighted) additive model is translation invariant in any variable.

This result can be easily proven by taking into account that if we translate each of the inputs and each of the outputs by adding a quantity (not necessarily the same for all the variables) then the set of constraints of the original model and that resulting from the translated variables coincide.

By virtue of this result, if any variable (input or output) has some negative value we can proceed in the following way: translate the values of the corresponding variable so that the resulting variable has all its values positive and then perform the analysis of the translated data set by means of the (weighted) additive model. What the translation invariance tells us is that the obtained results are exactly the same as if the original data set were analyzed. In fact, by means of a linear programming code we can solve the model resorting to the original data set. The reason for performing the translation so as to achieve non-negative data is that the DEA software packages typically require this condition. Moreover, in the presence of non-discretionary inputs or outputs (Banker and Morey, 1986) the same result holds, since the model formulation only skips the corresponding slack or excess variables from the objective function.

As already pointed out by Ali and Seiford (1990), the translation invariance property of the additive model may be very useful for dealing with the multiplicative model (Charnes et al., 1983), which is simply the additive model applied to the logarithms of the original data, in order to achieve a piecewise Cobb-Douglas envelopment. Let us now consider the case of the two basic radial models, starting with the VRS case.

Proposition 2. The envelopment form of the input (output)-oriented BCC model is translation invariant with respect only to outputs (inputs) and to non-discretionary inputs (outputs).

This result means that, for example, we can deal with any output variable in the input-oriented BCC model, even if all its data are negative, provided all the input variables have non-negative values. If non-discretionary inputs are present, they may also contain negative data. In that case we may consider each non-discretionary input as an output of the model, just by reversing its sign, and then the model can be solved (see Lozano-Vivas et al. (2002) for an application where all the environmental variables are located on the output side of an input-oriented BCC model, although some of them are originally non-discretionary inputs).

The presence of the efficiency score as a variable at, say, the input restrictions of an input-oriented BCC model prevents the equivalence between the set of original restrictions and the set of restrictions obtained after translating the inputs. Moreover, this equivalence holds if, and only if, the efficiency score of the unit under assessment equals 1, which means that the translated model identifies correctly the weak-efficient units. In other words, if we translate the inputs in an input-oriented BCC model, the classification of the units as weak-efficient or inefficient remains, but the efficiency score of each inefficient unit is distorted.

It should also be noted that both the BCC and the additive models are variable returns to scale (VRS) models, in contrast to the CCR model, which exhibits constant returns to scale (CRS). In fact, being a VRS model is the key for satisfying translation invariance or, in other words, the convexity constraint (the sum of the intensity variables equals 1, i.e., $\sum_{j=1}^{n} \lambda_j = 1$) is the clue. As said before, the fact that the BCC model is translation invariance only on one side is due to the presence of the efficiency score on the other side. That means that the efficiency score of an output-oriented BCC model may change if we perform a translation of outputs. Of course, the shape of the efficiency frontier will not change, since the VRS frontier is unique for all the VRS DEA models, such as the additive model. The classification of the points as (weak)-efficient or inefficient will not change either. But the ranking of the efficiency scores may change, as the following example shows.

Example 1.

Table 4-1 records the data corresponding to a sample of DMUs each consuming 2 inputs to produce 1 output.

Table 4-1. Example 1

DMU	A	B	C	D	E
X	1	1	0.5	1	0.5
Y1	2	3	1	4	4.2
Y2	4	3.6	4.95	4.8	0.8

The additive model shows that DMUs C, D and E are efficient, whereas A and B are inefficient. The BCC output-oriented DEA model yields 1.22 and 1.33 as the efficiency scores of A and B, respectively. However, if we translate the output Y1 by 10 units, the new efficiency scores for the inefficient DMUs are 1.17 for A and 1.08 for B, which shows that the efficiency ranking between these has changed.

Proposition 3. The envelopment form of the input (output)-oriented CCR model is not translation invariant with respect to either outputs or inputs.

The rationale behind this result can be easily explained as follows: In a constant returns to scale DEA frontier all the supporting hyperplanes of the production possibility set pass through the origin of coordinates. Consequently, if the data of any variable are translated, then the new supporting hyperplanes containing the facets of the frontier will change, since they all need to pass through the origin (these new hyperplanes will not be parallel to the original ones). Therefore, the targets for each inefficient unit will change accordingly, as well as the corresponding efficiency score.

Therefore, the CCR model cannot be used with any type of negative data. Translation of data is not allowed even if we deal with variables with positive data, as the next example shows.

Example 2.

Consider again the data in Table 4-1. The CCR output-oriented DEA model detects DMUs C and E as efficient and A, B and D as inefficient. The efficiency scores for these three latter DMUs are 1.89, 1.67 and 1.25, respectively. If we translate the input X by 0.5, then DMU D becomes efficient and the efficiency scores of A and B are now 1.47 and 1.33, respectively. Thus, we can see that the CCR model is not translation invariant, even in the case of translating inputs when using an output-oriented version of this model.

In any case, the assumption of constant returns to scale for the technology when negative data can exist is not possible (for a discussion on this issue see Portela el (2004)). In the last referenced paper, the authors show by means of a small example that, under CRS, any unit obtained as a proportion of an efficient unit is not necessarily efficient unless the data are non-negative. For this reason, when dealing with negative data, the technology is usually assumed to satisfy variable returns to scale which implies that the used DEA model includes the convexity constraint. It is nevertheless true that the convexity constraint can be relaxed, by changing the equality to an inequality, getting non-increasing returns to scale (NIRS, with $\sum_{j=1}^{n}\lambda_j \leq 1$) or non-decreasing returns to scale (NDRS, with $\sum_{j=1}^{n}\lambda_j \geq 1$) models with some weakened -although interesting- translation properties, as the next two (unknown) propositions state.

Proposition 4. The envelopment form of the output-oriented NIRS radial model is translation invariant with respect only to inputs. The same statement is true for the (weighted) additive NIRS model.

Proposition 5. The envelopment form of the input-oriented NDRS radial model is translation invariant with respect only to outputs. The same statement is true for the (weighted) additive NDRS model.

Observe that in the last two propositions the reference to nondiscretionary variables has disappeared in contrast to proposition 1.2. The loss is even deeper in the framework of the additive models. Nevertheless, the proofs of these two propositions are similar to proposition 1.2.

3. INTERVAL SCALE VARIABLES WITH NEGATIVE DATA AS A RESULT OF THE SUBTRACTION OF TWO RATIO SCALE VARIABLES

In many applications, the presence of negative data in the sample is a consequence of using derived variables obtained as the subtraction of two ratio scale variables, for instance, interval scale variables that represent profit or changes in other variables like sales or loans. Halme et al. (2002) point out the fact that the interval scale variables do not involve a true 0 and, consequently, cannot be used with the most widely used DEA models, i.e., the two basic radial models. As a matter of fact, the efficiency score of the CCR or of the BCC model, in the input (output) oriented case, corresponds to a ratio obtained through one of its inputs (outputs), taking as numerator the input (output) value of the projection and as denominator the original input (output) value. This ratio requires that the corresponding input (output) has a true 0, i.e., that the input (output) is a ratio scale variable. As a summary, the problem with interval scale variables arises from the fact that ratios of measurements on such a scale are meaningless, due to the absence of a true 0. Consequently, an input (output) oriented radial DEA model requires ratio scale inputs (outputs) and, at the same time, is able to deal with interval scale outputs (inputs). Moreover, non-discretionary variables do not cause any problem, since they are unaffected by the efficiency score, i.e., DEA models can deal directly with interval scale non-discretionary variables. For example, the number of accounts in a bank at a time point is non-discretionary in nature; we may consider as an interval scale output the increments in bank accounts with respect to that initial position. Summing up, in the framework of the basic radial DEA models the only variables that are not allowed to be interval scale variables are the ones affected by the efficiency score, e.g., the discretionary inputs in an input-oriented model.

To address this problem, the aforementioned authors propose an approach, which can be seen as an a priori approach to deal with negative data, based on the idea of replacing each of the interval scale variables with the two ratio scale variables that give rise to them. If, let us say, an input is replaced, one of the new variables is considered as an input and the one affected with the minus sign is considered as an output. It is well-known that the basic radial models are not symmetrical with respect to inputs and outputs, due to the presence of the efficiency score. Therefore, it does not seem a reasonable idea to decompose a discretionary interval scale input in a basic radial input-oriented model. What is clearly needed is a DEA model that is symmetrical in inputs and outputs, as the model proposed by Briec (1997). This is exactly the model considered by Halme et al. in their paper. To briefly describe their approach, let us consider a set of n DMUs each using m inputs to produce s outputs. (This assumption is mantained from now on). Assume also that all the variables are discretionary and that t among the m inputs and p among the s outputs have been measured on an interval scale. Assume further that we know the decomposition of each interval scale variable as the subtraction of two ratio scale variables. All these variables can be denoted as follows: Let $x_1,...,x_{m-t}$ be the ratio scale inputs and $\overline{x}_{m-t+1},...,\overline{x}_m$ the interval scale inputs. These latter t inputs can be expressed as $\overline{x}_{m-t+i} = x_{m-t+i} - y_{s+i}, i=1,...,t$. In a similar fashion, let $y_1,...,y_{s-p}$ be the ratio scale outputs and $\overline{y}_{s-p+1},...,\overline{y}_s$ the interval scale outputs. These latter p outputs can be expressed as $\overline{y}_{s-p+r} = y_{s-p+r} - x_{m+r}, r=1,...,p$. Halme et al. (2002) propose assessing the efficiency of a given DMU$_0$ by means of the next model

$$\text{Maximize } \sigma_0 + \varepsilon\left(\sum_{i=1}^{m} s_{i0}^- + \sum_{r=1}^{s} s_{r0}^+\right)$$

subject to:

$$\sum_{j=1}^{n} \lambda_j x_{ij} + \sigma_0 x_{i0} + s_{i0}^- = x_{i0}, \quad i=1,...,m+p$$

$$\sum_{j=1}^{n} \lambda_j y_{rj} - \sigma_0 y_{r0} - s_{r0}^+ = y_{r0}, \quad r=1,...,s+t$$

$$\sum_{j=1}^{n} \lambda_j = 1 \tag{1}$$

$$\lambda_j \geq 0 \quad\quad j=1,...,n$$
$$s_{i0}^- \geq 0 \quad\quad i=1,...,m$$
$$s_{r0}^+ \geq 0 \quad\quad r=1,...,s$$

Observe that the p added inputs correspond to the p original interval scale outputs, while the t added outputs correspond to the t original interval scale inputs.

One of the main findings of Halme et al. is that the units that are initially rated as efficient remain efficient when assessed by using the reformulated model. However, since the number of variables rise as a consequence of the decomposition of the interval scale variables, some of the inefficient units may become efficient, which is acknowledged by the authors as a drawback of the proposed approach. Finally, among the advantages, it should be noted that this approach can obviously be used with constant returns to scale technologies.

4. AVOIDING EFFICIENT UNITS WITH NEGATIVE OUTPUTS

Unlike the approach in the previous section, Seiford and Zhu (2002) propose an a posteriori approach to address the case when units with negative outputs are rated as efficient. These authors assume that units with positive output values, particularly the efficient ones, outperform those with negative values. In other words, although classifying the units with negative output values as efficient may be mathematically correct, this may not be managerially acceptable. For this reason, they propose to re-evaluate the units in a specific manner that we describe next. The approach by Seiford and Zhu is only concerned with the "classification invariance", and this is why it is restricted to the BCC model, although it can be directly applied to any weighted additive model. The term "classification invariance" is weaker than "translation invariance" since the former does not consider the efficiency scores. In fact, it means that the BCC efficiency classification of the units –as weak efficient or inefficient- is invariant to data transformation. The basic idea of this a posteriori model is to assess the efficient units with negative outputs with reference to the proper well-defined positive multiplier facets generated by the efficient units with positive data. To be precise, each efficient unit with negative output values is assigned to the facet that gives the highest efficiency score in the output-oriented model. The inefficient units are also re-evaluated with reference to the facets assigned to their efficient referent DMUs.

The model that provides the mentioned re-estimated efficiency score is formulated next, based on the data obtained after output data translation.

Minimize $\sum_{i=1}^{m} v_i x_{i0} + u_0$

subject to:

$$\sum_{r=1}^{s} \mu_r y_{rj} - \sum_{i=1}^{m} v_i x_{ij} - u_0 + d_j = 0, \quad j \in E^+ \quad (2)$$

$$\sum_{r=1}^{s} \mu_r y_{r0} = 1$$

$$d_j - Mb_j \leq 0, \qquad j \in E^+$$

$$\sum_{j \in E} b_j = |E^+| - k$$

$$b_j \in \{0,1\} \qquad j \in E^+$$
$$d_j \geq 0 \qquad j \in E^+$$
$$v_i \geq 0 \qquad i = 1,\ldots,m$$
$$\mu_r \geq 0 \qquad r = 1,\ldots,s$$
$$u_0 \text{ free}$$

Here E^+ denotes the set of extreme-efficient units within the set of DMUs with positive data and k, which is the dimension of the facet that will be used as referent for DMU$_0$, needs to be specified. This requires a previous analysis of the facets of the efficient frontier generated by the units in E^+ (see Olesen and Petersen (1996) and Cooper et al. (2006) to that end).

Model (2) includes the set of constraints of the dual multiplier formulation of the BCC model, together with another group of constraints that has the following purpose: $\sum_{j \in E} b_j = |E^+| - k$ implies that $b_j = 0$ for a subset of k DMU$_j$'s in E^+. Therefore, by virtue of the constraints $d_j - Mb_j \leq 0, j \in E^+$, $d_j = 0$ for these DMUs, and consequently, these will be the units that belong to the hyperplane and define the corresponding facet of the efficient frontier. Model (2) selects the subset of k DMU$_j$'s in E^+ that optimizes the objective function, i.e., that assigns the highest efficiency score to the unit being rated.

Concerning the specification of k, if we can select $k=m+s$ we will be resorting to a full dimensional efficient facet (FDEF) of the frontier. Otherwise, i.e., if $k<m+s$, then the considered facet will not be of full dimension. It is to be noted that, in the latter case, model (2) may become unbounded, as shown in our example in the last section.

For the reasons stated above, to proceed with this approach the users should first analyze the efficient frontier generated by the DMU$_j$'s in E^+. If an FDEF exists, then k should be set to $m+s$. Otherwise, k will be strictly lower than $m+s$ and the analyst must be aware of the possible difficulties when solving the resulting model.

5. THE DIRECTIONAL DISTANCE APPROACH

In this section we briefly describe the directional distance function approach by Portela et al. (2004). These authors address the problems of both setting targets and measuring relative efficiency in the presence of negative data.

5.1 Efficiency measurement

Portela et al. (2004) propose an approach to deal with negative data in DEA that provides efficiency scores that can be readily used without the need to transform the data. Moreover, these scores have a similar interpretation as the traditional efficiency scores of the radial DEA models. In fact, this approach is inspired by the directional distance model in Chambers et al. (1986). The so-called range directional model (RDM) is the following:

Maximize β_0

subject to:

$$\sum_{j=1}^{n} \lambda_j y_{rj} \geq y_{r0} + \beta_0 R_{r0}^+ \quad r=1,\ldots,s$$

$$\sum_{j=1}^{n} \lambda_j x_{ij} \leq x_{i0} - \beta_0 R_{i0}^- \quad i=1,\ldots,m \tag{3}$$

$$\sum_{j=1}^{n} \lambda_j = 1$$

$$\lambda_j \geq 0 \quad j=1,\ldots,n$$

where $R_{r0}^+ = \max_{j=1,\ldots,n}\{y_{rj}\} - y_{r0}$, r=1,...,s, and $R_{i0}^- = x_{i0} - \min_{j=1,\ldots,n}\{x_{ij}\}$, i=1,...,m, which can be seen as the ranges of possible improvement in each variable for DMU$_0$. With model (3) DMU$_0$ is projected onto the frontier along the direction of the vector $\begin{pmatrix} -R_0^- \\ R_0^+ \end{pmatrix}$, where $R_0^- = \left(R_{10}^-,\ldots,R_{m0}^-\right)'$ and $R_0^+ = \left(R_{10}^+,\ldots,R_{s0}^+\right)'$. This is why β_0^* has a similar geometric interpretation as the radial efficiency scores, but the efficiency assessment is now made with reference to an ideal point, with components $\left(\min_{j=1,\ldots,n}\{x_{1j}\},\ldots,\min_{j=1,\ldots,n}\{x_{mj}\},\max_{j=1,\ldots,n}\{y_{1j}\},\ldots,\max_{j=1,\ldots,n}\{y_{sj}\}\right)$, which plays the role of the origin in the classical radial case. In particular, it is to be noted that β_0^* can be interpreted as a percentage in terms of the ranges of possible improvement. If a unit is DEA efficient, its projection collapses with the unit and, consequently, $\beta_0^*=0$. Hence, β_0^* is a measure of inefficiency so that $1-\beta_0^*$ can be used as an efficiency measure.

This model is also translation invariant. This is due to the following two facts: it is a VRS model and the vector $\begin{pmatrix} -R_0^- \\ R_0^+ \end{pmatrix}$ is translation invariant. It is also units invariant, i.e., the rescaling of any variable does not change model (3), which is an important feature for a DEA model to deal with non-commensurable economic data.

It is also to be noted that the ranges of possible improvement are variables measured on a ratio scale and, consequently, the percentage in terms of the ranges of possible improvement (i.e., β_0^*) are meaningful quantities. For this reason, RDM can be used with variables measured on a ratio scale, such as profits or differences between variables.

Perhaps, the main drawback of the RDM model is that it cannot guarantee projections on the efficient frontier, as happens with the classical radial DEA models.

5.2 Target setting

Portela et al. (2004) point out that RDM provides targets giving priority to the variables in which the unit under assessment performs worse relative to other units, since these are the factors with the largest potential for improvement. Consequently, these targets may prove hard for the unit to achieve the efficiency in the short term.

To find closer targets the mentioned authors propose a different version of the RDM, called IRDM, by using in (3) the inverse of the ranges of possible improvement instead of the ranges themselves. The resulting model is proposed to be used only for target setting purposes, since it has some additional drawbacks. To be specific, IRDM is not so far units invariant and, in addition, it does not provide a measure with a straightforward interpretation (the "ideal" point varies from unit to unit). As a consequence of the use of the inverse of the ranges of possible improvement the obtained targets are expected to be near the closest ones. However, it cannot be assured closeness under any mathematical criterion.

6. EFFICIENCY MEASUREMENT AND TARGET SETTING BY MEANS OF WEIGHTED ADDITIVE MODELS

This section is devoted to the approach by Pastor and Ruiz (2005), which is probably the most recent contribution to deal with negative data in DEA. The idea of the mentioned approach is to relate the RDM model with a weighted additive model that, as it is well-known, provides efficient projections onto the efficient frontier. The mentioned authors consider first an already published weighted additive model whose objective function can be used as an efficiency measure. They propose a specification of that model that matches with the RDM. Finally, they propose a new refined version of the last model that determines the closest target for each DMU while considering the same efficiency measure.

6.1 Efficiency measurement

The approach by Pastor and Ruiz allows us also to deal directly with negative data and provides efficiency measures without the need to transform the data. In order to guarantee efficient projections, these authors propose using the following weighted-additive model, as suggested in Pastor (1994)

Maximize $\quad Z_0 = \dfrac{1}{m+s}\left(\sum_{i=1}^{m}\dfrac{s_{i0}^{-}}{R_{i0}^{-}} + \sum_{r=1}^{s}\dfrac{s_{r0}^{+}}{R_{r0}^{+}}\right)$

subject to:

$$\sum_{j=1}^{n}\lambda_j y_{rj} = y_{r0} + s_{r0}^{+} \qquad r=1,\ldots,s$$

$$\sum_{j=1}^{n}\lambda_j x_{ij} = x_{i0} - s_{i0}^{-} \qquad i=1,\ldots,m$$

$$\sum_{j=1}^{n}\lambda_j = 1 \qquad\qquad\qquad\qquad\qquad (4)$$

$$\lambda_j \geq 0 \qquad j=1,\ldots,n$$
$$s_{i0}^{-} \geq 0 \qquad i=1,\ldots,m$$
$$s_{r0}^{+} \geq 0 \qquad r=1,\ldots,s$$

To avoid problems with zeros, the slacks corresponding to variables with $R_{i0}^{-}=0$, for some $i=1,\ldots,m$, and/or $R_{r0}^{+}=0$, for some $r=1,\ldots,s$, are ignored.

Model (4) can be seen as the natural extension to the non-radial case of RDM. In fact, if we impose equiproportionate both input contractions and outputs expansions, which can be done by adding to (3) the constraints $s_{i0}^- = \beta_0 R_{i0}^-, i = 1,...,m,$ and $s_{r0}^+ = \beta_0 R_{r0}^+, r = 1,...,s,$ and, at the same time replacing the equalities by the corresponding inequalities in the constraints involving inputs and outputs, then this latter model becomes the RDM. In other words, RDM can be seen as a specification of model (4).

Furthermore, $\Gamma_0^* = 1 - Z_0^*$ can be used as an efficiency measure. In fact, this measure was proposed in Cooper et al. (1999) and satisfies some desirable properties such as $0 \leq \Gamma_0^* \leq 1$, $\Gamma_0^* = 1$ if, and only if, DMU_0 is efficient and Γ_0^* is strongly monotonic. In addition, (4) is also translation invariant and units invariant (see Lovell and Pastor, 1995).

For the same reason as in the case of RDM, (4) can also be used with interval scale data.

6.2   Target setting

One of the drawbacks of the additive model is that it yields the furthest targets on the efficient frontier in terms of the L_1-distance. This also happens with all of the additive-type DEA models, such as model (4), since they all somehow maximize the slacks (to be precise, they maximize a weighted sum of the slacks). This is why Pastor and Ruiz propose following the approach in Aparicio et al. (2005) in order to maximize the value of the efficiency measure Γ_0^*, which will be obtained with reference to the closest projection on the efficient frontier. With this approach the slacks will thus be minimized instead of maximized, and so, the resulting targets will be the closest ones according to the efficiency measure currently used, which is in addition a distance in a mathematical sense (a weighted L_1-distance). The formulation to be solved is the following:

Minimize $\quad z_0 = \dfrac{1}{m+s}\left(\displaystyle\sum_{i=1}^{m}\dfrac{s_{i0}^{-}}{R_{i0}^{-}} + \sum_{r=1}^{s}\dfrac{s_{r0}^{+}}{R_{r0}^{+}}\right)$

subject to:

$$\sum_{j\in E}\lambda_j x_{ij} = x_{i0} - s_{i0}^{-} \qquad i=1,\ldots,m \qquad (5.1)$$

$$\sum_{j\in E}\lambda_j y_{rj} = y_{r0} + s_{r0}^{+} \qquad r=1,\ldots,s \qquad (5.2)$$

$$\sum_{j\in E}\lambda_j = 1 \qquad (5.3)$$

$$-\sum_{i=1}^{m}v_i x_{ij} + \sum_{r=1}^{s}\mu_r y_{rj} + u_0 + d_j = 0 \qquad j\in E \qquad (5.4)$$

$$v_i \geq 1 \qquad i=1,\ldots,m \qquad (5.5)$$

$$\mu_r \geq 1 \qquad r=1,\ldots,s \qquad (5.6)$$

$$d_j \leq M b_j \qquad j\in E \qquad (5.7)$$

$$\lambda_j \leq 1 - b_j \qquad j\in E \qquad (5.8)$$

$$b_j \in \{0,1\} \qquad j\in E \qquad (5.9)$$

$$d_j, \lambda_j \geq 0 \qquad j\in E$$

$$s_{i0}^{-} \geq 0 \qquad i=1,\ldots,m$$

$$s_{r0}^{+} \geq 0 \qquad r=1,\ldots,s$$

$$u_0 \text{ free}$$

(5)

where M is a big positive quantity and E is the set of extreme efficient units.

Aparicio et al. (2005) prove that the set of constraints in the previous model provides a characterization of the set of points on the efficient frontier dominating DMU$_0$ so that the closest projection on the frontier will be obtained by minimizing the distance from this unit to these points. The rationale of this model is the following: with the constraints (5.1)-(5.3) we have that (X_0, Y_0) is projected onto a point, which dominates it, that is a combination of extreme efficient units, which in particular means that the resulting projection belongs to the production possibility set (PPS). This first subgroup of restrictions corresponds to the restrictions of the primal formulation. The second subgroup of constraints (5.4)-(5.6) is the one corresponding to the multiplier formulation of the additive DEA model, with the particularity that we have considered the restrictions (5.4) not for all the DMUs but only for the extreme efficient ones (see Ali and Seiford (1993) for the dual formulation of the additive model). With this second subgroup of constraints we allow for all the hyperplanes such that all the points of the PPS lie on one of the half-spaces determined by these hyperplanes. The third subgroup of constraints (5.7)-(5.8) includes the key restrictions that connect the two subgroups of constraints previously mentioned. Note that if $b_j = 0$ then (5.7) implies that $d_j = 0$, i.e., DMU$_j$ belongs to the hyperplane $-\sum_{i=1}^{m} v_i x_i + \sum_{r=1}^{s} \mu_r y_r = 0$. Moreover, $\lambda_j > 0$ only happens if $b_j = 0$, i.e., this model only considers the projections of (X_0, Y_0) that are a combination of extreme efficient units belonging to the same supporting hyperplane. In other words, with (5) we only allow for the dominating projection of (X_0, Y_0) that are on a facet of the frontier, which, in addition, is a facet of the efficient frontier since all the weights are strictly positive. Therefore, the minimum value of the distance $\frac{1}{m+s}\left(\sum_{i=1}^{m} \frac{s_{i0}^-}{R_{i0}^-} + \sum_{r=1}^{s} \frac{s_{r0}^+}{R_{r0}^+}\right)$ from (X_0, Y_0) to the efficient frontier is the optimal value of (5).

For the same reasons mentioned above, $\gamma_0^* = 1 - z_0^*$ can also be used as a DEA efficiency measure. And, in addition, it will provide the closest targets according to a weighted L$_1$-distance, which is a distance in a mathematical sense.

7. ILLUSTRATIVE EXAMPLE

In this section, we use the data recorded in Table 4-2 to illustrate how the different approaches revised in this work perform. These correspond to a small example in which the relative efficiency of 5 DMUs each consuming 2 inputs to produce 1 output is to be assessed. It is remarkable that some of the DMUs have negative data. In fact, this is a consequence of $\overline{Y}$ being the difference between the variables Y and X (whose values are recorded between the parentheses in Table 4-2)

Table 4-2. Data Set

DMU	X_1	X_2	$\overline{Y}$	(Y,X)
A	1	3	-1	(2,3)
B	3	1	1	(3,2)
C	2	3	1	(4,3)
D	3	2	-2	(1,3)
E	3	3	-1	(1,2)

First, we should note that DMUs A, B and C are the efficient units in the sample (in a Pareto-Koopmans sense). In fact, they are extreme efficient points.

In this situation, Halme et al. propose using model (1) with the variables Y and X in the manner explained in section 3 instead of using the output $\bar{Y}$, since this is a interval scale variable. The results of that analysis are recorded in Table 4-3

Table 4-3. Model (1) Results

DMU	σ_0^*	λ_1^*	λ_2^*	λ_3^*	λ_4^*	λ_5^*	slacks X_1	X_2	Y	X
A	0	1	0	0	0	0	0	0	0	0
B	0	0	1	0	0	0	0	0	0	0
C	0	0	0	1	0	0	0	0	0	0
D	0.2	0.3	0.7	0	0	0	0	0	1.9	0.1
E	0	0	0	0	0	1	0	2	1	0

We can see that DMUs A, B and C obviously remain efficient and that DMU D has an efficiency score of 0.2 (with associated slacks strictly positive in some variables). Finally, DMU E is rated as weak-efficient, since it has strictly positive slacks, in spite of having an inefficiency score equals 0.

The approaches presented in sections 5 and 6 in the present chapter can be used directly with the Y, since they allow us to deal with interval scale variables. Table 4-4 records the results of the efficiency assessment of this data set with RDM. This model detects DMUs A, B and C as efficient and DMUs D and E as inefficient. To be precise, it yields a score (of inefficiency) of 1/3 to D and 1/2 to E. However, it should be pointed out that in the case of DMU D the score 1/3 is not associated with a Pareto-efficient projection on the frontier, since its radial projection has a slack of 4/3 in the output Y. Thus, we can see that RDM might lead to targets that are not Pareto-efficient.

Table 4-4. RDM Results

DMU	β_0^*	λ_1^*	λ_2^*	λ_3^*	slacks X_1	X_2	Y
A	0	1	0	0	0	0	0
B	0	0	1	0	0	0	0
C	0	0	0	1	0	0	0
D	1/3	1/3	2/3	0	0	0	4/3
E	1/2	1/2	1/2	0	0	0	0

This drawback is avoided with model (4), which provides efficiency measures in the Pareto-sense (see Table 4-5). As could be expected (remember that, as explained in section 6, (4) extends RDM in the sense that

β_0^* can be obtained as the optimal value of a model that results from including a group of restrictions into its set of constraints, i.e., both set of constraints are nested), the inefficiency score of DMU D (2/3) is now higher than that of RDM (1/3), since the former inefficiency measure accounts for all the sources of inefficiency.

Table 4-5. Model (4) Results

DMU	Z_0^*	λ_1^*	λ_2^*	λ_3^*	s_{10}^{-*}	s_{20}^{-*}	s_{10}^{+*}
D	2/3	0	1	0	0	1	3
E	1/2	0	1	0	0	2	2

However, as explained before, model (4) provides efficiency measures obtained with reference to the furthest point on the efficient frontier. For this reason, we propose using model (5), which provides the closest targets associated with the inefficiency measure $\frac{1}{m+s}\left(\sum_{i=1}^{m}\frac{s_{i0}^-}{R_{i0}^-}+\sum_{r=1}^{s}\frac{s_{r0}^+}{R_{r0}^+}\right)$, which is also a distance in a mathematical sense (a weighted L_1-distance). We can see in Table 4-6 that, as could be expected, the inefficiency scores are now lower than those obtained with model (4), since they are associated with a closer efficient projection. In fact, (5) provides the closest targets. For instance, we can see in Table 4-5 that DMU E can achieve the efficiency by simply reducing X_1 by 2 units (which means an effort of 100% with respect to the range of possible improvement in this variable), whereas model (4) results for this DMU (see Table 4-5) would be indicating that it should both reduce X_2 by 100% and increase Y by 100% (with respect also to the range of possible improvement for Y) in order to achieve the efficiency, which is obviously a greater effort (this in real-life applications might prove hard to achieve, at least in the short term).

Table 4-6. Model (5) Results

DMU	z_0^*	λ_1^*	λ_2^*	λ_3^*	s_{10}^{-*}	s_{20}^{-*}	s_{10}^{+*}
D	7/18	1/2	1/2	0	1	0	2
E	1/3	1	0	0	2	0	0

Finally, we would like to point out that, if we follow the approach by Seiford and Zhu (2002), then DMU A should be re-evaluated since it is an efficient unit with a negative output value. However, model (2) is unbounded for this DMU, as a consequence of the fact that the efficient frontier has no FDEF (in this particular case, DMU A must be projected onto the facet generated by DMUs B and C). Let us now consider the situation in which the point (1,4,3.5) were added to the sample. This would generate a FDEF together with B and C, and then model (2) would have the value 1.5 (greater than 1) as the re-estimated efficiency score, which would be associated with the optimal solution $v_1 = 0.5$, $v_2 = 0.25$, $\mu = 0.5$ and $u_0 = 0.25$ (which are the coefficients of the hyperplane containing the mentioned FDEF).

8. CONCLUSIONS

In this chapter we have made a revision of the approaches dealing with negative data in DEA. The precedents were quite primitive, based on the idea of either transforming the data or eliminating the DMUs with negative data. The classical approach gets rid of the negative data just by translating them. Consequently, it studies the translation invariance of the different DEA models. In the most recent contributions the idea is to develop models that provide efficiency measures that can be readily interpreted without the need to transform the data. We have also paid attention to the issue of setting targets. To be specific, we have also been concerned with finding the closest target to the unit under assessment on the efficient frontier.

REFERENCES

1. Ali, A.I. and Seiford, L.M. (1990) "Translation invariance in data envelopment analysis", *Operations Research Letters*, 9, 403-405.
2. Ali, A.I. and Seiford, L.M. (1993) "The mathematical programming approach to efficiency analysis", in Fried, H. O., C. A. K. Lovell, S. S. Schmidt, eds. *The measurement of productive efficiency: techniques and applications*, New York, NY: Oxford University Press.
3. Aparicio, J., Ruiz, J.L. and Sirvent, I. (2005) "Closest targets and minimum distance to the efficient frontier in DEA", Trabajos I+D, I-2005-22, Centro de Investigación Operativa, Universidad Miguel Hernández, Spain.
4. Banker, R.D., Charnes, A. and Cooper, W.W. (1984) "Some Models for Estimating Technical and Scale Inefficiencies in Data Envelopment Analysis", *Management Science,* 30(9), 1078-1092.
5. Banker, R.D. and Morey, R.C. (1986) "Efficiency analysis for exogenously fixed inputs and outputs", *Operations Research*, 34(4), 513-521.
6. Briec, W. (1997) "A Graph-Type Extension of Farrell Technical Efficiency Measure", *Journal of Productivity Analysis*, 8 (1), 95-110.
7. Chambers, R.G., Chang, Y. and Färe, R. (1996) "Benefit and distance functions", *Journal of Economic Theory*, 70, 407-419.

8. Charnes, A., Cooper, W.W. and Rhodes, E. (1978) "Measuring the efficiency of decision making units", *European Journal of Operational Research,* 2/6, 429-444.
9. Charnes, A., Cooper, W.W. and S. Li (1989), "Using Data Envelopment Analysis to evaluate Efficiency in the Economic Performance of Chinese Cities", *Socio-Economic Planning Sciencies,* 23(6), 325-344.
10. Charnes, A., Cooper, W.W. and Thrall, R.M. (1991) "A structure for classifying and characterizing efficiency and inefficiency in Data Envelopment Analysis", *Journal of Productivity Analysis*, 2, 197-237.
11. Charnes, A., Cooper, W.W., Golany, B., Seiford, L. and Stutz, J. (1985) "Foundations of Data Envelopment Analysis for Pareto-Koopmans Efficient Empirical Productions Functions", *Journal of Econometrics*, 30, 91-107.
12. Charnes, A., Cooper, W.W. and R.M. Thrall (1986), "Classifying and characterizing efficiencies and inefficiencies in data envelopment analysis", *Operations Research Letters, 5(3), 105-110.*
13. Charnes, A., Cooper, W.W., Seiford, L. and Stutz, J. (1983), "Invariant multiplicative efficiency and piecewise Cobb-Douglas envelopments", *Operations Research Letters,* 2(3), 101-103.
14. Cooper, W.W., Park, K.S. and Pastor, J.T. (1999) "RAM: A range adjusted measure of inefficiency for use with additive models, an relations to other models and measures in DEA", *Journal of Productivity Analysis*, 11, 5-42.
15. Cooper, W.W., Ruiz, J.L. and Sirvent, I. (2006) "Choosing weights from alternative optimal solutions of dual multiplier models in DEA", *European Journal of Operational Research*, to appear.
16. Halme, M., Joro, T and Koivu, M. (2002), "Dealing with interval scale data in data envelopment analysis", *European Journal of Operational Research*, 137, 22-27.
17. Lovell, C.A.K. (1995), "Measuring the Macroeconomic Performance of the Taiwanese Economy", *International Journal of Production Economics*, 39, 165-178.
18. Lovell, C.A.K. and J.T. Pastor (1995), "Units invariant and translation invariant DEA models", *Operations Research Letters,* 18, 3, 147-151.
19. Lozano-Vivas, A., J.T. Pastor and J.M. Pastor (2002), "An Efficiency Comparison of European Banking Systems Operating under Different Environmental Conditions", *Journal of Productivity Analysis,* 18, 59-77.
20. Olesen, O. and Petersen, N.C. (1996) "Indicators of ill-conditioned data sets and model misspecification in Data Envelopment Analysis: An extended facet approach", *Management Science,* 42, 205-219.
21. Pastor J.T. and Ruiz, J.L. (2005), "On the treatment of negative data in DEA", Working Paper, Centro de Investigación Operativa, Universidad Miguel Hernandez, Spain.

22. Pastor, J.T. (1993), "Efficiency of Bank Branches through DEA: the attracting of liabilities", presented at the *IFORS93, Lisbon.*
23. Pastor, J.T. (1994), "New Additive DEA Models for handling Zero and Negative Data", Working Paper, Universidad de Alicante (out of print).
24. Pastor, J.T. (1996), "Translation invariance in data envelopment analysis: A generalization", *Annals of Operations Research*, 66, 93-102.
25. Seiford, L.M. and Zhu, J. (2002) "Classification invariance in data envelopment analysis", in Uncertainty and Optimality, Probability, Statistics and Operations Research, J.C. Misra (Ed.), World Scientific.
26. Silva Portela, M.C.A., Thanassoulis, E. and Simpson, G. (2004), "Negative data in DEA: a directional distance function approach applied to bank branches", *Journal of the Operational Research Society*, 55, 1111-1121.
27. Thore, S., G. Kozmetski and F. Philips (1994), "DEA of Financial Statements Data: The U.S. Computer Industry, later published in *Journal of Productivity Analysis,* 5(3), 229-248.
28. Thrall, R.M. (1996), "The lack of invariance of optimal dual solutions under translation", *Annals of Operations Research,* 66, 103-108.
29. Zhu, J. (1994), "Translation Invariance in Data Envelopment Analysis and its use", *Working Paper.*

Chapter 5

NON-DISCRETIONARY INPUTS

John Ruggiero
*Department of Economics and Finance, University of Dayton, Dayton, OH 45469-2251,
ruggiero@notes.udayton.edu*

Abstract: A motivating factor for Data Envelopment Analysis (DEA) was the measurement of technical efficiency in public sector applications with unknown input prices. Many applications, including public sector production, are also characterized by heterogeneous producers who face different technologies. This chapter discusses existing approaches to measuring performance when non-discretionary inputs affect the transformation of discretionary inputs into outputs. The suitability of the approaches depends on underlying assumptions about the relationship between non-discretionary inputs and outputs. One model treats non-discretionary inputs like discretionary inputs but uses a non-radial approach to project inefficient decision making units (DMUs) to the frontier holding non-discretionary inputs fixed. A potential drawback is the assumption that convexity holds with respect to non-discretionary inputs and outputs. Alternatively, the assumption of convexity can be relaxed by comparing DMUs only to other DMUs with similar environments. Other models use multiple stage models with regression to control for the effect that non-discretionary inputs have on production.

Key words: Data Envelopment Analysis (DEA), Performance, Efficiency, Non-discretionary inputs

1. INTRODUCTION

Data Envelopment Analysis is programming modeling approach for estimating technical efficiency assuming constant returns to scale introduced by Charnes, Cooper and Rhodes (1978) (CCR) and extended by Banker, Charnes and Cooper (1984) (BCC) to allow variable returns to scale. An underlying assumption necessary for the comparison of decision making units (DMUs) in both models is homogeneity – production possibilities of each and every DMU can be represented by the same piecewise linear technology. In many applications this assumption is not valid. For example, in education production depends not only on the levels of discretionary inputs like capital and labor but also on the socio-economic environment under which the DMU operates. Schools facing a harsher environment, perhaps represented by high poverty, will not be able to achieve outcome levels as high as other schools with low poverty levels even if they use the same amount of discretionary inputs.

Application of the standard DEA models to evaluate efficiency leads to biased estimates if the non-discretionary inputs are not properly controlled. Banker and Morey (1986) (BM) extended DEA to allow fixed, uncontrollable inputs by projecting inefficient DMUs to the frontier non-radially so that the resulting efficient DMU had the same levels of the non-discretionary inputs. The BM approach is consistent with short-run economic analyses with fixed inputs along the appropriate expansion path. As such, the model could be used for short-run analyses characterized by fixed inputs. However, the BM model assumes convexity with respect to outputs and non-discretionary inputs. In situations where such an assumption is not valid, the BM model will overestimate technical efficiency by allowing production impossibilities into the referent set.

An alternative approach, introduced by Ruggiero (1996) restricts the weights to zero for production possibilities with higher levels of the non-discretionary inputs. As a result, production impossibilities are appropriately excluded from the referent set. A limitation of this model is the increased information set necessary to estimate efficiency. In particular, excluding DMUs from the analysis eliminates potentially useful information and could lead to classification of efficient by default. However, simulation analyses in Ruggiero (1996), Syrjanen (2004) and Muñiz, Paradi, Ruggiero and Yang (2006) show that the model developed by Ruggiero (1996) performs relatively well in evaluating efficiency.

The problem of limited information in the Ruggiero model is magnified when there are multiple non-discretionary inputs. As the number of non-discretionary inputs increases, efficiency scores are biased upward and the number of DMUs classified as efficient by default increases. In such cases, multiple stage models have been incorporated. Ray (1991) introduced one such model where standard DEA models are applied in the first stage using only discretionary inputs and outputs. The resulting index, which captures not only inefficiency but also non-discretionary effects, is regressed in a second stage on the non-discretionary inputs to disentangle efficiency from environmental effects. The residual from the regression represents efficiency. Ruggiero (1998) showed that the Ray model works very well in controlling for non-discretionary inputs using rank correlation between true and estimated efficiency as the criterion. However, the approach overestimates efficiency and is unable to properly distinguish between efficient and inefficient DMUs.

An alternative multi-stage model was introduced by Ruggiero (1998) that extends Ray (1991) and Ruggiero (1996). Like the Ray model, this model uses DEA on discretionary inputs and outputs in the first stage and regression in the second stage. Unlike Ray, the residual from the regression is not used as the measure of efficiency. Instead, the regression is used to construct an overall index of environmental harshness, which is used in a third stage model using Ruggiero (1996). Effectively, this approach reduces the information required by using only one index representing overall non-discretionary effects. The simulation in Ruggiero (1998) shows that the multiple stage model is preferred to the original model of Ruggiero when there are multiple non-discretionary inputs.

Other models exist to control for environmental effects. Muñiz (2002) introduced a multiple stage model that focuses on excess slack and uses DEA models in all stages to control for the non-discretionary inputs. Yang and Paradi (2003) presented an alternative model that uses a handicapping function to adjust inputs and outputs to compensate for the non-discretionary factors. Muñiz, Paradi, Ruggiero and Yang (2006) compare the various approaches using simulated data and conclude that the multi-stage Ruggiero model performed well relative to all other approaches. Each approach has its own advantages. For purposes of this chapter, the focus will be on the BM model, Ray's model and the Ruggiero models.

The rest of the chapter is organized as follows. The next section presents the BM model to control for non-discretionary inputs. Using simulated data sets, the advantages and disadvantages are discussed. The models due to Ruggiero are then examined in the same way. The last section concludes.

2. PRODUCTION WITH NON-DISCRETIONARY INPUTS

We assume that DMU_j ($j = 1, \ldots, n$) uses m discretionary inputs x_{ij} ($i = 1,\ldots,m$) to produce s outputs y_{kj} ($k = 1,\ldots,s$) given r non-discretionary inputs z_{lj} ($l = 1,\ldots,r$). For convenience, we assume that in an increase in any non-discretionary input leads to an increase in at least one output, *ceteris paribus*. As a result, higher levels of the non-discretionary inputs lead to a more favorable production environment. Naïve application of the either the CCR or the BCC models, without properly considering the non-discretionary variables, leads to biased efficiency estimates. In this section, we consider such applications using the BCC model. The same arguments apply to the CCR model as well.

Consider first an application of the BCC model that only considers outputs and discretionary inputs. This DEA model can be written as

$$\theta_o^{1*} = \min \theta_o$$

subject to

$$\sum_{j=1}^{n} \lambda_j y_{kj} \geq y_{ko} \quad k = 1,...,s;$$

$$\sum_{j=1}^{n} \lambda_j x_{ij} \leq \theta_o x_{io} \quad i = 1,...,m; \quad (1)$$

$$\sum_{j=1}^{n} \lambda_j = 1$$

$$\lambda_j \geq 0 \quad j = 1, 2,..., n.$$

This formulation ignores the important role of the non-discretionary inputs. For illustrative purposes, we show how model (1) leads to biased efficiency estimates. In Figure 5-1, it is assumed that six $A - F$ produce one output using one discretionary input and one non-discretionary input. It is further assumed that all DMUs are operating efficiently. DMUs $A - C$ [$D - F$] all have the same non-discretionary input z_{10} (with $z_{1A} = z_{1B} = z_{1C} = z_{10}$) [$z_{11}$] where $z_{10} < z_{11}$. Because DMUs $D - F$ have a higher level of the non-discretionary input, each is able to produce a higher level of output y_1 with input usage x_1.

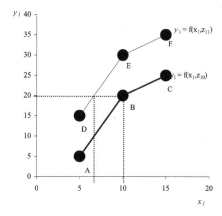

Figure 5-1. Production with Non-discretionary Inputs

DMUs $A - F$ are all on the frontier producing output with the minimum discretionary input level possible. Effectively, multiple frontiers arise depending on the level of the non-discretionary input. This set-up is consistent with microeconomic principles of short-run production with fixed inputs. Application of the BCC model results in incorrect classification of inefficiency for DMUs $A - C$ with projections to the frontier consistent with non-discretionary input level z_{11}. In the case of DMU A, "inefficiency" results from output slack and not radial contraction of the discretionary input. Efficient DMU C achieves an efficiency rating of 0.44 with a weighting structure of ⅔ (⅓) on DMU D (E). The resulting index is not capturing production inefficiency but rather an index of inefficiency and the effect that non-discretionary inputs have on production.

An alternative model would apply the BCC model with non-discretionary inputs treated the same as the discretionary inputs. This approach was adopted by Bessent, Bessent, Kennington and Reagan (1982) in an analysis of elementary schools in the Houston Independent District. The BCC version of this model is given by:

$$\theta_o^{2*} = \min \theta_o$$

subject to

$$\sum_{j=1}^{n} \lambda_j y_{kj} \geq y_{ko} \quad k = 1,...,s;$$

$$\sum_{j=1}^{n} \lambda_j x_{ij} \leq \theta_o x_{io} \quad i = 1,...,m; \quad (2)$$

$$\sum_{j=1}^{n} \lambda_j z_{ij} \leq \theta_o z_{lo} \quad l = 1,...,r;$$

$$\sum_{j=1}^{n} \lambda_j = 1$$

$$\lambda_j \geq 0 \quad j = 1, 2,..., n.$$

The problem with this model is the potential contraction of the non-discretionary inputs, which violates the assumption that these variables are exogenously fixed. The resulting projection for inefficient DMUs will be to a production possibility inconsistent with the DMUs' fixed level on the non-discretionary input. This is shown in Figure 5-2, where four DMUs $A - D$ are observed producing the same output with one discretionary input x_1 and one nondiscretionary input x_2. Application of (2) for DMU D leads to a peer comparison of DMU B. This is problematic because it implies that DMU D

is able to reduce the amount of exogenously fixed input x_2 from 20 units to 10.

There have been a few different approaches to measuring efficiency relative while controlling for the fixed levels of the nondiscretionary inputs. In the next section, we consider the Banker and Morey (1986) model that introduced the problem of inappropriate projections via equi-proportional reduction in all inputs. In particular, nondiscretionary variables are treated as exogenous and projections to the frontier are based on holding the nondiscretionary inputs fixed.

3. THE BANKER AND MOREY MODEL

3.1 Input-Oriented Model

Banker and Morey (1986) introduced the first DEA model that allowed for nondiscretionary inputs by modifying the input constraints to disallow input reduction on the fixed factor. The input-oriented model seeks the equi-proportional reduction in discretionary inputs while holding non-discretionary inputs and outputs constant.

The differences between this model and the standard BCC model are (i) the separation of input constraints for discretionary and non-discretionary inputs and (ii) the elimination of the efficiency index from the non-discretionary inputs. Importantly, this programming model provides the proper comparison for a given DMU by fixing the nondiscretionary inputs at the current level. Returning to Figure 5-2, the solution of the BM model for DMU *D* leads to a benchmark of DMU *A* which is observed producing the same level of output with the same level of the non-discretionary input.[1]

---

[1] On page 516 of the Banker and Morey (1986) paper, the CCR equivalent model is presented. Equation 32 multiplies the observed nondiscretionary input by the sum of the weights. However, decreasing returns to scale will lead to frontier projections to levels of the nondiscretionary inputs above the observed level. This appears to be at typo; in the paragraph above this model, the authors state that the CCR version of their model is obtained via the standard method of removing the convexity constraint. The simulation analysis in this section confirms this point.

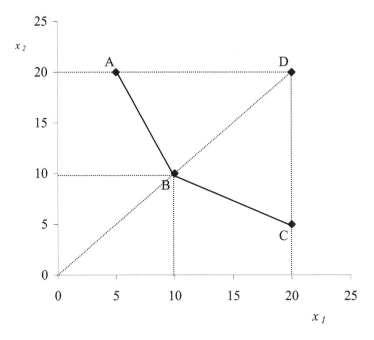

Figure 5-2. Radial Contraction

Assuming variable returns to scale, the BM model for evaluating the efficiency of DMU "o" is given by:

$$\theta_o^{3*} = \min \theta_o$$
subject to
$$\sum_{j=1}^{n} \lambda_j y_{kj} \geq y_{ko} \quad k = 1,\ldots,s;$$
$$\sum_{j=1}^{n} \lambda_j x_{ij} \leq \theta_o x_{io} \quad i = 1,\ldots,m; \quad (3)$$
$$\sum_{j=1}^{n} \lambda_j z_{lj} \leq z_{lo} \quad l = 1,\ldots,r;$$
$$\sum_{j=1}^{n} \lambda_j = 1$$
$$\lambda_j \geq 0 \quad j = 1,2,\ldots,n.$$

3.2 Illustrative Example Using Simulated Data

In this sub-section, the BM model is applied to simulated data for illustrative purposes. We assume that efficient production is characterized by constant returns to scale where two inputs are used to produce one output according to the following production function:

$$y_{1j} = x_{1j}^{0.4} x_{2j}^{0.6}.$$

Both inputs are generated from a normal distribution with mean 100 and standard deviation 25. It is assumed that input x_2 is fixed. Ln θ is generated from a |N(0,0.2)| distribution and observed production data is calculated in two ways:

Scenario 1: $y_{1j} = \theta_j x_{1j}^{0.4} x_{2j}^{0.6}$; and

Scenario 2: observed x_{1j} = efficient x_{1j}/θ_j.

The first scenario leads to inefficiency in the output, but since the production function is characterized by constant returns to scale, the input-oriented constant returns to scale model should be appropriate. The second scenario assumes inefficient usage of only the discretionary input. Tables 3-1 and 3-2 present the data and efficiency results using the BM model for scenarios 1 and 2, respectively.

The results in Table 5-1 show that the BM has a tendency to overestimate inefficiency given the data generating process in scenario 1. Notably, the mean absolute difference (MAD) between true and estimated efficiency is 0.10. However, the rank correlation is a high 0.95. The overestimation arises because efficiency is measured relative to the discretionary inputs only, ignoring the contribution to output from the nondiscretionary inputs. This problem was identified in Ruggiero (2005) which provided an alternative BM model consistent with this data generating process. To evaluate the performance of DMU "o" under variable returns to scale, the alternative model is:

$$\theta_o^{4*} = \min \frac{m}{m+r}\theta_o + \frac{r}{m+r}$$

subject to

$$\sum_{j=1}^{n} \lambda_j y_{kj} \geq y_{ko} \quad k = 1,\ldots,s;$$

$$\sum_{j=1}^{n} \lambda_j x_{ij} \leq \theta_o x_{io} \quad i = 1,\ldots,m; \quad (4)$$

$$\sum_{j=1}^{n} \lambda_j z_{lj} \leq z_{lo} \quad l = 1,\ldots,r;$$

$$\sum_{j=1}^{n} \lambda_j = 1$$

$$\lambda_j \geq 0 \quad j = 1, 2,\ldots,n.$$

Table 5-1. Simulated Data, Scenario 1

DMU	x_1	x_2	y_1	θ	BM	Alternative BM
1	92.49	68.06	67.48	0.88	0.73	0.86
2	106.11	131.91	113.06	0.94	0.88	0.94
3	129.96	143.33	133.88	0.97	0.96	0.98
4	45.41	94.15	62.37	0.89	0.86	0.93
5	127.38	72.83	88.31	0.97	0.98	0.99
6	82.74	57.74	42.66	0.64	0.36	0.68
7	53.83	75.56	64.24	0.97	0.97	0.99
8	80.66	47.05	44.75	0.77	0.53	0.76
9	85.80	89.90	85.52	0.97	0.95	0.97
10	103.37	90.86	86.52	0.90	0.80	0.90
11	91.83	90.74	70.06	0.77	0.55	0.77
12	133.57	97.87	97.92	0.88	0.74	0.87
13	95.35	87.17	77.80	0.86	0.71	0.86
14	149.31	121.64	128.22	0.97	0.95	0.98
15	159.39	83.63	103.35	0.95	0.95	0.98
16	141.54	59.69	80.87	0.96	1.00	1.00
17	113.47	122.55	86.69	0.73	0.51	0.76
18	147.97	97.89	114.61	0.99	1.00	1.00
19	86.91	116.88	100.83	0.97	0.97	0.98
20	90.47	118.94	73.93	0.69	0.48	0.74
21	63.90	78.82	63.88	0.88	0.76	0.88
22	61.96	90.93	60.49	0.78	0.61	0.81
23	99.19	100.70	99.79	1.00	1.00	1.00
24	91.93	154.86	124.10	0.99	1.00	1.00
25	56.44	81.59	51.02	0.72	0.53	0.77
26	35.56	136.19	74.03	0.93	1.00	1.00
27	68.01	83.66	70.56	0.92	0.84	0.92
28	118.94	111.67	109.26	0.95	0.90	0.95
29	121.87	114.89	102.61	0.87	0.74	0.87
30	65.70	72.11	68.53	0.99	0.99	0.99
31	117.35	108.07	96.40	0.86	0.72	0.86
32	76.50	93.98	69.50	0.80	0.63	0.82
33	103.29	113.94	105.18	0.96	0.93	0.97
34	103.47	77.23	71.34	0.82	0.63	0.82
35	147.12	112.18	84.83	0.68	0.41	0.70

All calculations by author.

Table 5-2. Simulated Data, Scenario 2

DMU	x_1	x_2	y_1	θ	BM	Alternative BM
1	105.47	68.06	76.94	0.88	0.89	0.95
2	113.47	131.91	120.91	0.94	0.96	0.98
3	133.78	143.33	137.82	0.97	0.98	0.99
4	51.21	94.15	70.33	0.89	0.94	0.97
5	131.37	72.83	91.08	0.97	1.00	1.00
6	129.34	57.74	66.68	0.64	0.65	0.82
7	55.28	75.56	65.97	0.97	1.00	1.00
8	105.23	47.05	58.37	0.77	0.79	0.89
9	88.53	89.90	88.24	0.97	0.98	0.99
10	114.31	90.86	95.67	0.90	0.92	0.96
11	119.50	90.74	91.17	0.77	0.77	0.89
12	151.18	97.87	110.83	0.88	0.90	0.95
13	110.73	87.17	90.35	0.86	0.87	0.94
14	153.74	121.64	132.03	0.97	0.99	1.00
15	166.94	83.63	108.24	0.95	0.99	1.00
16	147.56	59.69	84.31	0.96	1.00	1.00
17	155.56	122.55	118.84	0.73	0.74	0.87
18	149.09	97.89	115.48	0.99	1.00	1.00
19	89.48	116.88	103.81	0.97	1.00	1.00
20	130.46	118.94	106.61	0.69	0.71	0.86
21	72.49	78.82	72.47	0.88	0.90	0.95
22	79.90	90.93	77.99	0.78	0.80	0.90
23	99.49	100.70	100.09	1.00	1.00	1.00
24	93.12	154.86	125.71	0.99	1.00	1.00
25	77.88	81.59	70.41	0.72	0.74	0.87
26	38.24	136.19	79.59	0.93	1.00	1.00
27	74.22	83.66	77.01	0.92	0.94	0.97
28	124.67	111.67	114.52	0.95	0.97	0.98
29	139.71	114.89	117.63	0.87	0.88	0.94
30	66.61	72.11	69.47	0.99	1.00	1.00
31	135.96	108.07	111.69	0.86	0.88	0.94
32	95.28	93.98	86.55	0.80	0.82	0.91
33	107.58	113.94	109.56	0.96	0.97	0.99
34	125.91	77.23	86.81	0.82	0.84	0.92
35	216.84	112.18	125.03	0.68	0.69	0.85

All calculations by author.

The alternative BM model developed by Ruggiero (2005) provides a closer approximation to the true efficiency, achieving a MAD of 0.016, 84 percent lower than the original BM model. However, the rank correlation of 0.98 is the same.

The data generating process for scenario 2 assumes inefficiency only in the discretionary input. Efficient production was initially assumed and x_1 was rescaled by the inverse of the efficiency index. The BM models described above were then applied; the data and results are reported in Table 5-2. In contrast to scenario 1, the BM model outperforms the alternative BM model using the MAD criterion. The MAD for the standard BM model (0.019) is 72 percent lower than the MAD for the alternative BM model (0.069). The rank correlation between true and estimated efficiency for both models was 0.93. The simulations used in this section show that the choice of the standard or alternative BM model depends on the assumptions of the data generating process. The DGP for scenario 1 is consistent with the definition of efficiency as the ratio of observed to efficient production based on the underlying production function. The DGP for scenario 2 suggests that nondiscretionary inputs are held fixed at the associated efficient level.

4. ALTERNATIVE DEA MODELS

Alternative models have been developed to control for nondiscretionary inputs. Ray (1991) introduced a two-stage model that first applies DEA model (1) using only discretionary inputs and output and then uses regression to control for the effect nondiscretionary variables have on production. The other main model, due to Ruggiero (1996), adds constraints on the weights to ensure that any potential peer DMU has a level of nondiscretionary input no higher than the DMU under analysis.

4.1 Two-Stage Model Using DEA and Regression

The Ray (1991) model was developed as an alternative to the BM model. Rather than assuming convexity and treating the nondiscretionary input within the DEA model, DEA is applied using only discretionary inputs and outputs. This formulation is consistent with model (1). However, a second-stage regression is used to decompose the index from (1) into two components: the effect that nondiscretionary variables have on the transformation of discretionary inputs into outputs and (2) technical inefficiency. Give the results from (1), the following regression is performed:

$$\theta_j^{1*} = f(z_{1j},...,z_{rj},\varepsilon_j). \tag{5}$$

The use of regression to control for nondiscretionary variables is intuitive; the resulting error term represents unobserved technical efficiency. However, there are problems associated with regression that must be overcome. Firstly, a functional form must be assumed. Based on the nonparametric nature of the first stage DEA model, it is not clear what the proper functional form is. Normally, one could assume a Cobb-Douglas functional form and apply ordinary least squares (OLS) on the data in natural logarithm form. Secondly, the use of OLS may be inappropriate because the technical efficiency measure from the first stage is bounded above. McCarty and Yaisawarng (1993) applied the Ray model, using Tobit to estimate (5) instead of OLS. As part of the simulations below, this particular issue will be addressed.

The error term from the OLS regression represents technical efficiency. One potential limitation of the Ray approach is the two-sided mean zero residual. To get around this, one could apply the logic of the deterministic regression and adjust the intercept so that the adjusted residual is one-sided. For the purposes of the simulation in section 4.3, such an adjustment is made, leading to θ_j^{5*}.

4.2 Restricting Weights

A potential limitation of the BM model is the assumption of convexity with respect to the nondiscretionary inputs. This is not a problem, for example, when capital is fixed in the short-run and output levels are determined via an expansion path. Since capital is a discretionary input in the long run, the assumption of convexity is reasonable. For other applications, however, the assumption of convexity may be problematic. In education, output levels are determined by discretionary inputs (labor, capital, etc.) and environmental variables (parental education, income, etc.) In such an application, there is no theoretical justification for the assumption of convexity. Ruggiero (1996) introduced a model that modifies the weight structure to remove from the potential peer set any DMU that has a more favorable "environment."

Assuming variable returns to scale with respect to the discretionary inputs, the alternative model to evaluate the efficiency of DMU "o" is given by:

$$\theta_o^{6*} = \min \theta_o$$

subject to

$$\sum_{j=1}^{n} \lambda_j y_{kj} \geq y_{ko} \quad k = 1,\ldots,s;$$

$$\sum_{j=1}^{n} \lambda_j x_{ij} \leq \theta_o x_{io} \quad i = 1,\ldots,m; \qquad (6)$$

$$\sum_{j=1}^{n} \lambda_j = 1$$

$$\lambda_j \geq 0 \quad j = 1,2,\ldots,n;$$

$$\lambda_j = 0 \quad \forall l, j \ni z_{lj} > z_{lo}; \quad l = 1,\ldots,r; \quad j = 1,\ldots,n.$$

This model handles the nondiscretionary inputs differently than does BM. Notably, if a DMU has a higher level of any nondiscretionary input, the last constraint sets the weight to zero, disallowing the DMU from appearing in the referent set. In this case, the model seeks projection to the appropriate frontier determined by the level of the nondiscretionary inputs. Based on the assumptions of the model, there are potentially infinite output/discretionary frontiers that are nested. The CCR version of this model is easily obtained by deleting the convexity constraint.

Ruggiero (1998) showed that model (6) requires many DMUs because there tends to be classification of efficient by default as the number of nondiscretionary inputs increase. Ruggiero (1998) introduced a three-stage model to overcome this weakness. The first two stages are similar to Ray (1991), but the focus of the regression model (5) is to create an index of nondiscretionary effects. Instead of using the residual, the parameter estimates are used to construct the overall index. The index is then used as the only nondiscretionary input in model (6). Ruggiero (1998) showed using simulated data that the approach works well in estimating efficiency while maintaining the advantages of DEA.

4.3   Simulation Analysis

The simulation used in section 3.2 is insufficient for comparing the various models. First, the alternative models used above require more than 35 DMUs; in particular, restricting the weights via the Ruggiero model reduces the number of DMUs in the analysis of a given DMU. Secondly, the specification of the underlying production technology is inconsistent with the alternative models. The production function used for the BM model defines returns to scale with respect to the discretionary and nondiscretionary variables. This is inconsistent with models that assume that returns to scale are defined with respect to the discretionary inputs for given levels of the nondiscretionary inputs. The comparison to short vs. long run production provides the intuition. In the short run, capital is typically assumed fixed; however, in the long run all productive inputs are assumed variable. As a result, returns to scale are defined relative to long-run production. Nondiscretionary inputs are not variable even in the long-run. As a result, returns to scale should be considered relative to the discretionary inputs only.

For the simulation, we assume that output is produced using two discretionary inputs and one nondiscretionary input according to the following function:

$$y_{1j} = \theta_j z_{1j} x_{1j}^{0.4} x_{2j}^{0.6}.$$

Production is characterized by constant returns to scale with respect to the discretionary inputs for any given level of the nondiscretionary input. Note that the nondiscretionary input enters the production function multiplicatively. For this analysis, the number of DMUs was set at 1000. The discretionary inputs were generated normally with mean 100 and standard deviation 25; the nondiscretionary input was generated from a uniform distribution from 0 to 1. Finally, $Ln\ \theta$ was generated from a half-normal distribution with mean 0 and standard deviation 0.2. In this case, the data generating process is consistent with scenario 1 in section 3.2.

Table 5-3. Simulation Results

Model	Model Description	Correlation	Rank Correlation	MAD
1	Ray First Stage	0.20	0.17	0.40
2	Standard DEA	0.48	0.45	0.20
3	Banker and Morey VRS	0.50	0.48	0.13
4	Alternative BM	0.50	0.48	0.08
5	Ray Second Stage			
	OLS			
	Linear	0.83	0.89	0.22
	Log-Linear	0.93	0.97	0.27
	Tobit			
	Linear	0.83	0.89	0.22
	Log-Linear	0.93	0.97	0.27
6	Ruggiero Model	0.92	0.87	0.05

All calculations by author.

All models discussed above were used to measure efficiency. Performance of the estimators was based on correlation, rank correlation and MAD. In addition to the standard BM model, the alternative BM model is used. It is expected that the alternative model will achieve the same correlation coefficients but with a lower MAD. For the Ray model, linear and log-linear regression models were considered. OLS and Tobit regressions were both considered; the use of the Tobit model arises out of the truncation of the efficiency index at 1. The results of the analysis are reported in Table 5-3.

Some interesting results emerge from the analysis. Firstly, the log-linear specification is preferred to the linear specification for the second stage regression used by Ray. This may be a result of the assumed functional form for the production function. Secondly, there is no improvement in using Tobit over OLS in the second stage regression. For both the linear and log-linear models, the results between OLS and Tobit are essentially the same. Thirdly, the alternative BM model has the same (with rounding) correlation and rank correlation coefficients as the BM model but has a MAD that is 38 percent lower.

Other reported results are not as surprising. Given the data generating process, only the alternative models due to Ray (1991) and Ruggiero (1996) perform well with respect to the correlation and rank correlation measures. The log-linear specification for Ray's model performs the best achieving the highest correlations and rank correlations. However, the deterministic approach for adjusting the efficiency measure does not. Notably, while the correlations are the highest, the MADs are too. Hence, the Ray model does well in ranking performance but does not provide good approximations to the actual relative efficiency. Ruggiero's model, on the other hand, performs best in terms of the MAD and achieves relatively high correlation coefficients. These results confirm the results reported in Ruggiero (1996) and Ruggiero (1998).

5. CONCLUSIONS

Application of standard DEA models to analyze production technologies characterized by nondiscretionary inputs is problematic. Early applications treated the nondiscretionary inputs as discretionary, leading to improper frontier comparisons. Banker and Morey (1986) modified the standard DEA models to allow projection along expansion paths; this model correctly projects inefficient DMUs to a frontier point consistent with the fixed level of the nondiscretionary inputs. This model works well as long as production is convex with respect to the nondiscretionary inputs. Alternatively, in cases when the convexity assumption is too strong, one can employ the model developed by Ruggiero (1996). This model restricts the comparison set by removing all DMUs with more favorable levels of the nondiscretionary inputs.

In this chapter, various models for treating exogenously fixed nondiscretionary inputs were presented. As discussed above, a key consideration is the data generating process. Which model is preferred depends on the assumptions of the model. The BM model performs well assuming convexity with respect to the nondiscretionary inputs and observed discretionary outputs are scaled up due to inefficiency. However, if efficiency enters the production function via a more general output-oriented definition, the alternative BM model is preferred. Finally, the simulation results suggest that there are advantages to both the Ray and Ruggiero models in cases if the sample size is large enough.

REFERENCES

1. Banker, R, A. Charnes, W.W. Cooper (1984), "Some Models for Estimating Technical and Scale Inefficiencies in Data Envelopment Analysis", Management Science; 30; 1078-1092.
2. Banker, R, R. Morey (1986), "Efficiency Analysis for Exogenously Fixed Inputs and Outputs", Operations Research; 34; 513-521.
3. Bessent, A, W. Bessent, J. Kennington, B. Reagan (1982), "An Application of Mathematical-Programming to Assess Productivity in the Houston Independent School-District"; Management Science; 28; 1355-1367.
4. Charnes, A, W.W. Cooper, E. Rhodes (1978), "Measuring the Efficiency of Decision Making Units", European Journal of Operational Research; 2; 429-444.
5. McCarty, T, S. Yaisawarng (1993), "Technical Efficiency in New Jersey School Districts"; in H. Fried, C.A.K. Lovell, S. Schmidt (eds.), The Measurement of Productive Efficiency: Techniques and Applications; OxfordL Oxford University Press; 271-287.
6. Muñiz, M (2002), "Separating Managerial Inefficiency and External Conditions in Data Envelopment Analysis"; European Journal of Operational Research; 143; 625-643.
7. Muñiz, M, J. Paradi, J. Ruggiero, Z.J. Yang (2006), "Evaluating Alternative DEA Models Used to Control for Non-Discretionary Inputs", Computers & Operations Research; 33; 1173-1183.
8. Ray, S (1991), "Resource Use Efficiency in Public Schools. A Study of Connecticut Data", Management Science; 37; 1620-1628.

9. Ruggiero, J (1996), "On the Measurement of Technical Efficiency in the Public Sector", European Journal of Operational Research; 90; 553-565.
10. Ruggiero, J (1998), "Non-Discretionary Inputs in Data Envelopment Analysis"; European Journal of Operational Research; 111; 461-469.
11. Syrjanen, M.J. (2004), "Non-Discretionary and Discretionary Factors and Scale in Data Envelopment Analysis"; European Journal of Operational Research; 158; 20-33.

Chapter 6

DEA WITH UNDESIRABLE FACTORS

Zhongsheng Hua[1] and Yiwen Bian[2]
[1]*School of Management, University of Science and Technology of China, Hefei, Anhui 230026, People's Republic of China, zshua@ustc.edu.cn*

[2]*The Sydney Institute of Language & Commerce, Shanghai University, Shanghai 200072, People's Republic of China*

Abstract: The standard Data Envelopment Analysis (DEA) models rely on the assumption that inputs are minimized and outputs are maximized. However, when some inputs or outputs are undesirable factors (e.g., pollutants or wastes), these outputs (inputs) should be reduced (increased) to improve inefficiency. This chapter discusses the existing methods of treating undesirable factors in DEA. Under strongly disposable technology and weakly disposable technology, there are at least three approaches of treating undesirable outputs in the DEA literature. The first approach is the hyperbolic output efficiency measure that increases desirable outputs and decreases undesirable outputs simultaneously. Based on the classification invariance property, a linear monotone decreasing transformation is used to treat the undesirable outputs. A directional distance function is used to estimate the efficiency scores based on weak disposability of undesirable outputs. This chapter also presents an extended DEA model in which undesirable outputs and non-discretionary inputs are considered simultaneously.

Key words: Data Envelopment Analysis (DEA), Eco-efficiency, Undesirable factors, Non-discretionary inputs

1. INTRODUCTION

Data envelopment analysis (DEA) uses linear programming problems to evaluate the relative efficiencies and inefficiencies of peer decision-making units (DMUs) which produce multiple outputs by using multiple inputs.

Once DEA identifies the efficient frontier, DEA improves the performance of inefficient DMUs by either increasing the current output levels or decreasing the current input levels. However, both desirable (good) and undesirable (bad) output and input factors may be present. Consider a paper mill production where paper is produced with undesirable outputs of pollutants such as biochemical oxygen demand, suspended solids, particulates and sulfur oxides. If inefficiency exists in the production, the undesirable pollutants should be reduced to improve the inefficiency, i.e., the undesirable and desirable outputs should be treated differently when we evaluate the production performance of paper mills (see Seiford and Zhu, 2002). Cases where some inputs should be increased to improve efficiency can also occur (see Allen, 1999). For example, the aim of a recycling process is to use maximal quantity of the input waste. As pointed out by Simth (1990), undesirable factors (inputs/outputs) may appear in nonecological applications like health care (complications of medical operations) and business (tax payment).

We know that decreases in outputs are not allowed and only inputs are allowed to decrease in the standard DEA model (similarly, increases in inputs are not allowed and only outputs are allowed to increase). As pointed out by Seiford and Zhu (2002), although undesirable output (input) needs to be decreased (increased) in improving a DMU's efficiency, if one treats the undesirable outputs (inputs) as inputs (outputs), the resulting DEA model does not reflect the true production process.

Under two alternative reference technologies for undesirable outputs, i.e., strongly disposable technology (applicable if undesirable outputs are freely disposable) and weakly disposable technology (applicable when it is costly to dispose off undesirable outputs, e.g., due to regulatory actions), methods of treating undesirable outputs have been developed.

To treat desirable and undesirable outputs asymmetrically, Färe et al. (1989) used an enhanced hyperbolic output efficiency measure that evaluates DMUs' performance in terms of the ability to obtain an equiproportionate increase in desirable outputs and reduction in undesirable outputs. They developed their measures on a strongly disposable technology and on a weakly disposable technology. Their methodology has been applied to a sample of paper mills, and has been further developed by Färe et al. (1996) and Tyteca (1997).

Based on a strongly disposable technology of undesirable outputs, Seiford and Zhu (2002) developed a radial DEA model, in the presence of undesirable outputs, to improve the efficiency via increasing desirable outputs and decreasing undesirable outputs. By introducing a linear monotone decreasing transformation for undesirable outputs, the transformed undesirable outputs can then be treated just as desirable outputs in their model. In spirit of Seiford and Zhu's (2002) approach, Färe and Grosskopf (2004a) suggested an alternative approach in treating the undesirable factors by adopting a directional distance function to estimate DMUs' efficiencies based on weak disposability of undesirable outputs. As pointed out by Seiford and Zhu (2005) that, the approach of Färe and Grosskopf (2004a) can be linked to the additive DEA model. Following the approach of Seiford and Zhu (2002), Vencheh et al. (2005) developed a DEA-based model for efficiency evaluation incorporating undesirable inputs and undesirable outputs, simultaneously.

Recently, Hua et al. (2006) developed a DEA model in which undesirable outputs and non-discretionary inputs are considered simultaneously. They extended Seiford and Zhu's (2002) approach to incorporate non-discretionary inputs, and applied their method to evaluate eco-efficiencies of paper mills along the Huai River in China.

The reminder of the chapter is organized as follows. The next section describes the weak and strong disposability of undesirable outputs. We then present the hyperbolic output efficiency measure of undesirable outputs. Two approaches of treating undesirable outputs, developed by Seiford and Zhu (2002) and Färe and Grosskopf (2004a) are then introduced. We then show how to consider non-discretionary inputs and undesirable outputs simultaneously in an eco-efficiency evaluation problem. Discussions and conclusions are given at the last section.

2. WEAK AND STRONG DISPOSABILITY OF UNDESIRABLE OUTPUTS

When undesirable outputs are taken into consideration, the choice between two alternative disposable technologies (improvement technologies or reference technologies), i.e., strong and weak disposability of undesirable output, has an important impact on DMUs' efficiencies. A production process is said to exhibit strong disposability of undesirable outputs, e.g., heavy metals, CO_2, etc., if the undesirable outputs are freely disposable. The case of weak disposability refers to situations when a reduction in waste or emissions forces a lower production of desirable outputs, i.e., in order to meet some pollutant emission limits (regulations), reducing undesirable outputs may not be possible without assuming certain costs (see Zofio and Prieto, 2001).

For example, a power plant generates electricity by burning coal. Electricity is the plant's desirable output, and sulfur dioxide emission is its undesirable output. Weak disposability in this example implies that a 10% reduction in sulfur dioxide emission is possible if accompanied by a 10%

reduction in the output of electricity, holding the input vector constant (see Färe and Primont, 1995).

Suppose we have n independent decision making units, denoted by DMU_j ($j = 1, 2, \ldots, n$). Each DMU consumes m inputs x_{ij}, ($i = 1, 2, \ldots, m$) to produce s desirable outputs y_{rj}, ($r = 1, 2, \ldots, s$) and emits k undesirable outputs b_{tj}, ($t = 1, 2, \ldots, k$).

When the undesirable outputs are strong disposable, the production possibility set can be expressed as

$$T^s = \{(x,y,b) \mid \sum_{j=1}^{n} \eta_j x_j \leq x, \sum_{j=1}^{n} \eta_j y_j \geq y, \sum_{j=1}^{n} \eta_j b_j \geq b,$$
$$\eta_j \geq 0, \ j = 1, 2, \ldots, n\}.$$

When the undesirable outputs are weakly disposable, as suggested by Shephard (1970), the production possibility set may be written as

$$T^w = \{(x,y,b) \mid \sum_{j=1}^{n} \eta_j x_j \leq x, \sum_{j=1}^{n} \eta_j y_j \geq y, \sum_{j=1}^{n} \eta_j b_j = b,$$
$$\eta_j \geq 0, j = 1, 2, \ldots, n\}.$$

However, the concept of strong and weak disposability is relative and internal to each DMU's technology while the concept of regulations (reduction limits of undesirable outputs) is imposed externally upon DMUs.

For more detailed discussion on strong and weak disposability of undesirable outputs, the reader is referred to Färe and Primont (1995), and Färe and Grosskopf (2004a).

3. THE HYPERBOLIC OUTPUT EFFICIENCY MEASURE

This section presents the hyperbolic output efficiency measure developed in Färe et al. (1989). Hyperbolic efficiency explicitly incorporates the fact that the performance of production processes should be compared to an environmentally friendly standard, that is, reference facets for efficiency measurement should not be those defined by observed activities that produce both larger amounts of desirable production and waste but larger amounts of the former and smaller amounts of the latter (Zofio and Prieto, 2001)

Based on the assumption of strong disposability of undesirable outputs, DMUs' eco-efficiencies can be computed by solving the following programming problem:

Max ρ_0
subject to

$$\sum_{j=1}^{n} \eta_j x_{ij} + s_i^- = x_{i0}, \quad i = 1, 2, \ldots, m,$$

$$\sum_{j=1}^{n} \eta_j y_{rj} - s_r^+ = \rho_0 y_{r0}, \quad r = 1, 2, \ldots, s, \quad (1)$$

$$\sum_{j=1}^{n} \eta_j b_{tj} - s_t^+ = \frac{1}{\rho_0} b_{t0}, \quad t = 1, 2, \ldots, k,$$

$$\sum_{j=1}^{n} \eta_j = 1,$$

$$\eta_j, s_i^-, s_r^+, s_t^+ \geq 0, \text{ for all } j, i, r, t.$$

Obviously, model (1) is a non-linear programming problem, and increases desirable outputs and decreases undesirable outputs simultaneously. Similarly, based on the assumption of weak disposability of undesirable outputs, we have:

Max ρ_0
subject to

$$\sum_{j=1}^{n} \eta_j x_{ij} + s_i^- = x_{i0}, \quad i = 1, 2, \ldots, m,$$

$$\sum_{j=1}^{n} \eta_j y_{rj} - s_r^+ = \rho_0 y_{r0}, \quad r = 1, 2, \ldots, s,$$

$$\sum_{j=1}^{n} \eta_j b_{tj} = \frac{1}{\rho_0} b_{t0}, \quad t = 1, 2, \ldots, k, \quad (2)$$

$$\sum_{j=1}^{n} \eta_j = 1,$$

$$\eta_j, s_i^-, s_r^+ \geq 0, \text{ for all } j, i, r.$$

To solve models (1) and (2), Färe et al. (1989) suggested that the nonlinear constraints $\sum_{j=1}^{n} \eta_j b_{tj} \geq \frac{1}{\rho_0} b_{t0}$ approximated by linear expressions

$2b_{t0} - \rho_0 b_{t0} \leq \sum_{j=1}^{n} \eta_j b_{tj}$; similarly, the constraints $\sum_{j=1}^{n} \eta_j b_{tj} = \frac{1}{\rho_0} b_{t0}$ can be approximated by linear expressions $2b_{t0} - \rho_0 b_{t0} = \sum_{j=1}^{n} \eta_j b_{tj}$.

The above approaches of treating undesirable outputs can also be extended to treat undesirable inputs of a recycling process. Suppose DEA inputs data are $X_j = (X_j^D, X_j^U)$, where X_j^D and X_j^U represent desirable inputs to be decreased and undesirable inputs to be increased, respectively. Based on the assumption of strong disposability of undesirable inputs, we have

Min θ_0

subject to

$$\sum_{j=1}^{n} \eta_j x_j^D + s^- = \theta_0 x_0^D,$$

$$\sum_{j=1}^{n} \eta_j x_j^U + s^- = \frac{1}{\theta_0} x_0^U,$$

$$\sum_{j=1}^{n} \eta_j y_j - s^+ = y_0,$$

$$\sum_{j=1}^{n} \eta_j = 1,$$

$$\eta_j \geq 0, j = 1, 2, ..., n.$$

(3)

Obviously, model (3) decreases desirable inputs and increases undesirable inputs simultaneously. Similarly, based on the assumption of weak disposability of undesirable inputs, we have:

Min θ_0

subject to

$$\sum_{j=1}^{n} \eta_j x_j^D + s^- = \theta_0 x_0^D,$$

$$\sum_{j=1}^{n} \eta_j x_j^U = \frac{1}{\theta_0} x_0^U,$$

$$\sum_{j=1}^{n} \eta_j y_j - s^+ = y_0,$$

$$\sum_{j=1}^{n} \eta_j = 1,$$

$$\eta_j \geq 0, j = 1, 2, ..., n.$$

(4)

Models (3) and (4) are nonlinear programming problems, and can also be solved by applying similar linear approximations to the nonlinear constraints as those used in models (1) and (2).

4. A LINEAR TRANSFORMATION FOR UNDESIRABLE FACTORS

This section presents the work of Seiford and Zhu (2002). Under the context of the BCC model (Banker et al., 1984), Seiford and Zhu (2002) developed an alternative method in dealing with desirable and undesirable factors in DEA. They introduced a linear transformation approach to treat undesirable factors and then incorporated transformed undesirable factors into standard BCC DEA models.

Consider the treatment of undesirable outputs. In order to maintain translation invariance, there are at least two translation approaches to treat undesirable outputs: a linear monotone decreasing transformation or a nonlinear monotone decreasing transformation (e.g., $1/b$). As indicated by Lewis and Sexton (1999), the non-linear transformation will demolish the convexity relations. For the purpose of preserving convexity relations, Seiford and Zhu (2002) suggested a linear monotone decreasing transformation, $\overline{b}_j = -b_j + v \geq 0$, where v is a proper translation vector that makes $\overline{b}_j > 0$. That is, we multiply each undesirable output by (−1) and find a proper translation vector v to convert negative data to non-negative data.

Based upon the above linear transformation, the standard BCC DEA model can be modified as the following linear program:

Max h

subject to

$$\sum_{j=1}^{n} \eta_j x_{ij} + s_i^- = x_{i0}, \quad i = 1, 2, ..., m,$$

$$\sum_{j=1}^{n} \eta_j y_{rj} - s_r^+ = h y_{r0}, \quad r = 1, 2, ..., s,$$

$$\sum_{j=1}^{n} \eta_j \overline{b}_{tj} - s_t^+ = h \overline{b}_{t0}, \quad t = 1, 2, ..., k, \qquad (5)$$

$$\sum_{j=1}^{n} \eta_j = 1,$$

$\eta_j, s_i^-, s_r^+, s_t^+ \geq 0$, *for all* j, i, r, t.

Model (5) implicates that a DMU has the ability to expand desirable outputs and to decrease undesirable outputs simultaneously. A DMU_0 is efficient if $h_0^* = 1$ and all $s_i^- = s_r^+ = s_t^+ = 0$. If $h_0^* > 1$ and (or) some s_i^-, s_r^+ or s_t^+ is non-zero, then the DMU_0 is inefficient.

The above transformation approach can also be applied to situations when some inputs are undesirable and need to be increased to improve DMUs' performance. In this case, Seiford and Zhu (2002) rewrite the DEA input data as $X_j = (X_j^D, X_j^U)$, where X_j^D and X_j^U represent inputs to be decreased and increased, respectively. We multiply X_j^U by "-1" and then find a proper translation vector w to let all negative data to non-negative data, i.e., $\overline{x}_j^U = -x_j^U + w \geq 0$, then we have

Min τ

subject to

$$\sum_{j=1}^{n} \eta_j x_j^D + s^- = \tau x_0^D,$$

$$\sum_{j=1}^{n} \eta_j \overline{x}_j^U + s^- = \tau \overline{x}_0^U,$$

$$\sum_{j=1}^{n} \eta_j y_j - s^+ = y_0,$$

$$\sum_{j=1}^{n} \eta_j = 1,$$

$$\eta_j \geq 0, \ j = 1, 2, ..., n.$$

(6)

In model (6), X_j^D is decreased and X_j^U is increased for a DMU to improve its performance.

The linear transformation for undesirable factors can preserve the convexity and linearity of BCC DEA model. However, such approach can only be applied to the BCC model.

5. A DIRECTIONAL DISTANCE FUNCTION

Based on weak disposability of undesirable outputs, Färe and Grosskopf (2004a) provided an alternative approach to explicitly model a joint environmental technology. The key feature of the proposed approach is that it distinguishes between weak and strong disposability of undesirable outputs and uses a directional distance function.

Based on weak disposability of undesirable outputs, the production outputs set can be written as

$$P^w(x) = \{(x,y,b) \mid \sum_{j=1}^{n} \eta_j x_j \leq x, \sum_{j=1}^{n} \eta_j y_j \geq y, \sum_{j=1}^{n} \eta_j b_j = b,$$
$$\eta_j \geq 0, j = 1,2,...,n\}.$$

Null-jointness is imposed via the following restriction on the undesirable outputs: $b_{tj} > 0$, ($t = 1,2,...,k$, $j = 1,2,...,n$).

Since the traditional DEA model is a special case of directional output distance function, Färe and Grosskopf (2004a) adopted a more general directional output distance function to model the null-jointness production technology.

Let $g = (g_y, -g_b)$ be a direction vector. We have:

$\vec{D}(x_0, y_0, b_0; g) = Max\ \beta$

subject to

$$\sum_{j=1}^{n} \eta_j x_{ij} \leq x_{i0}, \quad i = 1,2,...,m,$$

$$\sum_{j=1}^{n} \eta_j y_{rj} \geq y_{r0} + \beta g_{y_r}, \quad r = 1,2,...,s, \quad\quad (7)$$

$$\sum_{j=1}^{n} \eta_j b_{tj} = b_{t0} - \beta g_{b_t}, \quad t = 1,2,...,k,$$

$$\eta_j \geq 0, \quad j = 1,2,...n.$$

Model (7) can also be rewritten as

$\vec{D}(x_0, y_0, b_0; g) = Max\ \beta$

subject to $\quad\quad (8)$

$(y_0 + \beta g_y, b_0 - \beta g_b) \in P^w(x).$

A DMU is efficient when $\vec{D}(x_0, y_0, b_0; g) = 0$ and inefficient when $\vec{D}(x_0, y_0, b_0; g) > 0$.

The key difference between this and the traditional DEA model is the inclusion of the direction vector $g = (g_y, -g_b)$. If $g_y = 1$ and $-g_b = -1$, then the solution represents the net performance improvement in terms of feasible increases in desirable outputs and decreases in undesirable outputs. Therefore, we have for desirable outputs $(y_{r0} + \vec{D}(x_0, y_0, b_0; g) \cdot 1) - y_{r0} = \vec{D}(x_0, y_0, b_0; g)$ and for undesirable outputs $(b_{t0} - \vec{D}(x_0, y_0, b_0; g) \cdot 1) - b_{t0} = \vec{D}(x_0, y_0, b_0; g)$.

Based on weak disposability of undesirable outputs and directional distance function approach, Färe and Grosskopf (2004b) also presented an index number approach to address eco-efficiency evaluation problem.

As pointed out by Seiford and Zhu (2005), model (7) can be linked to the additive DEA model (Charnes et al., 1985). Note in model (7), that g_y and g_b can be treated as user specified weights, and $\beta = 0$ is a feasible solution to model (7). Seiford and Zhu (2005) replace g_y and g_b by g_r^+ and g_t^-, respectively and obtain

$$\text{Max } \beta(\sum_{r=1}^{s} g_r^+ + \sum_{t=1}^{k} g_t^-)$$
subject to
$$\sum_{j=1}^{n} \eta_j x_{ij} \leq x_{i0}, \quad i = 1, 2, \ldots, m,$$
$$\sum_{j=1}^{n} \eta_j y_{rj} - \beta g_r^+ \geq y_{r0}, \quad r = 1, 2, \ldots, s, \qquad (9)$$
$$\sum_{j=1}^{n} \eta_j b_{tj} + \beta g_t^- = b_{t0}, \quad t = 1, 2, \ldots, k,$$
$$\eta_j \geq 0, \quad j = 1, 2, \ldots n,$$
$$\beta \geq 0.$$

If we use β_r^+ and β_t^- rather than a single β in model (9), then we have (Seiford and Zhu, 2005)

$$\text{Max } \sum_{r=1}^{s} \beta_r^+ + \sum_{t=1}^{k} \beta_t^-$$
subject to
$$\sum_{j=1}^{n} \eta_j x_{ij} \leq x_{i0}, \quad i = 1, 2, \ldots, m,$$
$$\sum_{j=1}^{n} \eta_j y_{rj} = y_{r0} + \beta_r^+ g_r^+, \quad r = 1, 2, \ldots, s, \qquad (10)$$
$$\sum_{j=1}^{n} \eta_j b_{tj} = b_{t0} - \beta_t^- g_t^-, \quad t = 1, 2, \ldots, k,$$
$$\eta_j \geq 0, \quad j = 1, 2, \ldots n,$$
$$\beta_r^+, \beta_t^- \geq 0.$$

As noted in Section 2, weak disposability is applied when inequalities are changed into equalities related to the constraints of undesirable outputs. In this case, weak disposability is automatically applied in model (10), because in optimality the equality always holds for the undesirable outputs (see Zhu (1996) and Seiford and Zhu (2005)).

Let $s_r^+ = \beta_r^+ g_r^+$ and $s_t^- = \beta_t^- g_t^-$, then model (10) can be written as the following additive DEA model:

$$\text{Max} \sum_{r=1}^{s} \frac{s_r^+}{g_r^+} + \sum_{t=1}^{k} \frac{s_t^-}{g_t^-}$$

subject to

$$\sum_{j=1}^{n} \eta_j x_{ij} \le x_{i0}, \quad i = 1, 2, ..., m,$$

$$\sum_{j=1}^{n} \eta_j y_{rj} - s_r^+ = y_{r0}, \quad r = 1, 2, ..., s, \tag{11}$$

$$\sum_{j=1}^{n} \eta_j b_{tj} + s_t^- = b_{t0}, \quad t = 1, 2, ..., k,$$

$$\eta_j \ge 0, \quad j = 1, 2, ... n,$$

$$s_r^+, s_t^- \ge 0.$$

Model (11) treats undesirable outputs as inputs, and decreases undesirable outputs and increases desirable outputs simultaneously.

6. NON-DISCRETIONARY INPUTS AND UNDESIRABLE OUTPUTS IN DEA

We here consider a performance evaluation problem provided in Hua et al. (2005) where the eco-efficiency evaluation of paper mills along the Huai River in China is discussed To deal with the increasingly serious environmental problem of the Huai River, State Environmental Protection Administration of China (EPAC) had passed an Act to limit the total quantities of main pollutants emission (e.g., biochemical/chemical oxygen demand (BOD/COD), nitrogen (N), etc) from industrial enterprises (e.g., paper mills) along the River. According to the EPAC's Act, an annual emission quota of BOD is allocated to each paper mill, and each paper mill should not emit more BOD than the allocated quota.

For these paper mills, emission of BOD is an undesirable output, and the corresponding emission quota is a non-discretionary input allocated by EPAC. The non-discretionary input, on the one hand, restricts pollutants emission. On the other hand, it also affects the production of desirable outputs (paper products). When we evaluate efficiencies of paper mills along the Huai River, undesirable outputs and non-discretionary inputs should be modeled in a single model.

Denote by $Z_j = (z_{1j}, z_{2j}, ..., z_{pj})^T$ the given p non-discretionary inputs, $X_j = (x_{1j}, x_{2j}, ..., x_{mj})^T$ the m discretionary inputs, $Y_j = (y_{1j}, y_{2j}, ..., y_{sj})^T$ the s desirable outputs and $B_j = (b_{1j}, b_{2j}, ..., b_{kj})^T$ the k undesirable outputs ($k \geq p$). We next follow Seiford and Zhu (2002) (or section 4) to make a linear monotone decreasing transformation, $\bar{b}_j = -b_j + v \geq 0$, where v is a proper translation vector that makes $\bar{b}_j > 0$. We then use the non-radial DEA model to generate the following model

$$\text{Max } \phi_0 = \frac{1}{s+k} \sum_{r=1}^{s} \alpha_r + \frac{1}{s+k} \sum_{t=1}^{k} \beta_t$$

subject to

$$\sum_{j=1}^{n} \eta_j x_{ij} + s_i^- = x_{i0}, \quad i = 1, 2, ..., m,$$

$$\sum_{j=1}^{n} \eta_j y_{rj} = \alpha_r y_{r0}, \quad r = 1, 2, ..., s,$$

$$\sum_{j=1}^{n} \eta_j \bar{b}_{tj} = \beta_t \bar{b}_{t0}, \quad t = 1, 2, ..., k, \tag{12}$$

$$\sum_{j=1}^{n} \eta_j = 1,$$

$$\alpha_r \geq 1, \beta_t \geq 1, \text{ for all } r, t,$$

$$\eta_j, s_i^- \geq 0, \text{ for all } j, i.$$

Model (12) is an output-oriented DEA model. Values of α_r and β_t are used to determine the amount of possible augmentation and reduction in each desirable and undesirable output, respectively. If some transformed undesirable outputs $\bar{b}_{t0}$ become zero, then we set $\beta_t = 1$ for those t with $\bar{b}_{t0} = 0$.

To incorporate the non-discretionary inputs into model (12), we let $z'_j = Q - z_j \geq 0$ ($j = 1, 2, ..., n$), where Q is a proper translation vector that makes $z'_j > 0$.

Banker and Morey (1986) used the following to model non-discretionary inputs

$$\sum_{j=1}^{n} \eta_j z'_{lj} \geq z'_{l0} \quad (l = 1, 2, ..., p). \tag{13}$$

In the presence of undesirable outputs, Eq.(13) is inappropriate because it considers only positive impacts of non-discretionary inputs on desirable outputs while ignoring impact of non-discretionary inputs on undesirable outputs.

Hua et al. (2005) modified (13) into

$$\sum_{j=1}^{n} \eta_j z'_{lj} = z'_{l0} \ (l=1,2,...,p). \tag{14}$$

Equation (14) requires that reference units utilize the same levels of the transformed non-discretionary inputs as that of the assessed DMU on average.

Based on the reference set defined in Equation (14), model (12) can be extended to the following output-oriented DEA model:

$$\text{Max } \varphi_0 = \frac{1}{s+k}\sum_{r=1}^{s}\alpha_r + \frac{1}{s+k}\sum_{t=1}^{k}\beta_t$$

subject to

$$\sum_{j=1}^{n} \eta_j x_{ij} + s_i^- = x_{i0}, \ i=1,2,...,m,$$

$$\sum_{j=1}^{n} \eta_j z'_{lj} = z'_{l0}, \ l=1,2,...,p,$$

$$\sum_{j=1}^{n} \eta_j y_{rj} = \alpha_r y_{r0}, \ r=1,2,...,s, \tag{15}$$

$$\sum_{j=1}^{n} \eta_j \bar{b}_{tj} = \beta_t \bar{b}_{t0}, \ t=1,2,...,k,$$

$$\sum_{j=1}^{n} \eta_j = 1,$$

$$\alpha_r \geq 1, \ \beta_t \geq 1, \ \text{for all } r,t,$$

$$\eta_j, s_i^- \geq 0, \ \text{for all } j,i.$$

For any given non-discretionary inputs, model (15) increases each desirable output and each transformed undesirable output simultaneously with different proportions.

Let $(\alpha_r^*, \beta_t^*, s_i^{-*}, \eta^*)$ be the optimal solution to model (15). A DMU_0 is efficient if $\varphi_0^* = 1$ (i.e., $\alpha_r^* = 1$ and $\beta_t^* = 1$) and all $s_i^{-*} = 0$. If $\varphi_0^* > 1$ (i.e., any of α_r^* or β_t^* is greater than one) and (or) some of s_i^{-*} are non-zero, then the DMU_0 is inefficient.

We now apply this approach to 32 paper mills. Table 6-1 reports characteristics of the data set with labor and capital as the two discretionary inputs, paper products as the desirable output, emitted BOD as the undesirable output, and the emission quota of BOD (BOD-Q) allocated to each paper mill as a non-discretionary input.

Table 6-1. Characteristics of the data set for 32 paper mills [a]

	Labor	Capital	BOD-Q (kg)	Paper (ton)	BOD (kg)
Max	1090	5902	33204	29881	28487.7
Min	122	1368.5	1504.2	5186.6	1453.3
Mean	630	3534.1	13898.6	17054	10698.5
Std. dev	326	1222.4	8782	7703.3	7969.2

[a] Note: Capital are stated in units of 10-thousand RMB Yuan. Labor is expressed in units of 1 person.

Table 6-2. Efficiency Results [1]

Mills	Model (12)	Model (15)	Mills	Model (12)	Model (15)
1	1.0000	1.0000	17	1.0000	1.0000
2	1.0000	1.0000	18	1.2318	1.2071
3	1.4028	1.0000	19	1.4529	1.3086
4	1.0814	1.0569	20	1.0000	1.0000
5	1.1644	1.1437	21	1.0000	1.0000
6	1.9412	1.5951	22	1.4731	1.1647
7	1.1063	1.0000	23	1.0000	1.0000
8	1.0000	1.0000	24	1.1157	1.0472
9	1.0000	1.0000	25	1.0000	1.0000
10	1.2790	1.0000	26	1.2321	1.2294
11	1.5171	1.0000	27	1.3739	1.3606
12	1.0000	1.0000	28	1.0000	1.0000
13	1.0431	1.0000	29	1.1365	1.0000
14	1.1041	1.0860	30	1.0000	1.0000
15	1.2754	1.2672	31	1.0000	1.0000
16	1.4318	1.0000	32	1.0000	1.0000
			Mean	1.1676	1.0771
			EDMU [a]	14	21

[a] EDMU represents the number of efficient DMUs

We set the translation vectors of $v = 50000$ and $w = 60000$ for undesirable output (BOD) and non-discretionary input (BOD-Q). Table 6-2 reports the efficiency results obtained from model (12) and model (15) for all paper mills.

Note in Table 6-2 that, when we ignore the non-discretionary input, for BCC models, 14 paper mills are deemed as efficient under model (12). However, we have only 11 BCC-inefficient paper mills under model (15), These results show that, when undesirable outputs are considered in performance evaluation and if the impacts of non-discretionary inputs on

---

[1] There is an error in the case study section in Hua et al. (2005). The translation invariance does not hold under the CCR model (15), i.e., the CCR model should not have been applied.

DMUs' efficiencies are not dealt with properly, the ranking of DMUs' performance may be severely distorted.

7. DISCUSSIONS AND CONCLUSION REMARKS

This chapter reviews existing approaches in solving DEA models with undesirable factors (inputs/outputs). These approaches are based on two important disposability technologies for undesirable outputs: one is strong disposal technology and the other is weak disposal technology.

In addition to the three methods discussed in the previous sections, there are some other methods for dealing with undesirable factors in the literature. Table 6-3 lists six methods for treating undesirable factors in DEA.

Table 6-3. Six methods for treating undesirable factors in DEA

Method	Definition
1	Ignoring undesirable factors in DEA models
2	Treating undesirable outputs (inputs) as inputs (outputs)
3	Treating undesirable factors in nonlinear DEA model (Färe et al., 1989)
4	Applying a nonlinear monotone decreasing transformation (e.g., $1/b$) to the undesirable factors
5	Using a linear monotone decreasing transformation to deal with undesirable factors (Seiford and Zhu, 2002)
6	Directional distance function approach (Färe and Grosskopf, 2004a)

To compare these six methods for treating undesirable factors, we use 30 DMUs with two inputs and three outputs as reported in Table 6-4. Each DMU has two desirable inputs (D-Input 1 and D-Input 2), two desirable outputs (D-Output 1 and D-Output 2) and one undesirable output (UD-Output 1).

We set the translation parameter $v = 1500$, and the direction vector $g = (500, 2000, 100)$. We then use six different methods to treat undesirable outputs in BCC DEA models, and the results (efficiency scores) are reported in Table 6-5.

When we ignore the undesirable output (method 1), 14 DMUs are deemed as efficient, and the mean efficiency is 1.2316. However, we have 17 efficient DMUs under method 2 and method 5, and 19 efficient DMUs under method 6. These results confirm the finding in Färe et al. (1989) that method 1 failing to credit DMUs for undesirable output reduction may severely distort DMUs' eco-efficiencies. Although there are 17 efficient under method 2, which is the same number as that from method 5, the mean efficiency is 1.1957, which is higher than that of method 5. This difference may be due to the fact that method 2 treats undesirable outputs as inputs, which does not reflect the true production process. As to method 3, the mean efficiency is 1.2242, which is even higher than that of method 2. The reason for this may be due to the use of approximation of the nonlinear programming problem. There are 15 efficient DMUs under method 4, and the corresponding mean efficiency is 1.1704, which is higher than that of method 5. This result may attribute to the nonlinear transformation adopted in method 4.

Table 6-4. Data set for the example

DMUs	D-Input 1	D-Input 2	D-Output 1	D-Output 2	UD-Output 1
1	437	1438	2015	14667	665
2	884	1061	3452	2822	491
3	1160	9171	2276	2484	417
4	626	10151	953	16434	302
5	374	8416	2578	19715	229
6	597	3038	3003	20743	1083
7	870	3342	1860	20494	1053
8	685	9984	3338	17126	740
9	582	8877	2859	9548	845
10	763	2829	1889	18683	517
11	689	6057	2583	15732	664
12	355	1609	1096	13104	313
13	851	2352	3924	3723	1206
14	926	1222	1107	13095	377
15	203	9698	2440	15588	792
16	1109	7141	4366	10550	524
17	861	4391	2601	5258	307
18	249	7856	1788	15869	1449
19	652	3173	793	12383	1131
20	364	3314	3456	18010	826
21	670	5422	3336	17568	1357
22	1023	4338	3791	20560	1089
23	1049	3665	4797	16524	652
24	1164	8549	2161	3907	999
25	1012	5162	812	10985	526
26	464	10504	4403	21532	218
27	406	9365	1825	21378	1339
28	1132	9958	2990	14905	231
29	593	3552	4019	3854	1431
30	262	6211	815	17440	965

Table 6-5. Results of DMUs' efficiencies [a]

DMUs	M 1	M 2	M 3	M 4	M 5	M 6
1	1.0000	1.0000	1.0000	1.0000	1.0000	0
2	1.0000	1.0000	1.0000	1.0000	1.0000	0
3	2.1076	2.0139	2.1076	1.7948	1.1773	1.9198
4	1.3079	1.3079	1.3079	1.3079	1.0686	0.8214
5	1.0063	1.0000	1.0063	1.0000	1.0000	0
6	1.0000	1.0000	1.0000	1.0000	1.0000	0
7	1.0137	1.0137	1.0137	1.0137	1.0136	0.1276
8	1.2540	1.2540	1.2358	1.2540	1.2540	1.7328
9	1.5604	1.5604	1.5578	1.5604	1.5604	3.0493
10	1.0678	1.0000	1.0678	1.0000	1.0000	0
11	1.3387	1.3236	1.3387	1.3387	1.2591	1.9199
12	1.0000	1.0000	1.0000	1.0000	1.0000	0
13	1.0000	1.0000	1.0000	1.0027	1.0000	0
14	1.0000	1.0000	1.0000	1.0000	1.0000	0
15	1.0000	1.0000	1.0000	1.0000	1.0000	0
16	1.0987	1.0721	1.0987	1.0830	1.0599	0.5328
17	1.7467	1.0000	1.7467	1.0405	1.0000	0
18	1.0575	1.0575	1.0000	1.0575	1.0575	0
19	1.6763	1.6763	1.6763	1.6117	1.6293	3.5787
20	1.0000	1.0000	1.0000	1.0000	1.0000	0
21	1.1444	1.1444	1.0000	1.1444	1.1444	0
22	1.0000	1.0000	1.0000	1.0000	1.0000	0
23	1.0000	1.0000	1.0000	1.0000	1.0000	0
24	2.2198	2.2198	2.2198	2.1810	2.1281	4.9951
25	1.9087	1.8110	1.9087	1.6759	1.2637	2.5453
26	1.0000	1.0000	1.0000	1.0000	1.0000	0
27	1.0000	1.0000	1.0000	1.0000	1.0000	0
28	1.4405	1.4160	1.4405	1.0463	1.0080	0.1012
29	1.0000	1.0000	1.0000	1.0000	1.0000	0
30	1.0000	1.0000	1.0000	1.0000	1.0000	0
Mean	1.2316	1.1957	1.2242	1.1704	1.1208	0.7108
EDMU [b]	14	17	16	15	17	19

[a] M 1-6 represent methods 1-6, respectively.
[b] EDMU represents the number of efficient DMUs.

It can be observed in Table 6-5 that the results of methods 5 and 6 are different. There are 17 efficient DMUs under method 5, while 19 efficient DMUs under method 6. One reason for this is the different reference technologies assumed in methods 5 and 6, i.e., method 5 assumes the strong disposability of undesirable outputs while method 6 assumes the weak disposability of undesirable outputs.

The ranking of DMUs determined by method 6 are strongly affected by the user specified weights (direction vector g). For example, if we set $g = (400, 1000, 500)$, results of method 6 will be different from those in Table 6-5.

The issue of dealing with undesirable factors in DEA is an important topic. The existing DEA approaches for processing undesirable factors have been focused on individual DMUs. Modeling other types of DEA models for addressing complicated eco-efficiency evaluation problems (e.g., network DEA models with undesirable factors, multi-component DEA models with undesirable factors, DEA models with imprecise data and undesirable factors) are interesting topics for future research.

REFERENCES

1. Allen, K. (1999), "DEA in the ecological context-an overview", In: Westermann, G. (Ed.), Data Envelopment Analysis in the Service Sector, Gabler, Wiesbaden, pp.203-235.
2. Ali, A.I., L.M. Seiford (1990), "Translation invariance in data envelopment analysis", Operations Research Letters 9, 403-405.
3. Banker, R.D., A. Charnes, W.W. Cooper (1984), "Some models for estimating technical and scale inefficiencies in data envelopment analysis", Management Science 30, 1078-1092.
4. Banker, RD., R. Morey (1986), "Efficiency analysis for exogenously fixed inputs and outputs", Operations Research 34, 513-521.
5. Charnes, A., W.W. Cooper, B. Golany, L.M. Seiford, J. Stutz (1985), "Foundations of data envelopment analysis for Paretop-Koopmans efficient empirical production functions", Journal of Econometrics 30, 91-107.
6. Färe, R., S. Grosskopf, C.A.K. Lovell, C. Pasurka (1989), "Multilateral productivity comparisons when some outputs are undesirable: a nonparametric approach", The Review of Economics and Statistics 71, 90-98.
7. Färe, R., S. Grosskopf, D. Tyteca (1996), "An activity analysis model of the environmental performance of firms—application to fossil-fuel-fired electric utilities", Ecological Economics 18, 161-175.
8. Färe, R., S. Grosskopf (2004a), "Modeling undesirable factors in efficiency evaluation: Comment", European Journal of Operational Research 157, 242-245.
9. Färe, R., D. Primont (1995), "Multi-output production and duality: Theory and Applications", Boston: Kluwer Academic Publishers.
10. Färe, R., S. Grosskopf (2004b), "Environmental performance: an index number approach", Resource and Energy Economics 26, 343-352.
11. Hua, Z.S., Y.W. Bian, L. Liang (2006), "Eco-efficiency analysis of paper mills along the Huai River: An extended DEA approach", OMEGA, International Journal of Management Science (in press).

12. Lewis, HF., TR. Sexton (1999), "Data envelopment analysis with reverse inputs", Paper presented at North America Productivity Workshop, Union College, Schenectady, NY.
13. Simth, P. (1990), "Data envelopment analysis applied to financial statements", Omega: International Journal of Management Science 18, 131-138.
14. Seiford, L.M., J. Zhu (2002), "Modeling undesirable factors in efficiency evaluation", European Journal of Operational Research 142, 16-20.
15. Seiford, L.M., J. Zhu (2005), "A response to comments on modeling undesirable factors in efficiency evaluation", European Journal of Operational Research 161, 579-581.
16. Seiford, L.M., J. Zhu (1998), "Identifying excesses and deficits in Chinese industrial productivity (1953-1990): A weighted data envelopment analysis approach", OMEGA, International Journal of Management Science 26(2), 269-279.
17. Seiford, L.M, J. Zhu (1999), "An investigation of returns to scale in data envelopment analysis", Omega 27, 1-11.
18. Scheel, H. (2001), "Undesirable outputs in efficiency valuations", European Journal of Operational Research 132, 400-410.
19. Tyteca D. (1997), "Linear programming models for the measurement of environment performance of firms—concepts and empirical results", Journal of Productivity Analysis 8, 183-197.
20. Thrall, R.M. (1996), "Duality, classification and slacks in DEA ", Annals of Operations Research 66, 109-138.
21. Vencheh, A.H., R.K. Matin, M.T. Kajani (2005), "Undesirable factors in efficiency measurement", Applied Mathematics and Computation 163, 547-552.
22. Word Commission on Environment and Development (WCED) (1987), "Our Common Future", Oxford University Press.
23. Zofio, J.L., A.M. Prieto (2001), "Environmental efficiency and regulatory standards: the case of CO_2 emissions form OECD industries", Resource and Energy Economics 23, 63-83.
24. Zhu, J. (1996), "Data envelopment analysis with preference structure", Journal of Operational Research Society 47, 136-150.

Part of the material in this chapter is adapted from Omega, International Journal of Management Science, Hua Z.S., Y.W., Bian, Liang L., Eco-efficiency analysis of paper mills along the Huai River: An extended DEA approach (in press), with permission from Elsevier Science.

Chapter 7

EUROPEAN NITRATE POLLUTION REGULATION AND FRENCH PIG FARMS' PERFORMANCE

Isabelle Piot-Lepetit and Monique Le Moing
INRA Economie, 4 allée Adolphe Bobierre, CS 61103, 35011 Rennes cedex, France, Isabelle.Piot@rennes.inra.fr, Monique.LeMoing@rennes.inra.fr

Abstract: This chapter highlights the usefulness of the directional distance function in measuring the impact of the EU Nitrate directive, which prevents the free disposal of organic manure and nitrogen surplus. Efficiency indices for the production and environmental performance of farms at an individual level are proposed, together with an evaluation of the impact caused by the said EU regulation. An empirical illustration, based on a sample of French pig farms located in Brittany in 1996, is provided. This chapter extends the previous approach to good and bad outputs within the framework of the directional distance function, by introducing a by-product (organic manure), which becomes a pollutant once a certain level of disposability is exceeded. In this specific case, the bad output is the nitrogen surplus - resulting from the nutrient balance of each farm – that is spread on the land. This extension to the model allows us to explicitly introduce the EU regulation on organic manure, which sets a spreading limit of 170kg/ha. Our results show that the extended model provides greater possibilities for increasing the level of production, and thus the revenue of each farm, while decreasing the bad product (nitrogen surplus) and complying with the mandatory standard on the spreading of organic manure.

Key words: Environmental regulation, Manure management, Farms' performance, Directional distance function, Data Envelopment Analysis (DEA).

1. INTRODUCTION

Agricultural activities are, in most cases, characterized by some kind of negative externalities. By negative externalities, we mean technological externalities, i.e., negative side effects from a particular farm's activity that reduce the production possibility set for other farms or the consumption set of individuals. The main environmental issues associated with pig production concern water and air pollution. One factor of water pollution arises from the inappropriate disposal of pig manure. These kinds of externalities or the production of "bad" outputs can be excessive simply because producers have no incentive to reduce the harmful environmental impact of their production. To influence farmers' behavior in a way that is favorable to the environment, a number of policy instruments have been introduced in the European Union (EU).

There are relatively few environmental policy measures relating specifically to the pig sector. Pig producers are affected by wider policies aimed at the livestock sector or the agricultural sector as a whole and do in fact face an array of regulations impacting on their production levels and farming practices. The major environmental objective of policy instruments affecting the pig sector has been to reduce the level of water pollution. The initial response by most governments in the European Union in addressing environmental issues in the pig sector has been to impose regulations, develop research programmes and provide on-farm technical assistance and extension services to farmers. These measures are predominately regulatory, are increasing in severity and complexity and involve a compulsory restriction on the freedom of choice of producers, i.e., they have to comply with specific rules or face penalties. Apart from payments to reduce the cost of meeting new regulations, economic instruments have rarely been used.

The European Union addresses issues of water management through the more broadly focused EU Water Framework directive and specific issues of water pollution from agriculture through the Nitrates directive (EU 676/91) and the Drinking Water directive. Each EU country is responsible for meeting the targets set by the Nitrates directive, and consequently, differences emerge at the country level. In particular, the Nitrates directive sets down precise limits on the quantity of manure that can be spread in designated areas. In addition to this regulation, technical assistance has been provided to assist the implementation of the Codes of Good Agricultural Practice required by the Nitrates directive. These codes inform farmers about practices that reduce the risk of nutrient pollution. Restrictions have also been brought in to control the way manure is spread, the type of facilities used for holding manure and the timing of the spreading. In France, the regulation concerning the management and disposal of manure has been in effect since 1993. Farmers have received subsidies to cover the costs of bringing buildings and manure storage facilities into line with environmental regulations.

The purpose of this chapter is to analyze the impact of the Nitrate directive on the performance of French pig farms, and in particular the mandatory standard on the spreading of organic manure. Using a recently developed technique, the directional distance function, we can explicitly treat the production of pollution from pig farms and introduce a standard affecting a by-product of pig production in the representation of the production possibility set.

In recent decades, there has been a growing interest in the use of efficiency measures that take undesirable or pollutant outputs into account (Tyteca, 1996; Allen, 1999). These measurements are based on the adjustment of conventional measures (Farrell, 1957) and most of the time, they consider pollution as an undesirable output. They develop efficiency measures that include the existence of undesirable or "bad" outputs in the production process and allow for a valuation of the impact of environmental regulations on farms' performance. Färe *et al*. (1989) established the basis for this extension by considering different assumptions on the disposability of bad outputs. Färe *et al*. (1996) develop several indicators of efficiency, considering that environmental restrictions on the production of waste can hamper the expansion of the production of goods. This approach is based on the use of Shephard's output distance functions (Shephard, 1970). Recently, a new representation of the technology based on Luenberger's benefit function (Luenberger, 1992; Färe and Grosskopf, 2000, 2004) has been developed. Chung *et al.* (1997) provide the basis for representing the joint production of desirable and undesirable outputs by extending Shephard's output distance function. A directional output distance function expands good outputs and contracts bad outputs simultaneously along a path defined by a direction vector. This directional distance function generalizes Shephard's input and output distance function (Chambers *et al.*, 1996; Chambers, 1998). It provides a representation of the technology, allowing an approach to production and environmental performance issues that may be useful in policy-oriented applications.

In this chapter, we make use of the directional distance function to evaluate the performance of pig farms, taking into account the presence of polluting waste (nitrogen surplus) in the pig sector. In our modeling, however, nitrogen surplus is not directly the by-product of pig production. The by-product is actually organic manure, while the bad output derived from the nutrient balance of the farm is manure surplus. The previous model based on the directional distance function has been extended so as to differentiate between organic manure and nutrient surplus. Furthermore, we provide an extension of the existing model of production technology, which explicitly integrates the individual constraint introduced by the EU Nitrates directive on the spreading of organic manure. This individual standard is considered as a right to produce allocated to each farmer. As regards the activity of each farm, some are highly constrained while others are not. The question is to consider how producers will individually adapt their production activity to not only comply with the regulation, but also maintain their activity at a good economic performance level. Our empirical application uses data from a cross-section of French farms located in Brittany in 1996.

The chapter begins in section 2 with a presentation of the methodology by which good and bad outputs are represented. Section 3 describes how this approach has been extended to introduce a regulatory constraint on the by-product of pig production and all mandatory restrictions resulting from the implementation of the Codes of Good Agricultural Practice. Section 4 describes the data and section 5 discusses the results. Finally, Section 6 concludes.

2. MODELLING TECHNOLOGIES WITH GOOD AND BAD OUTPUTS

When there exists a negative externality (technological externality), the production of desirable outputs is accompanied by the simultaneous or joint production of undesirable outputs. Here, we denote inputs by $x = (x_1,...,x_N) \in R_+^N$, good or desirable outputs by $y = (y_1,...,y_M) \in R_+^M$, and undesirable or bad outputs by $b = (b_1,...,b_S) \in R_+^S$. In the production context, good outputs are marketed goods, while bad outputs are often not marketed and may have a detrimental effect on the environment, thus involving a cost that is borne by society as a whole in the absence of any explicit regulations on the disposal of bad outputs.

The relationship between inputs and outputs is captured by the firm's technology, which can be expressed as a mapping $P(x) \subset R_+^{M+S}$ from an input vector x into the set of feasible output vectors (y,b). The output set may be expressed as:

$$P(x) = \{(y,b): x \text{ can produce } (y,b)\}, x \in R_+^N \qquad (2.1)$$

To model the production of both types of outputs, we need to take into account their characteristics and their interactions (Färe and Grosskopf, 2004). This implies modifying the traditional axioms of production to accommodate the analysis, by integrating the notions of null jointness and weak disposability of outputs. The former represents the joint production of desirable and undesirable outputs while the latter models the idea that bad outputs are costly to reduce.

The production technology is assumed to produce both desirable and undesirable outputs and it is assumed that it cannot produce one without the other (joint-production). This notion of null jointness has been defined by Shephard and Färe (1974). Formally, we have:

$$\text{if } (y,b) \in P(x) \text{ and } b = 0 \text{ then } y = 0 \qquad (2.2)$$

In words, if an output vector *(y,b)* is feasible and no bad outputs are produced, then under null jointness only zero output can be produced. Alternatively, if one wishes to produce good outputs then some bad outputs must also be produced. This hypothesis implies a reduction in the production possibility set.

In order to address the fact that bad outputs are costly to reduce, we impose a hypothesis of weak disposability of outputs, i.e.,

$$(y,b) \in P(x) \text{ then } (\theta y, \theta b) \in P(x) \text{ for } 0 \leq \theta \leq 1 \qquad (2.3)$$

In words, this states that a reduction in undesirable outputs can be achieved by reducing good outputs, given fixed input levels. This assumption models the idea that bad outputs are not freely disposable because certain regulatory restrictions exist. In this case, compliance with the regulation may require a reduction in the production of some of the good outputs or alternatively, an increase in some of the inputs, for example, when the cleaning up of undesirable outputs is mandatory.

In addition to impose weak disposability, we also assume that the desirable outputs are freely disposable, i.e.,

$$(y,b) \in P(x) \text{ and } y' \leq y \text{ imply } (y',b) \in P(x) \qquad (2.4)$$

In this hypothesis, by which any good outputs may be freely disposed of, we assume that there are no regulations on the disposability of these goods.

In Figure 7-1. the two assumptions on the disposability of outputs are illustrated. The output set built on the hypothesis that bad outputs are freely disposable is bounded by OECD while the output set built on the weak disposability hypothesis is bounded by OEBO. The difference between the two output sets may be viewed as a characterization of the effect of regulations on bad outputs. A regulation that is too weak may not prevent the free disposability of undesirable outputs. On the other hand, a regulation that is too strong may impose a drastic reduction in the production of desirable outputs.

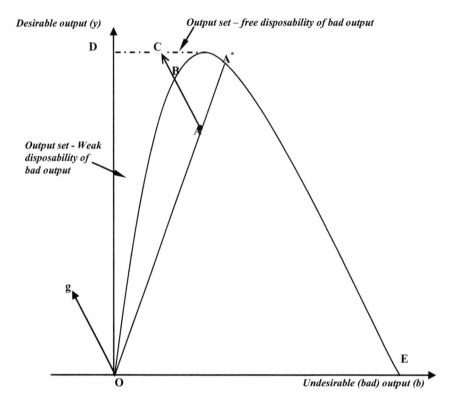

Figure 7-1. Weak disposability of outputs and the directional output distance function

An alternative representation of the technology, conveying the same information as the output set *P(x)*, is the directional output distance function, introduced by Chung et al. (1997). This distance function represents the technology of production and allows us to credit farms for reducing bad outputs while simultaneously increasing good outputs. Formally, it is defined as:

$$\vec{D}_o(x, y, b; g) = \sup\{\beta : ((y, b) + \beta.g) \in P(x)\} \quad (2.5)$$

where g is the vector of directions in which outputs are scaled. By construction, $D_o(x, y, b; g) \geq 0$ if and only if $(y, b) \in P(x)$. When $D_o(x, y, b; g) = 0$ the farm is on the boundary of the production set. The directional output distance takes *(y,b)* in the *g* direction and places it on the production frontier. In our case, $g=(y,-b)$, i.e., good outputs are increased and bad outputs are decreased.

In Figure 7-1., the directional distance function locates the productive plan A on the boundary of the output set built on the hypothesis of weak disposability of the bad output at point B. The distance AB measures the increase in production of the good output and the decrease in production of the bad output, resulting from a reduction in production inefficiency. Under the free disposability hypothesis, the directional distance function locates plan A at point C. A higher increase in the good output can be achieved due to the fact that A is unrestricted and can freely dispose of the bad output.

The directional distance function $\vec{D}_o(x,y,b;g)$ is of special interest in the measurement of production efficiency. According to the work of Debreu (1951), Farrell (1957) and Färe et al. (1985, 1994, 2004), production efficiency[1] is defined as the proportional rescaling of inputs or outputs that would bring the firm to the production frontier. Thus, we define the production efficiency index as:

$PE(x,y,b) = 1 + \vec{D}_o(x,y,b;g)$

In this chapter, we use Data Envelopment Analysis (DEA) to construct the technology. Following Färe et al. (2004), we assume that there are $k=1,..., K$ observations of inputs and outputs. For each observation, the directional distance function is computed as the solution to a linear programming problem. For example, for k',

$$\vec{D}_0(x_{k'}, y_{k'}, b_{k'}; y_{k'}, -b_{k'}) = \max \beta_{k'}$$

s.t.

$$\sum_{k=1}^{K} z_k y_{mk} \geq (1+\beta_{k'}) y_{mk'} \quad m = 1,..., M$$

$$\sum_{k=1}^{K} z_k b_{sk} = (1-\beta_{k'}) b_{sk'} \quad s = 1,..., S \quad (2.6)$$

$$\sum_{k=1}^{K} z_k x_{nk} \leq x_{nk'} \quad n = 1,..., N$$

$$z_k \geq 0 \quad k = 1,..., K$$

The z's are intensity variables which serve to construct the reference technology as convex combinations of the observed data. The inequalities for inputs and goods outputs make them freely disposable. The equality for

---

[1] We use the term of production efficiency, which is more general than technical efficiency, because in our case, the measurement of farm performance includes several components: technical efficiency, scale efficiency and environmental efficiency.

undesirable outputs makes them weakly disposable. Finally, the non-negativity constraints on the intensity variables z_k allow the model to exhibit constant returns to scale[2].

3. MODELLING TECHNOLOGIES WITH AN ENVIRONMENTAL STANDARD ON THE BY-OUTPUT

In response to high levels in water supplies, the European Union passed its Nitrate directive in 1991. This directive is the EU's main regulatory instrument for combating nitrogen-related pollution from agricultural sources. It covers all nitrogen of whatever origin (chemical fertilizers, effluents from animal farms or agro-food plants, sludge etc.) and all waters, regardless of their origin or use. The directive relates more particularly to agricultural production, and aims to limit the amount of residual nitrogen remaining in the soil after uptake by crops. The directive limits the spreading of organic manure per farm to 170 kilograms per hectare in vulnerable zones. Each European member state has organized its own implementation of this directive, defining a set of constraints relevant to its own country on the use of nitrogen fertilizers, the numbers of livestock, and the storage and disposal of manure. In France, the directive has been in effect since 1993; its aim is to rectify the most polluting practices and encourage them to change, so as to protect and even restore water quality.

In this context, organic manure is the by-product or joint production of animals. Formally, we define b_{Norg} as the level of organic manure produced jointly with the good output vector y. In order to represent the jointness of production between the two goods, we assume that b_{Norg} and y are null joint:

$$\text{if } (y, b_{Norg}) \in P(x) \text{ and } b_{Norg} = 0 \text{ then } y = 0 \tag{2.7}$$

The regulatory constraint introduced by the Nitrate directive is quantitative in nature, i.e., organic manure is restricted in quantity. The limit on the application of organic manure on land is set at a maximum rate of 170kg per hectare. It may be expressed as follows:

$$b_{Norg} \leq 170 \cdot x_{land} \tag{2.8}$$

---

[2] Alternative hypotheses concerning the return to scale can be introduced in the modelling (variable or non increasing return to scale). As we focus on the representation of bad outputs, we have chosen to use constant returns to scale at this stage.

where x_{land} is the total area of the farm which can be used for disposal of organic manure.

However, organic manure is not the bad output. In pig production, the undesirable output is the production of manure surplus. Up to a certain level, organic manure can be applied to fields for fertilization purposes. Furthermore, fertilizers are often added as well, especially to secure nitrate availability for the crops. The nutrients contained in manure and fertilizers are to some extent removed when the crops are harvested. However, surplus nutrients pose potential environmental problems. When the maximum level of nutrients needed for fertilizing purposes is reached, the surplus from spreading runs off on the land and contaminates surface and ground waters (OECD, 2003). These non-point source pollutions are difficult to manage and have a considerable impact on water quality. The level of nitrogen surplus is derived from the nutrient balance of the farm, which can be stated as follows:

$$b_{Nsurpl} = b_{Norg} + b_{N\min} - b_{N\exp} \tag{2.9}$$

where b_{Norg} is the level of organic manure, $b_{N\min}$ the level of mineral fertilizers and, $b_{N\exp}$ the level of nitrogen that is taken up by the crops on the land.

Thus, the quantitative standard resulting from the Nitrate directive only applies to one component of the nutrient balance, b_{Norg}, and not to b_{Nsurpl}. However, all the other constraints resulting from the implementation of this directive in European countries and in particular, in France, relate to the management of this nitrogen surplus and aim to reduce its diffusion in the environment. Most of the mandatory constraints - manure storage capacities, timing of spreading, etc. - apply to the undesirable output b_{Nsurpl}. Thus, we assume that b_{Nsurpl} is only weakly disposable.

$$(y, b_{Nsurpl}) \in P(x) \text{ and } y' \leq y \text{ imply } (y', b_{Nsurpl}) \in P(x) \tag{2.10}$$

Based on this new set of hypotheses, the directional distance function may be rewritten as follows. For the farm k' ($k'=1,...,K$), we have:

$$\vec{D}_0^{170}(x_{k'}, y_{k'}, b_{k'}; y_{k'}, -b_{k'}) = \max \beta_{k'}^{170}$$

s.t.

$$\sum_{k=1}^{K} z_k y_{mk} \geq (1 + \beta_{k'}^{170}) y_{mk'} \quad m = 1,\ldots,M$$

$$\sum_{k=1}^{K} z_k b_{sk} = (1 - \beta_{k'}^{170}) b_{sk'} \quad s = 1,\ldots,S-1$$

$$\sum_{k=1}^{K} z_k b_{Nsurpl\,k} = b_{Nsurpl}^{k'} \quad (2.11)$$

$$\sum_{k=1}^{K} z_k x_{nk} \leq x_{nk'} \quad n = 1,\ldots,N$$

$$z_k \geq 0 \quad k = 1,\ldots,K$$

$$b_{Nsurpl}^{k'} = b_{Norg}^{k'} + b_{N\min k'} - b_{N\exp k'}$$

$$b_{Norg}^{k'} \leq 170 \cdot x_{land\,k'}$$

where b_s are all other bad outputs resulting from the production of pigs (phosphorus, heavy metals, etc.). These are not directly considered by the specific regulation in the Nitrates directive. In the model, all the variables with an upper index k' are variable while the variables with a lower index k' are observed data.

4. DATA AND EMPIRICAL MODEL

The data used to conduct this research has been drawn from the Farm Accountancy Data Network (FADN) data set. The sample consists of farms specializing in pig farming activities in 1996 and located in Brittany, France's main pig producing region. The data set used here contains annual information from 181 farms. It includes information on quantities, both in physical and monetary terms, relating to the production of good outputs and the use of inputs. Data on the organic manure and nitrogen surplus of each farm are calculated. The level of organic manure is based on the number of animals on the farm. We have applied the coefficients provided by the CORPEN[3], which provide an approximation of the level of organic manure

---

[3] The CORPEN is an organisation responsible for defining the Codes of Good Agricultural Practice relating to the management of nitrogen and phosphorus. They also provide

produced by each type of animal. The level of manure surplus is derived from the individual nutrient balance of each farm. This balance is a tool used to provide estimates of flows of nitrogen across the farm boundary. Nutrient balances are defined as the difference between input and output flows. Input flows compare nitrogen from inorganic fertilizers, nitrogen from organic manure (not including losses due to emissions of ammonia), and nitrogen by deposition from the atmosphere. Output flows include uptake by harvested crops and livestock sold (Meisinger and Randall, 1991). These calculations are described in Table 7-1.

Table 7-1. Description of the data set

	#	b_{Nsurpl} (kg/ha)		b_{Norg} (kg/ha)	
		Mean	Std	Mean	Std
$b_{Norg} > 170$	71	229	177	312	186
$b_{Norg} \leq 170$	110	86	64	115	33
$b_{Nsurpl} > 0$	181	130	149	193	153

Table 7-2. Summary statistics for inputs and outputs

	Mean	St. Dev.	Minimum	Maximum
Total gross output (€)	268,441	171,051	38,717	893,510
Land (ha)	46	19.7	9	104
Livestock (Lu)	2,440	1,789	258	10,843
Labor (Awu)	1.8	0.7	0.8	4.2
Variable inputs (€)	167,252	112,826	16,491	606,868
Organic manure (kg)	7,592	3,943	882	21,805
Mineral manure (kg)	6,445	4,213	0	20,747
Exportation (kg)	8,068	1,967	1,012	25,615
Manure surplus (kg)	5,970	4,120	186	20,240

In the sample, all farms have a positive nitrogen surplus with an average level of 130kg/ha. However, this mean does not reflect the differences that exist between farms. 40% of the farms in the sample do not comply with the constraint imposed by the Nitrate directive, which requires a level of organic manure spread per hectare of less than 170kg. When evaluating farms' performance relative to their production and environmental efficiency, it seems important to integrate the standard derived from the EU regulation.

---

information on the average level of nitrogen produced by different types of animal and taken up by crops.

This restriction reduces the production frontier to a subset compatible with the environmental regulation.

To implement this approach, we have defined two empirical models by specifying a set of inputs and outputs. For the traditional model (section 2), which simply considers the presence of an undesirable by-output, we assume a production technology with one desirable output (total gross output), one bad output (manure surplus) and four inputs (land, livestock, labor and variable inputs). For the extended model (section 3), which considers the presence of an undesirable output and the regulation on organic manure from the EU Nitrate directive, we assume a production technology with one desirable output (total gross output), one by-output (organic manure), one bad output (manure surplus) and four inputs (land, livestock, labor and variable inputs). Summary statistics for these variables are reported in Table 7-2.

5. RESULTS

The outcomes obtained using the methodology presented above are shown in Table 7-3. As previously mentioned, a farm is efficient when its production efficiency measurement is 1. A farm with a score of over 1 can improve its performance by modifying its production process in order to reach the empirical production frontier along a path defined by the direction vector.

Table 7-3. Comparison of production efficiency measurement with and without the European Nitrate directive standard

	#	Model 1 - without standard		Model 2 – with standard	
		Mean	Std	Mean	Std
$b_{Norg} > 170$	71	1.103	0.098	1.074	0.104
$b_{Norg} \leq 170$	110	1.109	0.119	1.167	0.146
$b_{Nsurpl} > 0$	181	1.106	0.111	1.130	0.138

In the traditional model or Model 1, we only consider the production of one bad output, nitrogen surplus. The direction for the improvement of the production efficiency of each farm is defined in a way that allows an increase in the good output and a decrease in the bad output. As there exists an EU regulation on the management of manure through the Code of Good Agricultural Practice, we assume that farms cannot freely dispose of nitrogen surplus. Average results show that the farms in our sample can increase their production by 10%. This figure is equivalent to an average increase in the total gross output per farm of around €26,844 and a 597kg decrease in nitrogen surplus per farm, i.e., 13kg/ha. When the results are classified according to the initial situation of each farm in relation to the environmental restriction on organic manure, as set out in the EU Nitrate directive, no significant difference appears.

In the extended model or Model 2, we introduce the EU standard on organic manure into the modeling. We consider the production of a by-product - organic manure – and a bad output – nitrogen surplus. The direction for the improvement of the production efficiency of each farm is defined in a way that allows both an increase in the good output and compliance with the EU's mandatory limitation on organic manure. As in Model 1, we assume that farms cannot freely dispose of nitrogen surplus. The average results of this extended model show that farms from our sample can increase their production by 13% while complying with the EU Nitrate directive standard. This figure is equivalent to an average increase in the total gross output per farm of €34,897 and a 776kg decrease in nitrogen surplus per farm, i.e., 17kg/ha. Behind this figure, there is considerable disparity between farms above and below the standard. Farms above 170kg/ha have the lowest efficiency score. They are more "efficient". We can effectively consider that they are more bounded and thus, have fewer possibilities for reaching the production frontier, i.e., fewer possibilities for increasing production and complying with the EU Nitrate directive standard. On the other hand, farms below the limit have greater possibilities for adjustments to their level of production. If we consider their individual results, we can see that they increase their production level and thus the level of their by-product. Some are then bounded by the standard while others are still unrestricted. These results are presented in Table 7-4.

Table 7-4. Impact on organic manure and nitrogen surplus

	#	b^{Obs}_{Nsurpl}	b^{Opt}_{Nsurpl} Model 1	b^{Opt}_{Nsurpl} Model 2	b^{Obs}_{Norg}	b^{Opt}_{Norg} Model 2
		kg (kg/ha)	kg (kg/ha)	kg (kg/ha)	kg (kg/ha)	kg (kg/ha)
$b_{Norg} > 170$	71	8,245 (229)	7,396 (205)	4,199 (116)	10,061 (312)	6,016 (167)
$b_{Norg} \leq 170$	110	4,501 (86)	4,010 (76)	5,770 (110)	5,998 (116)	7,267 (139)
$b_{Nsurpl} > 0$	181	5,970 (130)	5,337 (116)	5,154 (112)	7,592 (193)	6,676 (152)

In Table 7-4, we can see that a lower production of nitrogen surplus is provided when we use the extended Model 2. When using the traditional

Model 1, we have an average decrease of around 14kg/ha of nitrogen surplus. However, we have no indication of the impact on the farm's manure management. Breaking down the results according to the initial situation of each farm with respect to the EU standard, shows that producers above the limit of 170kg/ha reduce their nitrogen surplus by 24kg/ha but we do not know if they are in conformity with the EU regulation. For farms below the EU standard, the reduction in nitrogen surplus is 10kg/ha on average.

Model 2 provides results that are very different in several ways. First, the average decrease in nitrogen surplus is 18kg/ha with each farm in the sample complying with the EU standard (152kg/ha on average). Second, producers below the standard reduce their initial nitrogen surplus by around 24kg/ha while producers above the standard decrease their nitrogen surplus by an average 113kg/ha. Third, figures relating to the level of organic manure (last two columns) show that there is an average decrease of 41kg/ha in the production of this by-product together with an increase in the production of the good output and compliance with the mandatory limit. Fourth, farms initially above the EU standard decrease their by-production by 145kg/ha per farm and fulfill the EU standard with an average organic manure production of 167kg/ha. Farms initially below the standard increase their production of organic manure by an average of 23kg/ha and still comply with the mandatory constraint (139kg/ha on average).

6. CONCLUSION

Agricultural activities take place in a setting characterized by the increasing presence of public regulations. Those that aim to reduce the emission of polluting wastes create a need for new methods of evaluating the impact of environmental regulations on farms' performance. This paper highlights the usefulness of the directional technology distance function in measuring the impact of the EU Nitrate directive, which prevents the free disposal of organic manure and nitrogen surplus. It provides efficiency indices for the production and environmental performance of farms at an individual level and an evaluation of the impact of the said EU regulation on this performance. This methodology is illustrated with an empirical application using a sample of French pig farms located in Brittany in 1996.

In specific relation to this empirical analysis, we extend the previous approach to good and bad outputs within the framework of the directional distance function, by introducing a by-product (organic manure), which becomes a pollutant only after a certain level of disposability is exceeded. In this specific case, the bad output is the nitrogen surplus - resulting from the nutrient balance of each farm - that is spread on the land. This extension to the model allows us to explicitly introduce the EU regulation on the spreading of organic manure, which sets a limit of 170kg/ha. Our results show that the extended model provides greater possibilities for increasing the level of production, and thus the revenue of each farm, while decreasing the bad product (nitrogen surplus) and complying with the mandatory standard on the spreading of organic manure.

This chapter provides an illustration that demonstrates the flexibility and usefulness of the directional distance function in evaluating the impact of environmental regulations on farms' performance. This approach may also be used to simulate other implementations of European regulations on nitrogen management. In particular, this kind of approach can be used to study the effect of replacing the management at the individual farm level of the constraint limiting the amount of manure spread per hectare, with the collective management of the said constraint by groups of farms, so as to determine whether it is possible to further reduce the negative impact of pig production on the environment and at the same time improve farms' economic performance.

REFERENCES

1. Allen, K., 1999, DEA in the ecological context: An overview, In: *Data Envelopment Analysis in the Service Sector*, G., Westermann, ed., Gabler Edition Wissenschaft, Weisbaden, pp. 203-235.
2. Chambers, R.G., Färe, R., and Grosskopf, S., 1996, Productivity growth in APEC countries, *Pacific Economic Review*, **1**(3):181-190.
3. Chambers, R.G., Chung, Y.H., and Färe, R., 1998, Profit, directional distance functions and Nerlovian efficiency, *Journal of Optimization Theory and Applications,* **98**(2):351-364.
4. Chung, Y.H., Färe, R., and Grosskopf, S., 1997, Productivity and undesirable outputs: A directional distance function approach, *Journal of Environmental Management*, **51**(3): 229-240.
5. Debreu, G., 1951, The coefficient of resource utilization, *Econometrica,* **19**(3):273-292.
6. Färe, R., and Grosskopf, S., 2000, Theory and application of directional distance functions, *Journal of Productivity Analysis*, **13**(2):93-103.
7. Färe, R., and Grosskopf, S., 2004, *New Directions: Efficiency and Productivity,* Kluwer Academic Publishers, Boston.
8. Färe, R., Grosskopf, S., and Lovell, C.A.K., 1985, *The Measurement of Efficiency of Production*, Studies in Productivity Analysis, Kluwer-Nijhoff Publishing, Boston.

9. Färe, R., Grosskopf, S., and Lovell, C.A.K., 1994, *Production Frontiers,* Cambridge University Press, Cambridge.
10. Färe, R., Grosskopf, S., Lovell, C.A.K., and Pasurka C., 1989, Multilateral productivity comparisons when some outputs are undesirable: A nonparametric approach, *The Review of Economics and Statistics*, **71**(1):90-98.
11. Färe, R., Grosskopf, S., and Tyteca, D., 1996, An activity analysis model of the environmental performance of Firms: Application to the Fossil Fuel Fired Electric Utilities, *Ecological Economics*, **18**(2):161-175.
12. Farrell, M. J., 1957, The measurement of productive efficiency, *Journal of the Royal Statistics Society,* Series A, **120**:253-282.
13. Luenberger, D., 1992, Benefit functions and duality, *Journal of Mathematical Economics*, **21**(5):461-481.
14. Meisinger, J.J., and Randall, G.W., 1991, Estimating Nitrogen Budgets of Soil-Crop Systems, In: *Managing Nitrogen for Groundwater Quality and Farm Profitability*, R.F., Follett, D.R., Keeney, and R.M., Cruse, ed., Soil Science Association of America, Madison, Wisconsin, pp. 85-124.
15. OECD, 2003, *Agriculture, trade and the environment: The pig sector*, OECD, Paris.
16. Shephard, R.W., and Färe, R., 1974, The law of diminishing returns, *Zeitschrift für Nationalökonomie,* **34**(1-2):69-90.
17. Shephard, R.W., 1970, *Theory of Cost and Production Functions,* Princeton University Press, Princeton.
18. Tyteca, D., 1996, On the measurement of the environmental performance of firms: A literature review and a productive efficiency perspective, *Journal of Environmental Management*, **46**(3):281-308.

ACKNOWLEDGEMENTS

This paper has benefited from comments and suggestions by participants at the IFORS (International Federation of Operational Research Societies) Triennial Conference, Honolulu, Hawaii, 2005 and at the AAEA (American Agricultural Economics Association) Annual Meeting, Providence, Rhode Island (USA), 2005.

Chapter 8

PCA-DEA
Reducing the curse of dimensionality

Nicole Adler[1] and Boaz Golany[2]
[1]*Hebrew University of Jerusalem, msnic@huji.ac.il*

[2]*Technion – Israel Institute of Technology*

Abstract: The purpose of this chapter is to present the combined use of principal component analysis (PCA) and data envelopment analysis (DEA) with the stated aim of reducing the curse of dimensionality that occurs in DEA when there is an excessive number of inputs and outputs in relation to the number of decision-making units. Various PCA-DEA formulations are developed in the chapter utilizing the results of principal component analyses to develop objective, assurance region type constraints on the DEA weights. The first set of models applies PCA to grouped data representing similar themes, such as quality or environmental measures. The second set of models, if needed, applies PCA to all inputs and separately to all outputs, thus further strengthening the discrimination power of DEA. A case study of municipal solid waste managements in the Oulu district of Finland, which has been frequently analyzed in the literature, will illustrate the different models and the power of the PCA-DEA formulation. In summary, it is clear that the use of principal components can noticeably improve the strength of DEA models.

Key words: Data envelopment analysis; Principal Component Analysis; Assurance Region Constraints

1. INTRODUCTION

The aim of this chapter is to present the combined use of principal component analysis and data envelopment analysis, namely PCA-DEA. The idea to combine these two methodologies was developed independently by Ueda and Hoshiai (1997) and Adler and Golany (2001, 2002). Zhu (1998) suggested principal component analysis of the output to input ratios as a complementary approach to DEA. The goal of the PCA-DEA model is to improve the discriminatory power within data envelopment analysis, which often fails when there is an excessive number of inputs and outputs in relation to the number of decision-making units (DMUs). When this occurs, the Pareto frontier may be defined by a large number of DMUs i.e. a substantial proportion of the units are considered efficient. As an illustration, we will analyze a municipal waste dataset from Finland (Hokkanen and Salminen (1997a, b), Sarkis (2000), Jenkins and Anderson (2003)) in which all DMUs are considered efficient under the standard BCC model (Banker et al. (1984)).

The following section describes both data envelopment analysis and principal component analysis. Subsequently, PCA-DEA constraint-modified models are developed, using objectively based, assurance region type constraints. The fourth section demonstrates the usefulness of this approach through the use of a dataset that has been substantially analyzed in the DEA and decision analysis literature. Finally, the last section presents a summary and conclusions drawing on these models.

2. DATA ENVELOPMENT ANALYSIS AND PRINCIPAL COMPONENT ANALYSIS

Data envelopment analysis is a technique that measures the relative efficiency of DMUs with multiple inputs and outputs but with no obvious production function to aggregate the data in its entirety. Since it may be unclear how to quantitatively combine the various inputs consumed and outputs produced, the DEA methodology computes the relative efficiency of each DMU individually through the use of weighted averages. PCA-DEA can be applied to the CCR (Charnes et al. (1978)), BCC (Banker et al. (1984)) and additive models (Charnes et al. (1985)). By comparing n units with q outputs denoted by Y and r inputs denoted by X, the efficiency measure for unit a is expressed as in models (1a, b and c), where both the envelopment side and multiplier side formulations are presented for the input oriented CCR, BCC and constant returns-to-scale additive models respectively.

$$\begin{array}{ll} \underset{\vartheta,\lambda}{\text{Min }} \vartheta & \underset{V,U}{\text{Max }} UY^a \\ \text{s.t. } Y\lambda \geq Y^a & \text{s.t. } VX^a = 1 \\ \vartheta X^a - X\lambda \geq 0 & VX - UY \geq 0 \\ \lambda, \vartheta \geq 0 & V, U \geq 0 \end{array} \qquad (1a)$$

$$\begin{aligned}&\underset{\vartheta,\lambda}{\text{Min }} \vartheta \\ &\text{s.t.} \quad Y\lambda \geq Y^a \\ &\quad \vartheta X^a - X\lambda \geq 0 \\ &\quad e\lambda = 1 \\ &\quad \lambda, \vartheta \geq 0\end{aligned} \qquad \begin{aligned}&\underset{V,U}{\text{Max }} UY^a - u^a \\ &\text{s.t.} \quad VX^a = 1 \\ &\quad VX - UY + u^a \geq 0 \\ &\quad V, U \geq 0, \ u^a \ \text{free}\end{aligned} \qquad (1b)$$

$$\begin{aligned}&\underset{s,\sigma}{\text{Max }} es + e\sigma \\ &\text{s.t.} \quad Y\lambda - s = Y^a \\ &\quad -X\lambda - \sigma = -X^a \\ &\quad \lambda, s, \sigma \geq 0\end{aligned} \qquad \begin{aligned}&\underset{V,U}{\text{Min }} VX^a - UY^a \\ &\text{s.t.} \quad VX - UY \geq 0 \\ &\quad V \geq e \\ &\quad U \geq e\end{aligned} \qquad (1c)$$

In (1), λ represents a vector of DMU weights chosen by the linear program, θ and u^a are constants, e a vector of ones, σ and s vectors of input and output slacks, respectively, and X^a and Y^a the input and output column vectors for DMU$_a$, respectively.

Principal component analysis (PCA) explains the variance structure of a matrix of data through linear combinations of variables. Consequently, it may be possible to reduce the data to a few principal components, which generally describe 80 to 90% of the variance in the data. If most of the population variance can be attributed to the first few components, then they can replace the original variables with minimum loss of information. As stated in Hair et al. (1995), let the random vector $X=[X_1, X_2, ..., X_p]$ (i.e. the original inputs/outputs chosen to be aggregated) have the correlation matrix C with eigenvalues $\lambda_1 \geq \lambda_2 \geq ... \geq \lambda_p \geq 0$ and normalized eigenvectors $l_1, l_2, ..., l_p$. Consider the linear combinations, where the superscript t represents the transpose operator:

$$X_{PC_i} = l_i^t X = l_{1i} X_1 + l_{2i} X_2 + ... + l_{pi} X_p$$

$$Var(X_{PC_i}) = l_i^t C l_i \qquad , \qquad i=1,2,...,p$$

$$Correlation(X_{PC_i}, X_{PC_k}) = l_i^t C l_k \qquad , \qquad i=1,2,...,p, k=1,2,...,p, i \neq k$$

The principal components are the uncorrelated linear combinations $X_{PC_1}, X_{PC_2}, ..., X_{PC_p}$ ranked by their variances in descending order. The PCA used here is based on correlation rather than covariance, as the variables used in DEA are often quantified in different units of measure. In

Section 4, PCA results are presented for the Finnish dataset utilized in this chapter.

3. THE PCA-DEA CONSTRAINED MODEL FORMULATION

This section first develops the PCA-DEA model formulation and then discusses the PCA-DEA constrained model.

3.1 PCA-DEA model

In order to use PC scores instead of the original data, the DEA model needs to be transformed to take into account the use of linear aggregation of data. Indeed, the formulation presented ensures that were we to use all the PC's, we would attain exactly the same solution as that achieved under the original DEA formulation. When using both original inputs and outputs and some aggregated variables, separate $X=\{X_o, X_{Lx}\}$ and $Y=\{Y_o, Y_{Ly}\}$, where $X_o(Y_o)$ represent inputs(outputs) whose original values will be used in the subsequent DEA. $X_{Lx}(Y_{Ly})$ represents inputs(outputs) whose values will be transformed through PCA. This separation of variables should be based on a logical understanding of the specific application under question. For example, in an application where outputs could be separated into direct explanations of resources, such as capital and labor, and indirect explanations such as quality or environmental measures, it would be natural to apply PCs to the indirect variables, which are often highly correlated and could be reduced with minimal to no loss of information. Another example can be drawn from an air transport application described in Adler and Golany (2001), in which the input variables could be logically split into three separate areas, namely sets of transportation, comfort and size variables. Each group could be handled individually as the correlation between variables within the group is naturally high.

Let $L_x = \{l_{ij}^x\}$ be the matrix of the PCA linear coefficients of input data and let $L_y = \{l_{st}^y\}$ be the matrix of the PCA linear coefficients of output data. Now, $X_{PC} = L_x X_{Lx}$ and $Y_{PC} = L_y Y_{Ly}$ are weighted sums of the corresponding original data, X_{Lx} and Y_{Ly}. We can consequently replace models (1) with models (2a, b and c), which refer to the input oriented CCR, BCC and additive models with PCA respectively.

$$\underset{\vartheta,\lambda}{\text{Min}} \ \vartheta$$

s.t. $Y_0\lambda - s_o = Y_0^a$

$Y_{PC}\lambda - L_y s_{PC} = Y_{PC}^a$

$\vartheta X_0^a - X_0\lambda - \sigma_0 = 0$ \hfill (2a)

$\vartheta X_{PC}^a - X_{PC}\lambda - L_x \sigma_{PC} = 0$

$L_x^{-1} X_{PC} \geq \sigma_{PC}$

$L_y^{-1} Y_{PC} \geq s_{PC}$

$\lambda, \vartheta, s_0, s_{PC}, \sigma_0, \sigma_{PC} \geq 0$

$$\underset{\vartheta,\lambda}{\text{Min}} \ \vartheta$$

s.t. $Y_0\lambda - s_o = Y_0^a$

$Y_{PC}\lambda - L_y s_{PC} = Y_{PC}^a$

$\vartheta X_0^a - X_0\lambda - \sigma_0 = 0$

$\vartheta X_{PC}^a - X_{PC}\lambda - L_x \sigma_{PC} = 0$ \hfill (2b)

$L_x^{-1} X_{PC} \geq \sigma_{PC}$

$L_y^{-1} Y_{PC} \geq s_{PC}$

$e\lambda = 1$

$\lambda, \vartheta, s_0, s_{PC}, \sigma_0, \sigma_{PC} \geq 0$

$$\underset{s_0, s_{PC}, \sigma_{PC}, \sigma_o, \lambda}{\text{Max}} \ es_o + es_{PC} + e\sigma_o + e\sigma_{PC}$$

s.t. $Y_o\lambda - s_o = Y_o^a$

$Y_{PC}\lambda - L_y s_{PC} = Y_{PC}^a$

$- X_o\lambda - \sigma_o = -X_o^a$

$- X_{PC}\lambda - L_x \sigma_{PC} = -X_{PC}^a$ \hfill (2c)

$L_x^{-1} X_{PC} \geq \sigma_{PC} \geq 0$

$L_y^{-1} Y_{PC} \geq s_{PC} \geq 0$

$s_o, \sigma_o, \lambda \geq 0$

The terms $L_x^{-1}(L_y^{-1})$ in (2) represent the inverse matrix of input(output) weights attained through the PCA.

It is worthwhile noting that the original additive model of Charnes et al. (1985) was extended by Lovell and Pastor (1995) to a weighted additive model whereby weights were introduced into the objective function in order to counter bias that might occur due to differences in magnitude of the values of the original variables. Along the same line, the relevant slacks in model (2) are weighted by the PC scores in order to counter the transformation and ensure an equivalent problem/solution to that of the original linear program. The PCA-DEA formulation is exactly equivalent to the original DEA model when the PCs explain 100% of the correlation in the original input and output matrices. Clearly, if we use less than full information, we will lose some of the explanatory powers of the data but we will improve the discriminatory power of the model. This means that we may mistakenly define a small number of efficient units as inefficient, however this error type is substantially smaller than its opposite whereby inefficient units are defined as efficient, because generally there are fewer efficient than inefficient units in the dataset (see Adler and Yazhemsky (2005) for further details). Furthermore, since DEA is a frontier-based methodology, it is potentially affected by data errors in measurement. To some extent, this may be avoided by removing some PCs.

The only other methodology for reducing dimensionality known to the authors is to ignore specific variables when doing the analysis. By doing so, all information drawn from those variables are automatically lost. By removing certain PCs, we will not lose an entire variable, unless the PC weight is placed entirely on the specific variable that is subsequently dropped, with a zero weight in all other PC combinations. In other words, choosing a limited number of variables is the private case of the PCA-DEA formulation.

Writing the dual of model (2) for DMU_a, we attain the following formulation, in which the effect of the PCA analysis can be seen directly through the weights $L_x(L_y)$.

$$\underset{V_0,V_{PC},U_0,U_{PC}}{Max} \quad U_0 Y_0^a + U_{PC} Y_{PC}^a$$

$$s.t. \quad V_0 X_0^a = 1$$

$$V_{PC} X_{PC}^a = 1$$

$$V_0 X_0 + V_{PC} X_{PC} - U_0 Y_0 - U_{PC} Y_{PC} \geq 0 \tag{3a}$$

$$V_{PC} L_x \geq 0$$

$$U_{PC} L_y \geq 0$$

$$V_0, U_0 \geq 0, \quad V_{PC}, U_{PC} \quad free$$

$$\underset{V_0,V_{PC},U_0,U_{PC}}{\text{Max}} \quad U_0 Y_0^a + U_{PC} Y_{PC}^a - u^a$$

$$\begin{aligned}
\text{s.t.} \quad & V_0 X_0^a = 1 \\
& V_{PC} X_{PC}^a = 1 \\
& V_0 X_0 + V_{PC} X_{PC} - U_0 Y_0 - U_{PC} Y_{PC} + u^a \geq 0 \\
& V_{PC} L_x \geq 0 \\
& U_{PC} L_y \geq 0 \\
& V_0, U_0 \geq 0, \quad V_{PC}, U_{PC}, u^a \text{ free}
\end{aligned}$$
(3b)

$$\underset{V_0,V_{PC},U_0,U_{PC}}{\text{Min}} \quad V_o X_o^a + V_{PC} X_{PC}^a - U_o Y_o^a - U_{PC} Y_{PC}^a$$

$$\begin{aligned}
\text{s.t.} \quad & V_o X_o + V_{PC} X_{PC} - U_o Y_o - U_{PC} Y_{PC} \geq 0 \\
& V_o \geq e \\
& U_o \geq e \\
& V_{PC} L_x \geq e \\
& U_{PC} L_y \geq e \\
& V_{PC}, U_{PC} \text{ free}
\end{aligned}$$
(3c)

Clearly, not only will U_o and V_o be computed, but also U_{PC} and V_{PC}, which can then be translated backwards to compute the 'original' weights. If all PCs are utilized, the weights should be the same as the original model, however given the potential in DEA for degeneracy, this may not always be the case. Even if all PCs are not accounted for, the weights can still be evaluated backwards on the original data, ensuring a complete DEA result that is interpretable for inefficient units in the standard manner.

3.2 PCA-DEA constrained model

The use of principal component analysis, whilst useful in reducing the problems of dimensions, may still be insufficient in providing an adequate ranking of the DMUs. Therefore, the introduction of constraints on the PC weights within DEA (V_{PC} and U_{PC}) may further aid the discrimination capabilities of the analysis. One of the major concerns with the assurance region and cone-ratio models is the additional preference information that needs to be drawn from the decision-makers in order to accomplish this task. This is particularly problematic given the less-than-opaque understanding of the DEA weight variables. Consequently, the objective use of weight constraints based on the output of the PCA may be considered helpful. This section discusses the addition of assurance region constraints to the PCA-DEA models, utilizing PCA weight information.

Several sets of weight constraints could be considered for this analysis, such as that of assurance regions, first developed in Thompson et al. (1986). Golany (1988) proposed a version of assurance regions with ordinal constraints on the weights. Other researchers followed with various extensions and the topic is reviewed in Cooper et al. (2000). Assurance region constraints introduce bounds on the DEA weights, for example to include all variables through a strictly positive value or to incorporate 'decision-makers' knowledge. In the light of the fact that PCA prioritizes the PCs in descending order of importance, additional constraints could simply require the weight of PC_1 to be at least that of PC_2, the weight of PC_2 to be at least that of PC_3 and so on, as specified in equation (4).

$$V_{PC_i} - V_{PC_{i+1}} \geq 0$$
$$U_{PC_i} - U_{PC_{i+1}} \geq 0 \quad \text{for } i=1,\ldots,m\text{-}1, \text{ where } m \text{ PCs are analyzed} \quad (4)$$

Equation (4) applies only to those weights associated with PC-modified data and only the PCs chosen to be included in the analysis.

In the second set of models, where PCA is applied to all inputs and outputs separately, the set of constraints described in equation (4) could then be applied to all PC variables, requiring only objective information and improving the discriminatory power of the original model. This is shown clearly in (5), where we present the standard additive model and the complete PCA-DEA constrained models as an example.

$$\underset{V,U}{\text{Min}} \ VX^a - UY^a$$
$$\text{s.t.} \ VX - UY \geq 0$$
$$V \geq e$$
$$U \geq e$$

$$\underset{VL_y,UL_x}{\text{Min}} \ V_{PC}X_{PC}^a - U_{PC}Y_{PC}^a$$
$$\text{s.t.} \ V_{PC}X_{PC} - U_{PC}Y_{PC} \geq 0$$
$$V_{PC}L_x \geq e$$
$$U_{PC}L_y \geq e \quad (5)$$
$$V_{PC_i} - V_{PC_{i+1}} \geq 0$$
$$U_{PC_i} - U_{PC_{i+1}} \geq 0$$

where $X_{PC}=L_xX$ implies that $V_{PC}X_{PC}=V_{PC}L_xX$, hence $V=V_{PC}L_x$ and equivalently $U=U_{PC}L_y$. PCs adding little to no information to the analysis could be dropped and the model reduced accordingly with minimal loss of information.

4. APPLICATION OF THE PCA-DEA MODELS

In order to illustrate the potential of these models, we will present the results of a dataset existing in the literature. The numerical illustration was first analyzed in Hokkanen and Salminen (1997a, b) in which 22 solid waste management treatment systems in the Oulu region of Finland were compared over 5 inputs and 3 outputs (defined in Sarkis (2000)). The original data is presented in Table 8-1 and the results of the principal components analysis in Table 8-2.

Table 8-1. Original data for location of solid waste management system in Oulu Finland

DMU	Inputs					Outputs		
	Cost	Global Effects	Health Effects	Acidificative releases	Surface water releases	Technical feasibility	Employees	Resource recovery
1	656	552,678,100	609	1190	670	5.00	14	13,900
2	786	539,113,200	575	1190	682	4.00	18	23,600
3	912	480,565,400	670	1222	594	4.00	24	39,767
4	589	559,780,715	411	1191	443	9.00	10	13,900
5	706	532,286,214	325	1191	404	7.00	14	23,600
6	834	470,613,514	500	1226	384	6.50	18	40,667
7	580	560,987,877	398	1191	420	9.00	10	13,900
8	682	532,224,858	314	1191	393	7.00	14	23,600
9	838	466,586,058	501	1229	373	6.50	22	41,747
10	579	561,555,877	373	1191	405	9.00	9	13,900
11	688	532,302,258	292	1191	370	7.00	13	23,600
12	838	465,356,158	499	1230	361	6.50	17	42,467
13	595	560,500,215	500	1191	538	9.00	12	13,900
14	709	532,974,014	402	1191	489	7.00	17	23,600
15	849	474,137,314	648	1226	538	6.50	20	40,667
16	604	560,500,215	500	1191	538	9.00	12	13,900
17	736	532,974,014	402	1191	489	7.00	17	23,600
18	871	474,137,314	648	1226	538	6.50	20	40,667
19	579	568,674,539	495	1193	558	9.00	7	13,900
20	695	536,936,873	424	1195	535	6.00	18	23,600
21	827	457,184,239	651	1237	513	7.00	16	45,167
22	982	457,206,173	651	1239	513	7.00	16	45,167

In this example, it could be argued that two PCs on the input side and two PCs on the output side explain the vast majority of the variance in the original data matrices, since they each explain more than 95% of the correlation, as shown in Table 8-2. It should be noted here that the results of a PCA are unique up to a sign and must therefore be chosen carefully to ensure feasibility of the subsequent PCA-DEA model (Dillon and Goldstein (1984)).

Table 8-2. Principal Component Analysis

Inputs (L_x)	PCx_1	PCx_2	PCx_3	PCx_4	PCx_5
% correlation explained	66.877	28.603	4.0412	0.32335	0.15546
Lx_1	-0.51527	0.079805	0.70108	-0.48201	0.065437
Lx_2	0.51843	-0.2507	-0.10604	-0.6928	0.42091
Lx_3	-0.43406	-0.47338	-0.47478	-0.36942	-0.47499
Lx_4	-0.5214	0.18233	-0.44219	0.040172	0.70551
Lx_5	-0.07394	-0.82064	0.2762	0.38679	0.30853

Outputs (L_y)	PCy_1	PCy_2	PCy_3
% correlation explained	79.011	16.9	4.0891
Ly_1	-0.55565	-0.69075	-0.46273
Ly_2	0.62471	0.020397	-0.78059
Ly_3	0.54863	-0.72281	0.42018

Table 8-3. PCA-DEA Results

Normalized data		Model A (Cost + 2 PC inp., original out)			Model B (2 PC inp., original out)	
DMU	CCR	CCR-PCA	constrained CCR-PCA	BCC-PCA	CCR-PCA	BCC-PCA
1	0.84	0.84	0.84	0.96	0.57	0.75
2	0.87	0.87	0.87	0.88	0.65	0.79
3	1.00	1.00	1.00	1.00	0.88	1.00
4	1.00	0.98	0.98	0.98	0.97	0.97
5	0.99	0.96	0.96	0.98	0.96	0.97
6	0.99	0.98	0.93	0.99	0.97	0.98
7	1.00	1.00	1.00	1.00	1.00	1.00
8	1.00	0.98	0.98	1.00	0.98	0.99
9	1.00	1.00	1.00	1.00	1.00	1.00
10	1.00	1.00	1.00	1.00	1.00	1.00
11	1.00	1.00	1.00	1.00	1.00	1.00
12	1.00	1.00	0.93	1.00	1.00	1.00
13	1.00	1.00	1.00	1.00	0.91	1.00
14	1.00	0.98	0.98	1.00	0.92	0.93
15	0.98	0.95	0.93	0.96	0.82	0.83
16	1.00	0.99	0.99	1.00	0.91	1.00
17	1.00	0.95	0.95	0.97	0.92	0.92
18	0.98	0.93	0.91	0.94	0.82	0.82
19	1.00	1.00	0.97	1.00	0.84	0.84
20	1.00	1.00	1.00	1.00	0.82	0.89
21	1.00	1.00	0.90	1.00	0.90	1.00
22	1.00	0.90	0.79	1.00	0.88	0.99

Table 8-3 presents the results of various DEA models, including the standard CCR, partial PCA-CCR and PCA-BCC models and a constrained PCA-CCR model. Table 8-4 presents the results of the complete PCA-CCR and PCA-BCC models, and the benchmarks resulting from it for further comparison. First it should be noted that all the original data was normalized by dividing through the elements in each vector by its mean, as was discussed in the previous literature (Sarkis (2000)), otherwise the linear programs may become infeasible due to the substantial differences in measurement (see Table 8-1). The standard CCR results are presented in the second column of Table 8-3 and the BCC results do not appear as all the DMUs were deemed efficient. In Model A, there are 3 inputs, the original cost and 2 PCs, based on the remaining 4 environmental inputs. The 2 PCs explain 98% of the variance of the 4 inputs. The 3 original outputs are used on the output side. This formulation is representative of the first model type explained in the previous section. In the CCR-PCA model, 6 DMUs previously efficient become inefficient and in the BCC-PCA model, 8 DMUs are now considered inefficient. We also present a constrained CCR-PCA model in which the weight on PC_1 must be at least equal to that of the weight on PC_2. Under these conditions, there is a greater differentiation between the units, with 3 additional DMUs becoming inefficient and another 4 reducing their efficiency levels. In Model B, 2 PC inputs are introduced to explain the original 3 outputs. This reduces the number of efficient DMUs in the CCR model from 73% to 23% and in the BCC model from 100% to 41%, resulting in 9 efficient waste management systems.

In Table 8-4, the two PC input, two PC output constant and variable returns-to-scale models results are presented, alongside the benchmarks for inefficient units and the number of times an efficient DMU appears in the peer set of the inefficient DMUs. In the PCA-CCR, whereby 2 PCs represent the input and output side respectively, only 2 DMUs remain efficient, namely DMUs 9 and 12. Interestingly, when we peruse the various results presented in Hokkanen and Salminen (1997b), using multi-criteria decision analysis based on decision-makers preferences, we see that DMU 9 is always ranked first and under their equal weights scenario, DMU 12 is then ranked second. There is also general agreement at to the least efficient units, including DMUs 1, 2, 15 and 18. In the PCA-BCC model, 5 DMUs are deemed efficient, namely 23% of the sample set. Clearly, DMUs 9 and 12 remain efficient and DMU 3, which would appear to be an outlier, is BCC-PCA efficient but not CCR-PCA efficient. Interestingly, DMU 3 does not appear in the reference set of any other inefficient DMU. The DMU would appear to be particularly large both in resource usage and outputs, especially with reference to the number of employees. The benchmarks were computed in a three stage procedure, whereby after computing the radial efficiency score and then the slacks, benchmarks were chosen only from the subset considered both radially efficient and without positive slacks i.e. Pareto-Koopmans efficient. The point of this exercise is to show that the results of a standard DEA procedure are also relevant when applying PCA-DEA models.

Table 8-4. PCA-BCC Benchmarking Results

	2 PC input, 2 PC output		
DMU	CCR-PCA	BCC-PCA	Benchmarks for BCC model
1	0.45	0.73	10(0.52), 11(0.48)
2	0.54	0.76	11(0.76), 9(0.24)
3	0.81	1.00	1
4	0.93	0.94	11(0.47), 10(0.3), 12(0.23)
5	0.90	0.96	11(0.95), 9(0.05)
6	0.97	0.97	12(0.57), 9(0.33), 11(0.10)
7	0.95	0.96	11 (0.47), 10(0.30), 12(0.23)
8	0.92	0.98	11 (0.95), 9(0.05)
9	1.00	1.00	10
10	0.98	0.99	7
11	0.96	1.00	17
12	1.00	1.00	10
13	0.82	0.84	11(0.65), 12(0.31), 10(0.04)
14	0.81	0.88	11(0.81), 9(0.19)
15	0.82	0.82	9(0.71), 12(0.21), 11(0.08)
16	0.82	0.84	11(0.65), 12(0.31), 10(0.04)
17	0.81	0.88	11(0.81), 9(0.19)
18	0.82	0.82	9(0.71), 12(0.21),11(0.08)
19	0.83	0.84	10(0.68), 11(0.22), 12(0.1)
20	0.70	0.85	11(0.76), 9(0.24)
21	0.90	1.00	2
22	0.88	0.99	21(1)

5. SUMMARY AND CONCLUSIONS

This chapter has shown how principal components analysis (PCA) may be utilized within the data envelopment analysis (DEA) context to noticeably improve the discriminatory power of the model. Frequently, an extensive number of decision making units are considered relatively efficient within DEA, due to an excessive number of inputs and outputs relative to the number of units. Assurance regions and cone-ratio constraints are often used in order to restrict the variability of the weights, thus improving the discriminatory power of the standard DEA formulations. However, use of these techniques requires additional, preferential information, which is often difficult to attain. This research has suggested an objective solution to the problem.

The first model presented uses PCA on sets of inputs or outputs and reduces the number of variables in the model to those PCs that explain at least 80% of the correlation of the original data matrix. Some information will be lost, however this is minimized through the use of PCA and does not require the decision maker to remove complete variables from the analysis or specify which variables are more important. Consequently, the adapted formulation avoids losing all the information included within a variable that is then dropped, rather losing a principal component explaining by definition very little of the correlation between the original grouped data. Additional strength can be added to the model by including constraints over the PC variables, whereby the weight on the first PC (that explains the greatest variation in the original dataset) must be at least equal to that of the second PC and so on. These assurance region type constraints are based on objective information alone. Furthermore, it is possible to work backwards and compute values for DEA weights and benchmarks, providing the same level of information that exists in the standard DEA models.

The second model applies PCA to all inputs and separately to all outputs. The PCA-DEA model can then be constrained in the same manner as discussed previously, this time over all PCs, potentially improving the discriminatory power of the model with no loss of information. If required, PCs can be dropped and the remaining PCs constrained to further improve discriminatory power.

The models are demonstrated using an illustration analyzing the efficiency of waste treatment centers in a district of Finland. The original DEA models suggest that 73% and 100% of the centers are constant and variable returns-to-scale efficient respectively, because of the nature of the data and the fact that 8 variables were identified to describe 22 centers. The PCA-DEA formulations presented here lead to the identification of DMU_9 and DMU_{12} as the most efficient departments amongst the small dataset.

REFERENCES

1. Adler N and Golany B (2001). Evaluation of deregulated airline networks using data envelopment analysis combined with principal component analysis with an application to Western Europe. European J Opl Res 132: 18-31.
2. Adler N and Golany B (2002). Including Principal Component Weights to improve discrimination in Data Envelopment Analysis, J Opl Res, 53 (9), pp. 985-991.
3. Adler N and Yazhemsky E (2005). Improving discrimination in DEA: PCA-DEA versus variable reduction. Which method at what cost? Working Paper, Hebrew University Business School.
4. Banker RD, Charnes A and Cooper WW (1984). Models for estimating technical and returns-to-scale efficiencies in DEA. Management Sci 30: 1078-1092.
5. Charnes A, Cooper WW and Rhodes E (1978). Measuring the efficiency of decision making units. European J Opl Res 2: 429-444.

6. Charnes A, Cooper WW, Golany B, Seiford L and Stutz J (1985). Foundations of Data Envelopment Analysis for Pareto-Koopmans Efficient Empirical Production Functions. J Econometrics 30: 91-107.
7. Cooper WW, Seiford LM and Tone K (2000). Data Envelopment Analysis: A Comprehensive Text with Models, Applications, References and DEA-Solver Software. Kluwer Academic Publishers, Boston.
8. Dillon WR and Goldstein M (1984). Multivariate Analysis: Methods and Applications. John Wiley & Sons, New York.
9. Golany B (1988). A note on including ordinal relations among multipliers in data envelopment analysis. Management Sci 34: 1029-1033.
10. Hair JF, Anderson RE, Tatham RL and Black WC (1995). Multivariate Data Analysis. Prentice-Hall: Englewood Cliffs, NJ.
11. Hokkanen J and Salminen P (1997a). Choosing a solid waste management system using multicriteria decision analysis. European J Opl Res 90: 461-72.
12. Hokkanen J and Salminen P (1997b). Electre III and IV methods in an environmental problem. J Multi-Criteria Analysis 6: 216-26.
13. Jenkins L and Anderson M (2003). A multivariate statistical approach to reducing the number of variables in DEA. European J Opl Res 147: 51-61.
14. Lovell CAK and Pastor JT (1995). Units invariant and translation invariant DEA models. Operational Research Letters 18: 147-151.
15. Sarkis J (2000). A comparative analysis of DEA as a discrete alternative multiple criteria decision tool. European J Opl Res 123: 543-57.
16. Thompson RG, Singleton FD, Thrall RM and Smith BA (1986). Comparative site evaluations for locating a high-energy lab in Texas. Interfaces 16:6 35-49.
17. Ueda T and Hoshiai Y (1997). Application of principal component analysis for parsimonious summarization of DEA inputs and/or outputs. J Op Res Soc of Japan 40: 466-478.
18. Zhu, J. (1998). Data envelopment analysis vs. principal component analysis: An illustrative study of economic performance of Chinese cities. European Journal of Operational Research 111 50-61.

Some of the material in this chapter draws on the following two papers, which are being reproduced with kind permission of Elsevier and Palgrave Macmillan respectively. Adler N. and Golany B., (2001). Evaluation of deregulated airline networks using data envelopment analysis combined with principal component analysis with an application to Western Europe, European Journal of Operational Research, 132 (2), pp. 260-73. and Adler N. and Golany B., (2002). Including Principal Component Weights to improve discrimination in Data Envelopment Analysis, Journal of the Operational Research Society, 53 (9), pp. 985-991. Nicole would like to thank Ekaterina Yazhemsky for generous research support and the Recanati Fund for partial funding of this research.

Chapter 9

MINING NONPARAMETRIC FRONTIERS

José H. Dulá[1]
Department of Management, Virginia Commonwealth University, Richmond, VA 23284

Abstract: Data envelopment analysis (DEA) is firmly anchored in efficiency and productivity paradigms. This research claims new application domains for DEA by releasing it from these moorings. The same reasons why efficient entities are of interest in DEA apply to the geometric equivalent in general point sets since they are based on the data's magnitude limits relative to the other data points. A framework for non-parametric frontier analysis is derived from a new set of first principles. This chapter deals with the extension of data envelopment analysis to the general problem of mining oriented outliers.

Key words: Data Envelopment Analysis (DEA), Linear Programming.

1. INTRODUCTION

Data Envelopment Analysis (DEA) was derived from fundamental principles of efficiency and productivity theory. The first principle for the methodology is the familiar definition of efficiency based on the ratio of an output and an input. The original contribution of the paper that introduced and formalized the topic by Charnes, *et al.* (1978) was the generalization of the notion of this ratio to the case when multiple incommensurate outputs and inputs are to be used to evaluate and compare the efficiency of a group of entities.

An entity in DEA is called a Decision Making Unit (DMU) and the entities in a study are characterized by a common set of attributes which are partitioned into inputs and outputs. A calculation to combine the magnitudes of these incommensurate attributes involves a decision about how each will

---

[1] This work was partially funded by a VCU Summer Research Grant.

be weighted. The premise behind DEA is that weights can be evaluated so that each entity attains an optimized efficiency score. The fundamental DEA principle is that a DMU cannot be classified as efficient if it cannot attain the overall best efficiency score among all entities even when its attributes are optimally weighted. A variety of linear programs (LPs) can be formulated to obtain optimal weights. The first DEA LP formulated for this purpose was by Charnes, *et al.* (1978) and applied to the case of constant returns (CR) to scale. Soon after, Banker, *et al.* (1984) formulated LPs for the variable, increasing, and decreasing returns to scale (VR, IR, and DR, respectively).

The solution to a DEA LP for a particular entity resolves the question of whether weights for its attributes exist such that an efficiency ratio is possible that dominates all the other entities' ratios using the same weights. If the answer is affirmative, the entity is classified as efficient. The original LP formulations provided benchmarking information about how inefficient entities can attain efficiency by decreasing inputs or increasing outputs. Other formulations followed such as the slack-based "additive" LP of Charnes, *et al.* (1985) and others, e.g.: Tone (2001), Bougnol, *et al.* (2005).

The fundamental geometrical object in DEA is the production possibility set. This set is defined by the data and its elements are all possible output vectors that can be transformed from all allowable input combinations under convexity, returns to scale, and free-disposability assumptions. This ends up being an unbounded convex polyhedron that, somehow, minimally "wraps around" the data points. It can be analytically described by constrained linear operations on the data points. There is a geometric equivalence between data points located on a subset of the boundary of the production possibility set known as the *frontier* and the DEA efficiency classification. The production possibility set is closely related to the LPs' primal and dual feasible regions. It is, however, neither of these regions although its polyhedral nature is the reason LP plays a major role on DEA.

DEA is about extremes. The extreme elements of the production possibility set are efficient DMUs, and the converse is prevalent. The identification of efficient DMUs in DEA generates interesting geometric problems involving polyhedral sets and their boundaries. This problem can be studied and interpreted for its own sake independently of the original DEA efficiency paradigms.

The mechanisms that give the production possibility set and its boundary a primary role in DEA and the procedures that have evolved to identify a data point's relation to it are the motivation for this chapter. What results is a generalization that permits new applications for nonparametric frontiers as a tool for mining point sets for elements that possess particular properties. The generalization is based on the analogy between an efficient DMU and points in general point sets that exhibit extreme properties relative to the rest of the data.

2. THE DEA PARADIGM AND THE PRODUCTION POSSIBILITY SET

The data domain for a DEA study is n data points; one for each entity. Each data point has m components partitioned into two types, those pertaining to m_1 inputs, $0 \neq x^j \geq 0$, and those corresponding to m_2 outputs, $0 \neq y^j \geq 0$.

DEA generalizes the conventional efficiency calculation of the ratio of an output and an input to compare n entities to the multivariate case given multiple incommensurate inputs and outputs:

$$Efficiency = \frac{\text{Weighted sum of outputs of DMU j}}{\text{Weighted sum of inputs of DMUj}}; j = 1,\ldots,n. \qquad (1)$$

The immediate issue that arises about weight selection is answered by "allowing each DMU to select its own weights" such that:
- The weights are optimal in the sense that the efficiency score is the best possible;
- A DMU's optimal score is compared to the scores attained by the other DMUs using the same weights;
- Efficiency scores are normalized usually to make a value of 1 the reference; and
- No zero weights are allowed. This prevents the situation where two identical DMUs except for, say, one output, end up having the same score simply by choosing zero for the weight used for that output. This situation results in what are called "weak efficient" DMUs in DEA.

These conditions translate directly to the following fractional program:

$$\max_{\omega>0,\sigma>0} \frac{\langle \sigma, y^{j^*} \rangle}{\langle \omega, x^{j^*} \rangle}; \quad \text{s.t.}: \frac{\langle \sigma, y^j \rangle}{\langle \omega, x^j \rangle} \leq 1; \; j = 1, \ldots, n; \qquad (2)$$

where ω, σ are the weights for the inputs and outputs, respectively, $\langle \sigma, z \rangle = \sum_i \sigma_i z_i$ is the inner product of the vectors in the argument, and j^* is the index of the DMU being scored; i.e., one of the n DMUs the efficiency status of which is under inquiry. This mathematical program has two LP equivalents, depending on whether the transformation fixes the numerator or denominator of the objective function (Charnes and Cooper (1962)):

$$\text{CR1:} \begin{cases} \max_{\omega>0,\sigma>0} & \langle \sigma, y^{j^*} \rangle \\ \text{s.t.} & -\langle \omega, x^j \rangle + \langle \sigma, y^j \rangle \leq 0, \quad \forall j; \\ & \langle \omega, x^{j^*} \rangle = 1. \end{cases} \quad \text{CR2:} \begin{cases} \min_{\omega>0,\sigma>0} & \langle \omega, x^{j^*} \rangle \\ \text{s.t.} & -\langle \omega, x^j \rangle + \langle \sigma, y^j \rangle \leq 0, \quad \forall j; \\ & \langle \sigma, y^{j^*} \rangle = 1. \end{cases} \quad (3)$$

These DEA LPs model a constant returns to scale environment since the original efficiency ratios are not affected by any positive scaling applied to both input and output vectors. CR1 is said to be *input* oriented since it is based on fixing the denominator of the fractional program's objective function which involves the input data; CR2's orientation is, therefore, on *outputs*. The two LPs are called *multiplier* forms since the decision variables are the weights. The restriction that $\omega > 0$, $\sigma > 0$ can be relaxed to simple nonnegativity at the risk of misclassifying some weak efficient entities as efficient. We will note this risk and work henceforth with the relaxed LPs.

Other returns to scale are modeled by introducing a new variable, β. Depending on the restriction on β, we obtain the different environments: variable returns – no restriction – while $\beta \geq 0$ and $\beta \leq 0$ model the increasing and decreasing returns. The two multiplier forms for the variable returns to scale become:

$$\text{VR1:} \begin{cases} \max_{\omega\geq 0,\sigma\geq 0,\beta} & \langle \sigma, y^{j^*} \rangle + \beta \\ \text{s.t.} & -\langle \omega, x^j \rangle + \langle \sigma, y^j \rangle + \beta \leq 0, \quad \forall j; \\ & \langle \omega, x^{j^*} \rangle = 1; \end{cases} \quad \text{VR2:} \begin{cases} \min_{\omega\geq 0,\sigma\geq 0,\beta} & \langle \omega, x^{j^*} \rangle + \beta \\ \text{s.t.} & -\langle \omega, x^j \rangle + \langle \sigma, y^j \rangle + \beta \leq 0, \quad \forall j; \\ & \langle \sigma, y^{j^*} \rangle = 1. \end{cases} \quad (4)$$

The dual of the multiplier form is an *envelopment* LP. The envelopment LPs corresponding to VR1 and VR2 are:

$$\text{DVR1} \begin{cases} \max_{\lambda_j\geq 0,\theta} & \theta \\ \text{s.t.} & \sum_j \begin{bmatrix} -x^j \\ y^j \end{bmatrix} \lambda_j \geq \begin{bmatrix} -x^{j^*}\theta \\ y^{j^*} \end{bmatrix}; \\ & \sum_j \lambda_j = 1; \end{cases} \quad \text{DVR2} \begin{cases} \min_{\lambda_j\geq 0,\phi} & \phi \\ \text{s.t.} & \sum_j \begin{bmatrix} -x^j \\ y^j \end{bmatrix} \lambda_j \geq \begin{bmatrix} -x^{j^*} \\ y^{j^*}\phi \end{bmatrix}; \\ & \sum_j \lambda_j = 1. \end{cases} \quad (5)$$

The production possibility set is the set of all output vectors $y \geq 0$ that can be produced from input vectors $x \geq 0$. Using Banker's, et al. (1984) "postulates," four such sets may be generated by a DEA data set; one for each returns to scale assumption: (see also Seiford and Thrall (1990) and

Dulá and Thrall (2001)):

$$\mathcal{P}^{CR} = \{z | z \leq \sum_{j=1}^{n} \begin{bmatrix} -x^j \\ y^j \end{bmatrix} \lambda_j, \lambda_j \geq 0;\ \forall j\};$$

$$\mathcal{P}^{VR} = \{z | z \leq \sum_{j=1}^{n} \begin{bmatrix} -x^j \\ y^j \end{bmatrix} \lambda_j, \sum_j \lambda_j = 1, \lambda_j \geq 0;\ \forall j\}; \qquad (6)$$

$$\mathcal{P}^{IR} = \{z | z \leq \sum_{j=1}^{n} \begin{bmatrix} -x^j \\ y^j \end{bmatrix} \lambda_j, \sum_j \lambda_j \geq 1, \lambda_j \geq 0;\ \forall j\};$$

$$\mathcal{P}^{DR} = \{z | z \leq \sum_{j=1}^{n} \begin{bmatrix} -x^j \\ y^j \end{bmatrix} \lambda_j, \sum_j \lambda_j \leq 1, \lambda_j \geq 0;\ \forall j\}.$$

Each set is a *hull* of the data which, in some sense, tightly "envelopes" them in the form of an unbounded polyhedron. All four production possibility sets are unbounded since the m "surpluses" in the main constraints provide independent directions of recession. This is a direct consequence of the "free disposability" postulate in DEA; that is, the premise that inefficient production is always a possibility either as additional inputs for a given output vector or fewer outputs for a given input vector. This also implies that the four production possibility sets in DEA have dimension m since the surpluses provide a full basis of unit vectors, defining their recession cone. P^{CR} is a cone whereas P^{VR}, P^{IR}, P^{DR}, are polyhedrons with, possibly, many extreme points.

We will focus on the VR case and its production possibility set, P^{VR}. This case is always interesting since it is, quite possibly, the most widely used in practice and because the polyhedron has the most extreme points. The set of extreme points for P^{VR} is a superset for the set of extreme elements (points and rays) of the other three hulls (Dulá and Thrall, (2001)). Our equivalences will relate to this particular hull.

A variety of LP formulations can be used to identify the efficient DMUs. Their purpose differs depending on the benchmarking recommendations or measures of efficiency they provide. It is safe to state that DEA LPs follow a general structure we will call "standard" (without concerning ourselves with deleted domain analyses). Such formulations have easily recognizable multiplier and envelopment versions where the multiplier LP will be the one to include a set of n constraints algebraically equivalent to:

$$\langle \pi, a^j \rangle - \beta \begin{pmatrix} \leq \\ \geq \end{pmatrix} 0; \ j = 1, \ldots, n; \tag{7}$$

$$\pi_i \begin{cases} \geq 0; & \text{if attribute } i \text{ is an output;} \\ \leq 0; & \text{if attribute } i \text{ is an input;} \end{cases}$$

where

$$a^j = \begin{bmatrix} x^j \\ y^j \end{bmatrix}; \ \forall j. \tag{8}$$

We can see this standard form in the formulations presented above: CR1 & CR2 ($\beta = 0$ in the CR case) and VR1 & VR2. In the envelopment LP this translates dually to a system where the left hand side will include

$$\Sigma_j \begin{bmatrix} -x^j \\ y^j \end{bmatrix} \lambda_j \tag{9}$$

for $\lambda \geq 0$ with a condition on $\Sigma_j \lambda_j$ depending on the restriction on β. We can recognize these elements in DVR1 & DVR2 but also in the definition of the production possibility set; e.g.: P^{VR}.

With this structure, basic LP theory determines that an optimal solution, (π^*, β^*), when a^{j^*} is efficient, is such that

$$\langle \pi^*, a^{j^*} \rangle - \beta^* = 0. \tag{10}$$

The required feasibility of the rest of the constraints at optimality allows the interpretation of (π^*, β^*) as parameters of a hyperplane, $H(\pi^*, \beta^*) = \{z \mid \langle \pi^*, z \rangle = \beta^*\}$ containing a^{j^*} and such that the rest of the data points belong to it or to one of the halfspaces it defines. This provides insight to understand a familiar result in DEA.

Result 1. *Let (π^*, β^*) be an optimal solution for a standard multiplier DEA LP to score DMU a^{j^*}. If a^{j^*} is efficient then $H(\pi^*, \beta^*)$ is a supporting hyperplane for the production possibility set, P^{VR}, and a^{j^*} belongs to the support set.*

Geometrically, the frontier of P^{VR} is defined by its extreme points. Its frontier is the union of all bounded faces each of which is a convex combination of a subset of the extreme points. These extreme points correspond to actual DMUs; that is, a subset of the DEA data set will be all the extreme points that define the faces of the frontier. Any data point which is extreme to the production possibility set is necessarily on the frontier. Other data points may belong to the frontier, and we will want to identify them, but experience shows that these tend to be rare.

Efficient DMUs are what we will call *oriented outliers* based on the sense that they exhibit values and their combinations that reflect the relative limits taken on by the data with a certain orientation criteria. The orientation aspect of these limits comes from the fact that only a specific region of the boundary is of interest. This is defined by the assignment of attributes to inputs and outputs and the fact that the interest is always on DMUs that exhibit smaller input requirements to produce larger quantities of outputs.

The notion of an oriented outlier is purely geometric and applies to any finite multivariate point set. Any point set defines hulls which depend on the points' locations and attribute assignments analogous to DEA's inputs and outputs. A frontier to any hull is defined by the emphasis on the components' magnitudes: the interest can be on greater magnitudes for some (e.g., outputs in DEA) and the opposite for others (e.g., inputs in DEA). The same reasons why these outliers are of interest in DEA are valid in other applications since they are based on the data's oriented limits relative to the rest of the data. In the next section we will motivate the generalization of the fundamental geometric concepts in DEA to general data mining.

3. FRONTIER MINING

DEA derives from fundamental efficiency and productivity theory principles. Its basic premise is that it identifies entities in an empirical frontier that approximates a true but unavailable production frontier. DEA's production possibility set is a proxy for an actual production frontier in the absence of any knowledge of the parameters that define it. New application domains can be claimed for DEA if it is released from its original moorings.

The following definition generalizes the concept of an oriented outlier.

Definition. *A data point, a^{j*}, is an oriented outlier if it is on the support set of a supporting hyperplane of a hull of a point set.*

A DEA efficient DMU is, by this definition, an oriented outlier. The shape of a hull of a point set depends on assessments as to whether only larger or smaller magnitudes for the attribute's value are prioritized. When larger is better in all dimensions the hull is equivalent to production possibility sets in DEA where all attributes are outputs. The presence of "undesirable" attributes (e.g., DEA inputs), that is, attributes where the focus is on smaller magnitudes, define a different hull. Each attribute can take on two orientations leading to 2^m different hulls. It may even be the case in general that there is a component for which there is an interest in the values at both ends. We will see how our generalizations will increase the number of potential hulls in a point set to 3^m by introducing the possibility of a focus on both larger and smaller magnitudes for attributes.

There is a natural interest in identifying oriented outliers in multivariate point sets in contexts outside of DEA. Consider, for example, the particular tax return where the total in charitable contributions or employee deductions is the largest among all returns within a given category. We can imagine that a government revenue agency would consider such a return interesting. In the same way, a security agency may focus on the individual who has made the largest number of monetary transfers, or the largest magnitude transfer, to a problematic location on the globe. Such records in a point set are oriented outliers in one of the dimensions in the sense that they attain one of two extreme values (largest rather than smallest) there. Finding them reduces to a sorting of records based on the value in that dimension.

Entities operating under strenuous circumstances push the limits in several key dimensions even though no single one may attain an extreme value. Such entities may be identified by generating extreme values when dimensions are combined. In the situations above, the tax return with the largest sum of the charitable contributions and employee deductions may prove interesting; or the individual whose money transfer events plus total monetary value of the transfers is the largest when added up may merit closer scrutiny. The record that emerges as an outlier based on this two-dimensional analysis using these simple criteria is just one of, possibly, many that can emerge if the two values are weighted differently. All such points are, in the same sense, oriented outliers and all would be interesting for different reasons.

Now consider the general case in which the point set consists of points $a^1,...,a^n$ the components of which are values without an input or output designation. Each data point has m components: $a^j = (a^j_1,...,a^j_m)$. We are, however, interested in a "focus" or orientation on either larger or smaller magnitudes for each component.

Using the simple sum of the attribute values to identify an entity means we place equal importance on each attribute in the identification criterion. Modifying attributes' weights results in different weighted sums and may be used to reflect different priorities or concerns. Negative weights shift the emphasis from larger magnitudes, as in the example above, to the case where the focus is on smaller extremes. Weighted sums are maximized by different entities depending on the weights. The question that arises in this context is:

Given a specific entity, j^*, are there a set of m weights that will make its weighted sum of the attribute values the maximum from among all entities' weighted sums?

An affirmative answer means that entity j^* has component magnitudes that can be somehow combined to attain a maximum value or "extreme". A negative answer to the question means the entity is never extreme and that other entities manage to attain larger weighted sums no matter what weights are used.

The question above can be answered by verifying the feasibility of a linear system. The motivation for identifying oriented outliers from the discussion above provides the first principles for its formulation. Let $\pi_1,...,\pi_m$ be the weights to be applied to the attributes. We will work only with nonnegative attribute values; that is: $a^j \geq 0$; $\forall j$. Consider the problem of answering the question above algebraically; i.e., are there values for m weights, π_i, $i = 1,...,m$ such that entity j^* attains a weighted sum that is greater than, or equal to, the weighted sums of the other entities using the same weights? Thus:

$$\text{M\_Feas-1} \begin{cases} \sum_i a_i^j \pi_i \leq \sum_i a_i^{j^*} \pi_i, & \forall j \\ \pi \neq 0, \end{cases} \quad (11)$$

where

$$\pi_i \begin{cases} \geq 0; & \text{if the focus is on larger magnitudes for attribute } i; \\ \leq 0; & \text{if the focus is on smaller magnitudes for attribute } i. \end{cases} \quad (12)$$

and $\pi \neq 0$ means the zero vector is excluded (to obviate the trivial solution).

If a solution, $\pi^* = (\pi^*_i)$, $i = 1,...,m$, exists, the answer to the question is affirmative and entity j^* is a point that would interest the analyst looking for entities that exhibit extreme characteristics. If no solution is possible then the system is infeasible and the answer to the question is negative. To understand this better let us rewrite the system by introducing a new variable, $\beta = \sum_i a^{j^*}_i \pi_i$, and substituting:

$$\text{M\_Feas-2} \begin{cases} \sum_i a_i^j \pi_i - \beta \leq 0; & \forall j \\ \sum_i a_i^{j^*} \pi_i - \beta = 0; \\ \pi \neq 0; \end{cases} \quad (13)$$

where π is restricted as in (12). This representation allows a more direct geometric insight and interpretation. A solution (π^*, β^*) provides the parameters for a hyperplane, $H(\pi^*, \beta^*)$ in $\mathcal{R}^m$. This hyperplane contains the point a^{j^*} and the rest of the data belongs to one of the two halfspaces it defines. However, there is more happening as we can see from the following result:

Result 2. Let (π^*, β^*) be a feasible solution for the system (M_Feas-2) to score DMU a^{j*}. Then, $H(\pi^*, \beta^*)$ is a supporting hyperplane for the VR hull of the point set and the point a^{j*} belongs to the support set.

The result follows from an application of *Slater's Theorem of Alternative* (see Mangasarian (1969), p. 27) to the feasibility system (M_Feas-2). Result 2 states that the halfspace defined by the hyperplane, $H(\pi^*, \beta^*)$, contains more than just the point set cloud – identically the VR hull of the data is located there including, of course, its entire recession. When system (M_Feas-2) has a feasible solution with a^{j*}, it means this point is on the boundary of the VR hull making it an oriented outlier according to our definition. If a^{j*} is not on the boundary of the VR hull of the data, this support is impossible and no feasible solution exists.

DEA requires that attributes be either inputs or outputs. This creates the potential of 2^m different production possibility sets just for the VR model. The framework for identifying oriented outliers developed here permits a generalization where the orientation focus extends to large and small magnitudes simultaneously for any attribute. This is achieved with a new feasibility system, (M_Feas-3), obtained by modifying the restrictions in (M_Feas-2) (see expression (12)) of the corresponding multipliers as follows:

$$\pi_i \begin{cases} \geq 0; & \text{if the focus is on larger magnitudes for attribute } i; \\ \leq 0; & \text{if the focus is on smaller magnitudes for attribute } i; \\ \text{unrestricted}; & \text{if both extremes for attribute } i \text{ are interesting.} \end{cases} \quad (14)$$

The effect on the geometry of the hull defined by the point set is to remove recession directions in the dimensions where the multipliers are unrestricted. The analytical description of the hull is a modification of the characterization of P^{VR} where the corresponding relation becomes an equality. The resultant hulls are unknown in DEA. The possibility of a third category for the attributes means that the number of VR-type hulls that can be created from the same point set is as many as 3^m. If all multipliers, π, are unrestricted, any oriented outlier that emerges from the solution to the LP is somewhere on the boundary of the convex hull, P^{CH}, of the data. The convex hull of a point set is a bounded polyhedron defined as follows:

$$\mathcal{P}^{CH} = \{z | z = \sum_j a^j \lambda_j, \sum_j \lambda_j = 1, \lambda_j \geq 0\}. \quad (15)$$

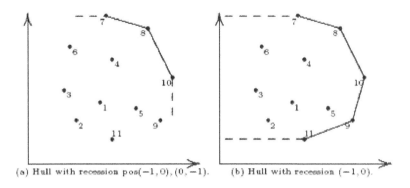

(a) Hull with recession pos$(-1,0),(0,-1)$.   (b) Hull with recession $(-1,0)$.

Figure 9-1. Two hulls of the same set of 11 points.

A focus on both the highest and lowest magnitudes for a single attribute can be visualized in two dimensions. Figure 9-1 depicts two hulls of the same two-dimensional point set. The hull on the left is a standard VR production possibility set with two outputs. The second hull results from orienting the first attribute so that the focus is on larger magnitudes while the second is focused on both upper and lower limits. In this case the hull corresponds to a feasibility system (M_Feas-3) with $\pi_1 \geq 0$ and π_2, unrestricted. A characterization of this two-dimensional hull is as follows

$$\left\{ \begin{bmatrix} z_1 \\ z_2 \end{bmatrix} \middle| \begin{bmatrix} z_1 \leq \sum_j a_1^j \lambda_j \\ z_2 = \sum_j a_2^j \lambda_j \end{bmatrix} ; \sum_j \lambda_j = 1;\ \lambda \geq 0 \right\}. \quad (16)$$

Notice how data points 9 and 11 emerge as outliers in the second hull reflecting the emphasis on both higher and lower magnitude limits for the second attribute.

A note about the role of weak efficiency. Weak efficient DMUs in DEA are boundary points of the production possibility set that are not efficient. They present several problems for DEA. They confound analyses because, although their score indicates efficiency, they necessarily generate zero weights in the relaxed DEA multiplier LPs. This is not, however, sufficient to conclude they are not efficient; efficient DMUs can also generate zero weights. Charnes, *et al.* (1978) resolved this by imposing the strict positivity condition on the weights referred to earlier which, of course, cannot be implemented in linear programming directly requiring the use of a "non-Archimedean" constant, $\varepsilon \geq 0$, large enough to prevent weak efficient DMUs from having the same scores as efficient ones, but not so large that it begins to misclassify efficient DMUs. This is a difficult balance making the use of the non-Archimedean constant problematic – and unpopular. An alternative is to use the additive formulation (Charnes *at al.* (1985)). This LP provides necessary and sufficient conditions for identifying truly efficient DMUs.

The same situation occurs in our development of the concept of oriented outliers. A data point located on the boundary of a receding face of a hull is, by our original definition, an oriented outlier, but when it is not an actual extreme point of the hull, it is equivalent to a weakly efficient DMU. This could be considered problematic in a mining analysis since these points are actually in the category where one, or a combination of some, of the dimensions could yield dominating values if some of the weight(s) used were zero. The feasibility systems derived above suffers from the same flaw as many relaxed DEA LPs in that they cannot distinguish between interesting oriented outliers (efficient DMUs) and other less interesting boundary data points (weak efficient DMUs). This problem can be remedied in our model by applying the "additive" notion of Charnes *at al.* (1985); that is, restrictions (14) on the variables of feasibility system (M_Feas-3) are modified as follows:

$$\pi_i \begin{cases} \geq 1; & \text{if the focus is on larger magnitudes for attribute } i; \\ \leq 1; & \text{if the focus is on smaller magnitudes for attribute } i; \\ \text{unrestricted}; & \text{if both extremes for attribute } i \text{ are interesting.} \end{cases} \quad (17)$$

We will refer to the resultant feasibility system as (M_Feas-4). The same reasons that do not permit weak efficient DMUs to be classified as efficient, exclude non-extreme point boundary entities on receding faces from being feasible in this modification of System (M_Feas-4).

The concept of an oriented outlier emerges naturally from an interest in special data points in a point set that reflect limits of the magnitudes of the attributes either individually or optimally combined. There is a direct correspondence between these oriented outliers in general point sets and efficient DMUs in DEA data sets under variable returns to scale assumptions. Oriented outliers, however, are a purely geometric concept totally detached from any notion of production or efficiency paradigms. The model permits an extension where a simultaneous focus on larger and smaller magnitudes can be modeled, introducing new shapes for the traditional VR production possibility set. This defines a third category of entities in addition to the standard desirable (outputs) and undesirable (inputs) designations. These generalizations increase modeling flexibility by allowing the application of this technique to mine any multidimensional point set.

4. COMPUTATIONAL TESTS

It is not difficult to imagine mining several hundreds of thousands of credit card records using models to detect potentially profitable, or problematic, customers, or for unusual activity that might reveal fraud or theft, by identifying the different oriented outliers that emerge. It is useful to understand the computational requirements for mining massive data sets for oriented outliers.

The feasibility of systems such as (M_Feas-2), (M_Feas-3), (M_Feas-4), can be ascertained using LP. The systems simply require an objective function and any LP solver will conclusively resolve its feasibility. A complete implementation to classify all n points in a data set requires solving as many LPs. A procedure for this has the following structure:

|LP-Based Oriented Outlier Detection Procedure.|

```
For j = 1 to n do:
  Step 1. j* ← j.
  Step 2. Solve appropriate LP.
  Step 3. LP Feasible?
    Yes: Entity j* is an oriented outlier.
    No: Entity j* is not an oriented outlier.
Next j*.
```

This simple procedure can mask an onerous computational task. The size of the LPs solved are, more or less, the size of the data matrix: $m \times n$. Every iteration requires the solution of a dense, high aspect ratio, LP. The LP is slightly modified each iteration. In a small scale, the procedure is easy to implement manually and can be coded directly in a language such as Visual Basic for Applications (VBA) when the data is in a spreadsheet. Large scale applications require more powerful tools and involve more implementation issues.

The procedure was implemented using an LP formulation which is essentially the dual of the LP that results from using the objective function max $\sum_i a^{j*}_i \pi_i - \beta$ on the system (M_Feas-4). The dual is an envelopment type LP which has the advantage of many fewer rows than columns.

Enhancements and improvements for the procedure are possible. Besides all that is known about improving LP performance (e.g., multiple pricing, product forms, hot starts, etc.), there are techniques that exploit the special structure of LPs based on systems such as (M_Feas-4). The most effective is Reduced Basis Entry (RBE) (see Ali (1993)). The idea works on the property that only extreme elements are necessary for basic feasible solutions in any

LP based on an envelopment hull such as P^{VR}. Any other entity plays no role in defining the optimal solution and it makes no difference whether its data are present or absent from the LP coefficient matrix. The idea of RBE is to omit inefficient entities from subsequent LP formulations as they are identified. This idea is easy to implement as LPs are iteratively formulated and solved. The systematic application of this approach progressively reduces the size of the LPs as the procedure iterates.

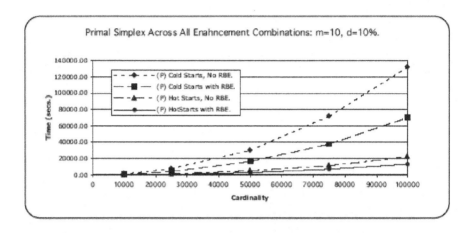

Figure 9-2. Experiments on point sets with 10K, 25K, 50K, 75K and 100K points ("cardinality") all with 10 dimensions and 10% extreme points.

Figure 9-2 summarizes the results of an implementation designed to understand the performance of the procedure in large scale applications and to test the roles of improvements and enhancements. The procedure was coded in Fortran 90 and the LP solver was the primal simplex in the CPLEX 8.0 *Callable Library* (ILOG (2002)). Runs were executed on an Intel P4-478 processor running at 2.53 MGz with 1024 MB of core memory. From the figure we see it takes 132,580 seconds (more than 36 hours) to solve the largest, $n = 100K$, problem in a "naive" implementation without any enhancements or improvements. More important information, though, is the relation between time and number of entities. Insights into this relation are obtained when we look at the time consumed per LP solved. This detail was investigated in Dulá (2005) and it turns out the time per LP behaves consistently linear with the cardinality (number of points) of the data set when using the primal simplex algorithm. Since the procedure requires the solution of n LPs, a quadratic relation between cardinality and computational times would be expected and this is verified graphically in the figure.

Also important is to learn that simple actions can greatly reduce computation times. Recall that the sequence of LPs solved differ little from one iteration to the next. A "hot starts" advanced basis feature would benefit from this and the experiments reveal that this is indeed worth it if the LP solver has one as with CPLEX 8.0 and others. For the case of $n = 100,000$, the impact of hot starts was a reduction of more than an 80% in computational time. Interestingly, RBE has an almost additive effect on the reduction of yet another 40% reduction for this particular problem suggesting that, in combination, the two features can produce a full order of magnitude reduction in times compared to the naive implementation. These dramatic effects are evident in the figure.

As a data mining tool, the identification of oriented outliers has modest computational requirements. These experiments suggest that it can be practically applied to massive point sets with well beyond 100,000 points.

5. CONCLUSIONS

Efficient DMUs represent entities that exhibit extreme characteristics of the data. The same reasons that compel their identification can be expected to apply to other point sets not necessarily connected to productivity and efficiency analyses.

Efficient DMUs correspond to points on the boundary of a specific polyhedral set known as the production possibility set constructed by special linear combinations of the data. The process of efficiency identification in DEA is essentially a geometrical problem and efficient DMUs are a form of data outlier made interesting by applying an orientation criterion based on whether larger or smaller magnitudes are desirable for each attribute. Any m-dimensional point set where each dimension is a magnitude contains such oriented outliers. Their identification is an interesting geometrical problem in itself with a variety of modeling implications and applications.

This chapter has developed from first principles a geometrical framework for the problem of identifying oriented outliers in general magnitude point sets. This was motivated by interesting problems unconnected with issues of productivity and efficiency. The resultant model resembles a variable returns to scale analysis in DEA. The new framework leads to new extensions that define geometric objects unknown in DEA such as when the focus is on both higher and lower values in some dimensions. Since computational requirements for this kind of analysis are quite manageable, it is possible to apply it to massive data sets. This methodology can be one more analytical tool in the toolbox of management scientist to extract additional information from data.

REFERENCES

1. Ali, A.I., (1993), Streamlined computation for Data Envelopment Analysis. *European Journal of Operational Research.* **64**, 61-67.
2. Banker, R.D., A. Charnes, and W.W. Cooper, (1984), Some models for estimating technological and scale inefficiencies in Data Envelopment Analysis. *Management Science*, **30**, 1078-1092.
3. Bougnol, M.-L., J.H. Dulá, D. Retzlaff-Roberts and N.K. Womer, (2005), Nonparametric frontier analysis with multiple constituencies. *Journal of the Operational Research Society*, **56**, No. 3, pp. 252–266.
4. Charnes, A. and W.W. Cooper, (1962), Programming with linear fractional functionals. *Naval Research Logistics Quaterly.* **9**, 181-185.
5. Charnes, A., W.W. Cooper, B. Golany, L. Seiford, and J. Stutz, (1985), Foundations of Data Envelopment Analysis for Pareto-Koopmans efficient empirical production functions. *Journal of Econometrics.* **30**, 91-107.
6. Charnes, A., W.W. Cooper, and E. Rhodes, (1978). Measuring the efficiency of decision making units. *European Journal of Operational Research,* **2**, 429-444.
7. Cooper, W.W., K.S. Park, and J.T. Pastor, (1999), RAM: A range adjusted measure of inefficiency for use with additive models and relations to other models and measures in DEA. *Journal of Productivity Analysis*, **11**, 5-42.
8. Dulá, J.H. and R.M. Thrall, (2001), A computational framework for accelerating DEA. *Journal of Productivity Analysis*, **16**, 63-78.
9. Dulá, J.H., 2005, "A computational study of DEA with massive data sets. To appear in *Computers & Operations Research.*
10. ILOG (2002), *ILOG CPLEX 8.0 User's Manual*, ILOG S.A., Gentilly, France.
11. Mangasarian, O.L., (1969), *Nonlinear Programming*, McGraw-Hill Book Company, New York.
12. Seiford, L.M. and R.M. Thrall (1990). Recent developments in DEA: the mathematical programming approach to frontier analysis, *J. of Econometrics*, **46**, 7-38.
13. Tone, K., (2001). A slacks-based measure of efficiency in Data Envelopment Analysis. *European Journal of Operational Research*, **130**, 498-509.

Chapter 10

DEA PRESENTED GRAPHICALLY USING MULTI-DIMENSIONAL SCALING

Nicole Adler, Adi Raveh and Ekaterina Yazhemsky
Hebrew University of Jerusalem, msnic@huji.ac.il

Abstract: The purpose of this chapter is to present data envelopment analysis (DEA) as a two-dimensional plot that permits an easy, graphical explanation of the results. Due to the multiple dimensions of the problem, graphical presentation of DEA results has proven somewhat elusive up to now. Co-Plot, a variant of multi-dimensional scaling, places each decision-making unit in a two-dimensional space in which the location of each observation is determined by all variables simultaneously. The graphical display technique exhibits observations as points and variables (ratios) as arrows, relative to the same center-of-gravity. Observations are mapped such that similar decision-making units are closely located on the plot, signifying that they belong to a group possessing comparable characteristics and behavior. In this chapter, we will analyze 19 Finnish Forestry Boards using Co-Plot to examine the original data and then to present the results of various weight-constrained DEA models, including that of PCA-DEA.

Key words: Data envelopment analysis; Co-Plot; Multi-dimensional scaling.

1. INTRODUCTION

The purpose of this chapter is to suggest a methodology for presenting data envelopment analysis (DEA) graphically. The display methodology utilized, Co-Plot, is a variant of multi-dimensional scaling (Raveh, 2000a, 2000b). Co-Plot locates each decision-making unit (DMU) in a two-dimensional space in which the location of each observation is determined by all variables simultaneously. The graphical display technique exhibits DMUs as points and variables as arrows, relative to the same axis and origin or center-of-gravity. Observations are mapped such that similar DMUs are closely located on the plot, signifying that they belong to a group possessing comparable characteristics and behavior. The Co-Plot method has been applied in multiple fields, including a study of socioeconomic differences among cities (Lipshitz and Raveh (1994)), national versus corporate cultural fit in mergers and acquisitions (Weber et al. (1996), computers (Gilady et al. (1996)) and the Greek banking system (Raveh (2000a)).

Presenting DEA graphically, due to its multiple variable nature, has proven difficult and some have argued that this has left decision-makers and managers at a loss in interpreting the results. Consequently, several papers have been published in the literature discussing graphical depictions of DEA results over the past fifteen years. Desai and Walters (1991) argued *"representation of multiple dimensions in a series of projections into a plane of two orthogonal axes is inadequate because of the inability to represent relative spatial positions in two dimensional projections"*. However, Co-Plot manages to do precisely this through a representation of variables (arrows) and observations (circles) simultaneously. Belton and Vickers (1993) proposed a visual interactive approach to presenting sensitivity analysis, Hackman et al. (1994) presented an algorithm that permits explicit presentation of a two-dimensional section of the production possibility set and Maital and Vaninsky (1999) extended this idea to include an analysis of local optimization, permitting an inefficient DMU to gradually reach the full-efficiency frontier. Albriktsen and Forsund (1990), El-Mahgary and Lahdelma (1995), Golany and Storbeck (1999) and Talluri et al. (2000) presented a series of two-dimensional charts that analyze the DEA results pictorially, including plots, histograms, bar charts and line graphs. The combination of Co-Plot and DEA represents a substantial departure from the existing techniques presented in the literature to date and has proven statistically useful in Adler and Raveh (2006).

Co-Plot may be used initially to analyze the data collected graphically. The method, presented in section 2, can be used to identify observation outliers, which may need further scrutiny, as they are likely to affect the envelope frontier. It can also be used to identify unimportant variables that could be removed from the analysis with little effect on the subsequent DEA results. In Section 3, we will describe how Co-Plot and DEA can be used to their mutual benefit. The value of this cross-fertilization is illustrated in Section 4 through the analysis of Finnish Forestry Boards, first investigated in Viitala and Hänninen (1998) and Joro and Viitala (2004), a particularly difficult sample set due to the substantial number of variables (16 in total) compared to the total number of DMUs (a mere 19). This section will present the various pictorial displays that could be employed to present the DEA results convincingly to relevant decision-makers. It will also present the usefulness of PCA-DEA (see Chapter 8) in reducing the dimensionality of the sample and presenting the results. Section 4 presents conclusions and possible future directions for this cross-fertilization of methodologies.

2. CO-PLOT

Co-Plot (Raveh (2000b)) is based on the integration of mapping concepts, using a variant of regression analysis that superimposes two graphs sequentially. The first graph maps n observations, in our case DMUs, and the second graph is conditioned on the first one, and maps k arrows (variables or ratios) portrayed individually. Given a data matrix X_{nk} of n rows (DMUs) and k columns (criteria), Co-Plot consists of four stages, two preliminary adaptations of the data matrix and two subsequent stages that compute two maps sequentially that are then superimposed to produce a single picture. First we will describe the stages required to produce a Co-Plot and then we will demonstrate the result using a simple illustration.

Stage 1: Normalize X_{nk}, in order to treat the variables equally, by removing the column mean $(\bar{x}_{.j})$ and dividing by the standard deviation (S_j) as follows:
$Z_{ij} = (x_{ij} - \bar{x}_{.j}) / S_j$.

Stage 2: Compute a measure of dissimilarity, $D_{il} \geq 0$, between each pair of observations (rows of Z_{nk}). A symmetric $n \times n$ matrix (D_{il}) is produced from the $\binom{n}{2}$ different pairs of observations. For example, compute the city-block distance i.e. the sum of absolute deviations to be used as a measure of dissimilarity, similar to the efficiency measurement in the additive DEA model (Charnes et al. (1985)).

$$D_{il} = \sum_{j=1}^{k} |Z_{ij} - Z_{lj}| \geq 0, (1 \leq i,l \leq n).$$

Stage 3: Map D_{il} using a multi-dimensional scaling method. Guttman's (1968) smallest space analysis provides one method for computing a graphic presentation of the pair-wise interrelationships of a set of objects. This stage yields $2n$ coordinates (X_{1i}, X_{2i}), $i=1,...,n$ where each row $Z = (Z_{i1},...,Z_{ik})$ is mapped into a point in the two-dimensional space (X_{1i}, X_{2i}). Thus, observations are represented as n points, P_i, $i=1,...,n$ in a Euclidean space of two dimensions. Smallest space analysis uses the coefficient of alienation, θ, as a measure of goodness-of-fit.

Stage 4: Draw k arrows $(\tilde{X}_j, j = 1,..., k)$ within the Euclidean space obtained in Stage 3. Each variable, j, is represented by an arrow emerging from the center of gravity of the points P_i. Each arrow $\tilde{X}_j$ is chosen such that the correlation between the values of variable j and their **projections** on the arrow is **maximal**. Therefore, highly correlated criteria will point in approximately the same direction. The cosines of angles between the arrows are approximately proportional to the correlations between their associated criteria.

Table 10-1. Example Dataset

DMU	x_1	x_2	x_3	x_4
A	1	3	5	8
B	2	6	15	5
C	3	9	28	4
D	4	7	14	2
E	5	8	4	1
F	6	11	-1	-3

Two types of measure, one for Stage 3 and another for Stage 4, assess the goodness-of-fit of Co-Plot. In Stage 3, a single coefficient of goodness-of-fit for the configuration of n observations is obtained from the smallest space analysis method, known as the coefficient of alienation θ. (For more details about θ, see Raveh (1986)). A general rule-of-thumb states that the picture is statistically significant if $\theta \leq 0.15$. In stage 4, k individual goodness-of-fit measures are obtained for each of the k variables separately. These represent the magnitudes of the k maximal correlations, r_j^* $j=1,...,k$, that measure the goodness-of-fit of the k regressions. The correlations r_j^*, can be helpful in deciding whether to include or eliminate criteria. Criteria that do not fit the graphical display, namely those that have low r_j^*, may be eliminated since the plot has not taken the variable into significant account (the length of the arrow will be relatively short). The higher the correlation r_j^*, the better $\widetilde{X}_j$ represents the common direction and order of the projections of the n points along the rotated axis $\widetilde{X}_j$ (arrow j). Hence, the length of the arrow is proportional to the goodness-of-fit measure r_j^*.

In order to demonstrate the approach, a simple 6-observation, 4-variable example is presented in Table *10-1*. The first two variables (x_1 and x_2) are highly correlated, the third variable (x_3) is orthogonal to the first two and the fourth variable (x_4) has an almost pure negative correlation to the first variable.

As a result of the correlations between variables, the Co-Plot diagram presented in Figure *10-1* shows x_1 and x_2 relatively close to each other, x_4 pointing in the opposite direction and x_3 lying in-between. Consequently, x_2 (or x_1) is identified as providing little information and may be removed from the analysis with little to no effect on the results. With respect to the observations, *A* is separated from the other units because it has the lowest x_1 value and the highest x_4 value. *A* could be considered an outlier and appears as such in Figure *10-1*, where the circle representing the observation appears high on the x_4 axes and far from x_1. *F* was given negative x_3 and x_4 values and is depicted far from both these variable arrows, whilst *C* is a specialist in

x_3 with a particularly elevated value hence appears high on the x_3 arrow. Finally, D is average in all variables and therefore appears relatively close to the center-of-gravity, the point from which all arrows begin.

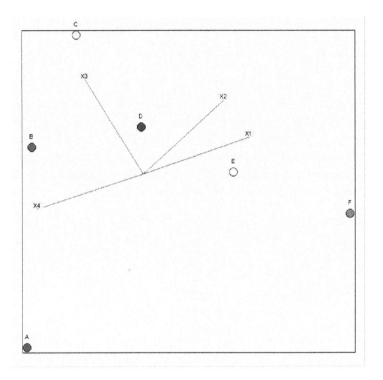

Figure 10-1. Co-Plot of illustrative data presented in Table 10-1

3. CO-PLOT AND DEA

It is being suggested in this chapter that Co-Plot can be used to present the results of a DEA graphically, by replacing the original variables with the ratio of variables (output$_l$ / input$_k$). Consequently, the efficient units in a DEA appear in the outer ring or sector of observations in the plot and it is easy to identify over which ratios specific observations are particularly good. This idea is in the spirit of Sinuany-Stern and Friedman (1998) and Zhu (1998) who used ratios within a discriminant analysis and principal component analysis respectively and subsequently compared the results to that of DEA. Essentially, the higher the ratio a unit receives, the more efficient the DMU is considered over that specific attribute.

In order to demonstrate this point, we will first analyze the example presented in Section 2. Let us assume that the first three variables are inputs and that x_4 represents a single output. In the next step, we define the following 3 ratios, r_{lk}, as follows:

$$r_{lk} = \frac{Output_l}{Input_k} \qquad \forall k = 1,2,3 \ \ l = 4.$$

The results of Co-Plot over the three ratios are presented in Figure *10-2*[1]. The efficient DMUs, according to the constant returns-to-scale additive DEA model, are of a dark color and the inefficient DMUs are light colored. The additive DEA model (Charnes et al. (1985)) was chosen in this case simply because some of the data was negative. It should also be noted that the use of color substantially improves the presentation of the results.

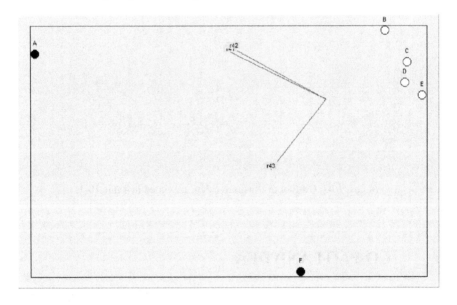

Figure 10-2. Co-Plot presenting the ratio of variables and DEA results pictorially

The constant returns-to-scale (CRS), additive DEA results for this illustration are presented in Table *10-2*[2]. *A* and *F* are identified as the

---

[1] A software program for producing Visual Co-Plot 5.5 graphs is freely available at http://www.cs.huji.ac.il/~davidt/vcoplot/.
[2] All DEA results presented in this chapter were computing using Holger Scheel's EMS program, which can be downloaded at http://www.wiso.uni-dortmund.de /lsfg/or/scheel/ems.

efficient DMUs, receiving zero slack efficiency scores, and appear in Co-Plot to the left of the variable arrows. The remaining points represent inefficient DMUs and lie behind the arrows. Were the arrows (variables) to cover all four quadrants of the plot, the efficient DMUs would appear in a ring around the entire plot.

Table 10-2. Additive CRS DEA results for Table 1 Example

DMU	Score	Benchmarks	Slacks x_1	x_2	x_3	score without x_2
A	0		4			0
B	17.38	A (0.63)	1.37	4.12	11.87	13.25
C	35.50	A (0.50)	2.5	7.5	25.5	28.00
D	22.75	A (0.25)	3.75	6.25	12.75	16.50
E	15.88	A (0.12)	4.87	7.62	3.37	8.25
F	0		0			0

An additional point to be noted in Figure *10-2* is the closeness of ratios r_{41} and r_{42}, strongly suggesting that one ratio would be sufficient to accurately present the dataset. The DEA was re-analyzed without the x_2 variable and the efficiency results, in terms of dichotomous classification, did not change, as demonstrated in the last column of Table *10-2*. The Co-Plot diagram did not change either, except for the removal of one of the arrows.

4. FINNISH FORESTRY BOARD ILLUSTRATION

We now turn to a real dataset, in order to demonstrate some of the wide-ranging uses Co-Plot may be deployed in a DEA exercise. The 19 Finnish Forestry Boards presented in Viitala and Hänninen (1998) is an interesting case study, among other things because of the issues of dimensionality that frequently occur. In this case study, there were 19 DMUs and 16 variables; one input and 15 outputs, as presented in *Table 10-3*.

In Joro and Viitala (2004), various weighting schemes were applied in order to incorporate additional managerial information and values that would overcome the discrimination issues that arose. We will discuss two of the different types of weight restrictions that were applied. The first weight scheme (OW) weakly ordered all the output weights and required strict positivity. The second weight scheme, an assurance region type of constraint (AR), restricted the relations between output weights according to a permitted variation from average unit cost. In this section, we will first analyze the original dataset, then we present a Co-plot of the ratios, present the effects of the weight restriction schemes pictorially and discuss the path over which inefficient DMUs need to cross in order to become efficient. Finally, we will compare the weight schemes with PCA-DEA (see Chapter 8), and view the results graphically.

Table 10-3. Variables over which Forestry Boards were analyzed

Input	Output	
Total Costs		**Forest Management Planning**
	1.	Woodlot-level forest management plans
	2.	Regional forest management planning
		Forest Ditching
	3.	Forest ditches planned
	4.	Forest ditching supervised
	5.	Forest ditches inspected and improved
		Forest Road Construction
	6.	Forest roads planned
	7.	Forest road building supervised
	8.	Forest roads inspected and approved
		Training and Extension
	9.	Forest owners attending group extension meetings
	10.	Forest owners offered face-to-face assistance
	11.	Training offered to Forestry Board personnel
		Handling Administrative Matters of Forest Improvement
	12.	Forest improvement projects approved
	13.	Regeneration plans approved
		Overseeing the Applications of Forestry Laws
	14.	Forest sites inspected for quality control
	15.	Forest sites inspected for tax relief

In *Figure 10-3,* the raw data is presented graphically using a plot with $\theta=0.10$ and the average of correlations equal to 0.83. We can immediately see that DMU 1 is the smallest unit and that DMUs 11, 13, 17 and 19 are the largest and, by definition, will be BCC efficient (Banker et al. (1984)) due to their individual specializations. Furthermore, DMU 18 would appear to be a special unit in that it does not belong to any cluster.

Co-Plot can also be used to analyze each variable individually and we have chosen variable 9, the number of forest owners attending group extension meetings, randomly. *Figure 10-4* presents a Co-Plot in which the data was sorted automatically, with DMU 9 to the far right of the arrow producing the smallest amount of this output and DMUs 10 and 12 producing the most. We can also see the large gap between the two most prolific forestry boards and the next in line, namely DMU 2.

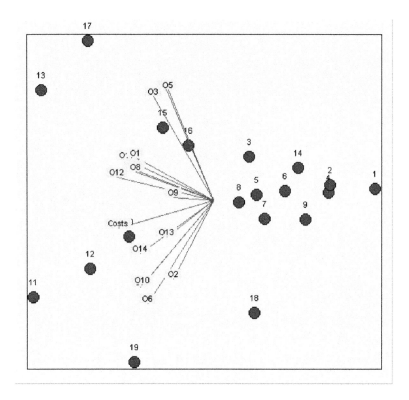

Figure 10-3. Presentation of Raw Forestry Data using Co-Plot

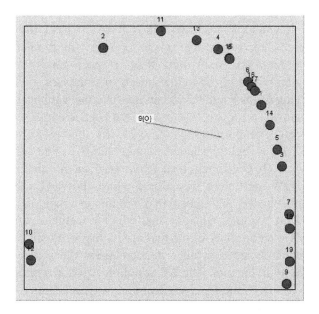

Figure 10-4. Co-Plot Presentation of Output 9

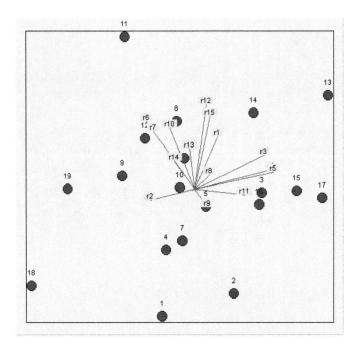

Figure 10-5. Co-Plot of Finnish Forestry Boards over Ratios of Variables

In *Figure 10-5,* the Co-Plot display presents the ratios of data for 19 forestry boards over 15 ratios, with a coefficient of alienation θ=0.19 and an average of correlations 0.647. The text output of this plot is presented in *Table 10-*4. It would appear that ratios 8 and 9 have little value in this analysis, hence their respective arrows are very short relatively and their correlation values are 0.23 and 0.24 respectively. *r8* and *r9* were consequently eliminated from in the subsequent analyses.

By removing ratios 8 and 9, we can improve the statistical significance of the plot, as shown in *Figure 10-6,* where the coefficient of alienation is 0.16 and the average of correlations is 0.72. In this plot, the single BCC inefficient DMU is lightly shaded, namely DMU3, and all the remaining observations are BCC efficient (see *Table 10-6* for full details). The DMUs that become OW inefficient are colored white, the DMUs that become AR inefficient are shaded and those that remain efficient in all models are colored black. It becomes apparent that the OW weight restriction approach removes DMUs on an inner circle that are neither specialists in any specific area nor reasonably good at all criteria, namely those placed close to the center-of-gravity in the plot. The AR weight restriction approach removes all the inner DMUs from the envelope frontier, as demonstrated in *Figure 10-6*.

This is true for all observations except DMU 18, which is efficient in all four models apart from the assurance region model, where it is the closest unit to the Pareto frontier with an efficiency score of 0.95.

Table 10-4. Correlation of Ratios

Coefficient of Alienation: 0.188 Center of Gravity: (55.20,40.70)			Average of Correlations: 0.647		
Observation	X_{1i}	X_{2i}	Variable	Degree	Correlation
1	44.31	0	1	66	0.62
2	68.43	7.35	2	-167	0.51
3	77.78	39.6	3	25	0.85
4	45.41	21.14	4	14	0.88
5	58.91	35.22	5	12	0.9
6	51.66	50.66	6	127	0.9
7	50.98	24.2	7	127	0.77
8	49.24	62.49	**8**	**44**	**0.23**
9	30.68	44.88	**9**	**-54**	**0.24**
10	50.18	451.18	10	113	0.71
11	31.67	89.48	11	-6	0.56
12	38.43	56.93	12	82	0.92
13	100	70.74	13	98	0.43
14	75.04	65.14	14	128	0.38
15	89.58	40.06	15	78	0.8
16	76.84	35.75			
17	98.1	37.94			
18	0	9.69			
19	12.34	40.85			

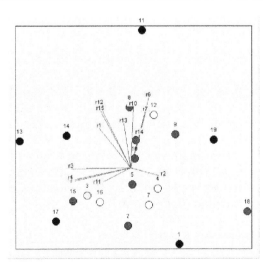

Figure 10-6. Co-plot graphic display for 19 forestry boards with 13 ratios

Co-plot could also be used to demonstrate the path over which an inefficient DMU needs to travel in order to be considered efficient. For example DMU 3 is considered inefficient in all models and in *Figure 10-7* the inefficient DMU and its hypothetical, efficient counterpart can be viewed and explained. Under the OW weight restricted model, the DMUs that belong to the benchmark group appear dark and DMU 3 is lightly shaded. The weight on DMU 11 is minimal, in other words the facet defined for hypothetical DMU 3* includes DMUs 5 (17%), 13 (20%) and 14 (63%).

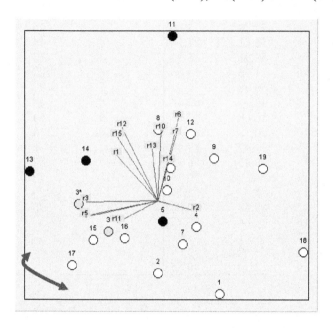

Figure 10-7. Co-Plot displaying the path to efficiency for DMU 3

An additional point that we would like to discuss in this section is the potential use of PCA-DEA (described in detail in Chapter 8) to improve the discriminatory power of DEA with such datasets. After running a principal component analysis of the 15 outputs, we can see that three principal components will explain more than 80% of the original data variance and should be sufficient to analyze this dataset. The three components and their powers of explanation are presented in *Table 10-5* and a Co-Plot of the PCA-DEA results is presented in *Figure 10-8*.

Table 10-5. Principal Component Analysis of Outputs

	% information
PC1	56.6
PC2	15.5
PC3	12.8
Total Variance Explained	84.8

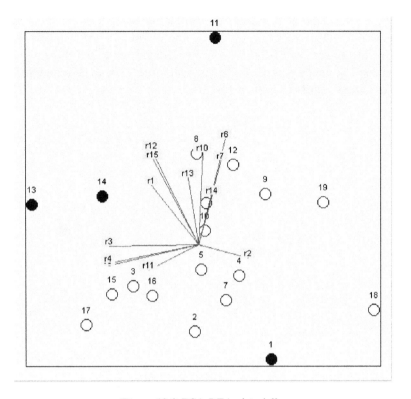

Figure 10-8. PCA-DEA pictorially

Table 10-6. DEA scores for various models

DMU	BCC	CCR	OW (VRS)	AR (VRS)	3 PCA-DEA (VRS)
1	1	1	1	1	1
2	1	1	1	0.85	0.92
3	0.98	0.92	0.86	0.8	0.83
4	1	0.92	0.88	0.81	0.90
5	1	1	1	0.77	0.84
6	1	1	1	0.8	0.91
7	1	1	0.86	0.77	0.72
8	1	1	1	0.85	0.97
9	1	1	1	0.83	0.98
10	1	1	1	0.84	0.82
11	1	1	1	1	1
12	1	1	0.98	0.81	0.91
13	1	1	1	1	1
14	1	1	1	1	1
15	1	1	1	0.86	0.89
16	1	1	0.96	0.81	0.84
17	1	1	1	1	0.94
18	1	1	1	0.95	0.98
19	1	1	1	1	**0.99**

Figure *10-8* clearly presents the remaining 4 DMUs that are efficient under the PCA-DEA model in the outer ring of data. In order to compare this model with the results of the weight-constrained models, OW and AR, we present the efficiency scores in the last column of *Table 10-6* and a Co-Plot of the different results in *Figure 10-9*.

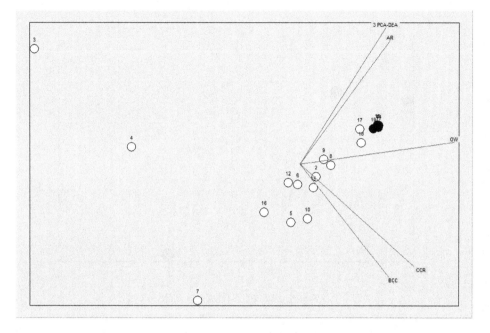

Figure 10-9. Co-Plot displaying the results of various DEA models

Figure 10-9 presents a Co-Plot with value of alienation equal to 0.04 and average of correlations equal to 0.91. PCA-DEA has provided the greatest level of discrimination, with results very similar to the other models, but removes an additional two DMUs from the efficient set of the AR model. Indeed, one could argue that the most efficient set of DMUs, painted a dark color in *Figure 10-9*, are likely to be truly relatively efficient, since all the models are in agreement over this cluster. Other DMUs, such as DMU 3, 4 and 7 are likely to be either very small or strictly inefficient and DMUs 17 and 18 would appear to be very close to the efficient frontier, based on the findings of all the models combined.

5. CONCLUSIONS

In this chapter we have discussed two methodologies that can be used to analyze multiple variable data. Data envelopment analysis (DEA) is frequently used to analyze the relative efficiency of a set of homogeneous units with multiple inputs and outputs. The results group the data into two sets, those units that are considered efficient and define the Pareto frontier and those that are not. Co-Plot is a graphical technique, which reduces each unit to an observation in two dimensions, enabling a plot to be drawn of the entire dataset. After applying Co-Plot over the set of variable ratios (each output divided by each input), in order to align the technique to the idea of efficiency as utilized in DEA, we can then use Co-Plot to graphically display the DEA results.

This chapter has illustrated the approach using a problematic dataset drawn from the literature, in which 19 Finnish Forestry Boards are compared over one input and fifteen outputs. The Co-Plot presents the variable direction in the form of arrows, DMUs in the form of circles and colors separate the efficient from inefficient units. The further an observation appears along a particular ray, the more efficient that DMU is with respect to that ratio. The DEA efficient units can be clearly identified in the Co-Plot as the (partial) outer-rings. Thus, each methodology can be used as an aid in understanding the relative efficiency of the observations. Whilst DEA provides a detailed analysis of how inefficient units can be improved to attain relative efficiency, a series of Co-Plots can be used as graphical aids to explain the results to management. Furthermore, Co-Plot is also useful in an exploratory data analysis to identify extreme outliers, which at the least need to be considered closely for potential data measurement errors, if not lack of homogeneity amongst observations. It can also be used to identify unnecessary variables that contribute little to the analysis, which may help with the problem of excess dimensionality that occurs in DEA.

REFERENCES

1. Adler, N., A. Raveh. 2006. Presenting DEA graphically. forthcoming in *Omega*.
2. Albriktsen, R., F. Forsund. 1990. A productive study of the Norwegian building industry. *Journal of Productivity Analysis* **2** 53-66.
3. Banker, R.D., A. Charnes, W.W. Cooper. 1984. Some models for estimating technical and scale inefficiencies in data envelopment analysis. *Management Science* **30/9** 1078-1092.
4. Belton, V., S.P. Vickers. 1993. Demystifying DEA – A virtual interactive approach based on multiple criteria analysis. *Journal of the Operational Research Society* **44** 883-896.

5. Charnes, A., W.W. Cooper, B. Golany, L. Seiford, J. Stutz. 1985. Foundations of data envelopment analysis for Pareto-Koopmans efficient empirical production functions. *Journal of Econometrics* **30** 91-107.
6. Desai, A., L.C. Walters. 1991. Graphical presentations of data envelopment analysis: Management implications from parallel axes representations. *Decision Sciences* **22** 335-353.
7. El-Mahgary, S., R. Lahdelma. 1995. Data envelopment analysis: Visualizing the results. *European Journal of Operational Research* **85** 700-710.
8. Gilady, R., Y. Spector, A. Raveh. 1996. Multidimensional scaling: An analysis of computers 1980-2990. *European Journal of Operational Research* **95** 439-450.
9. Golany, B., J.E. Storbeck. 1999. A data envelopment analysis of the operational efficiency of bank branches. *Interfaces* **29** 14-26.
10. Guttman, L. 1968. A general non-metric technique for finding the smallest space for a configuration of points. *Psychometrica* **33** 469-506.
11. Hackman, S.T., U. Passy, L.K. Platzman. 1994. Explicit representation of the two-dimensional section of a production possibility set. *The Journal of Productivity Analysis* **5** 161-170.
12. Joro, T., E-J Viitala. 2004. Weight-restricted DEA in action: From expert opinions to mathematical tools. *Journal of the Operational Research Society* **55** 814-821.
13. Lipshitz, G., A. Raveh. 1994. Applications of the Co-Plot method in the study of socioeconomic differences among cities: A basis for a differential development policy. *Urban Studies* **31** 123-135.
14. Maital, S., A. Vaninsky. 1999. Data envelopment analysis with a single DMU: A graphic projected-gradient approach. *European Journal of Operational Research* **115** 518-528.
15. Raveh, A. 1986. On measures of monotone association. *The American Statistician* **40** 117-123.
16. Raveh, A. 2000a. The Greek banking system: Reanalysis of performance. *European Journal of Operational Research* **120** 525-534.
17. Raveh, A. 2000b. Co-Plot: A graphic display method for geometrical representations of MCDM. *European Journal of Operational Research* **125** 670-678.
18. Scheel, H. 2001. Efficiency Measurement System. http://www.wiso.uni-dortmund.de/lsfg/or/scheel/ems.
19. Sinuany-Stern, Z., L. Friedman. 1998. DEA and the discriminant analysis of ratios for ranking units. *European Journal of Operational Research* **111** 470-78.

20. Talby, D. 2005. Visual Co-Plot 5.5. http://www.cs.huji.ac.il/~davidt/vcoplot/.
21. Talluri, S., M.M. Whiteside, S.J. Seipel. 2000. A nonparametric stochastic procedure for FMS evaluation. *European Journal of Operational Research* **124** 529-538.
22. Viitala, H.J., H. Hänninen. 1998. Measuring the efficiency of public forestry organizations. *Forestry Science* **44** 298-307.
23. Weber, C.A., A. Desai. 1996. Determination pf paths to vendor market efficiency using parallel coordinates representation: A negotiation tool for buyers. *European Journal of Operational Research* **90** 142-155.
24. Weber, Y., O. Shenkar, A. Raveh. 1996. National versus corporate cultural fit in mergers and acquisitions: An exploratory study. *Management Science* **42** 1215-1227.
25. Zhu, J. 1998. Data envelopment analysis versus principal component analysis: An illustrative study of economic performance of Chinese cities. *European Journal of Operational Research* **111** 50-61.

The authors would like to thank the Recanati Fund for partial funding of this research.

Chapter 11

DEA MODELS FOR SUPPLY CHAIN OR MULTI-STAGE STRUCTURE

Wade D. Cook[1], Liang Liang[2], Feng Yang[2], and Joe Zhu[3]

[1]Schulich School of Business, York University, Toronto, Ontario, Canada, M3J 1P3, wcook@shulich.yorku.ca

[2]School of Business, University of Science and Technology of China, He Fei, An Hui Province, P.R. China 230026 lliang@ustc.edu.cn

[3]Department of Management, Worcester Polytechnic Institute, Worcester, MA 01609, jzhu@wpi.edu

Abstract: Standard data envelopment analysis (DEA) models cannot be used directly to measure the performance of a supply chain and its members, because of the existence of the intermediate measures connecting those members. This observation is true for any situations where DMUs contain multi-stage processes. This chapter presents several DEA-based approaches in a seller-buyer supply chain context. Some DEA models are developed under the assumption that the relationship between the seller and buyer is treated as leader-follower and cooperative, respectively. In the leader-follower (or non-cooperative) structure, the leader is first evaluated using the standard DEA model, and then the follower is evaluated by a new DEA-based model which incorporates the DEA efficiency information for the leader. In the cooperative structure, one maximizes the joint efficiency that is modeled as the average of the seller's and buyer's efficiency scores, and both supply chain members are evaluated simultaneously.

Key words: Supply chain; Efficiency; Best practice; Performance; Data envelopment analysis (DEA); Buyer; Seller.

1. INTRODUCTION

It has been recognized that lack of appropriate performance measurement systems has been a major obstacle to effective management of supply chains (Lee and Billington, 1992). While there are studies on supply chain performance using the methodology of data envelopment analysis (DEA), the research has been focused on a single member of the supply chain. For example, Weber and Desai (1996) employed DEA to construct an index of relative supplier performance.

Note that an effective management of the supply chain requires knowing the performance of the overall chain rather than simply the performance of the individual supply chain members. Each supply chain member has is own strategy to achieve efficiency, however, what is best for one member may not work in favor of another member. Sometimes, because of the possible conflicts between supply chain members, one member's inefficiency may be caused by another's efficient operations. For example, the supplier may increase its raw material price to enhance its revenue and to achieve an efficient performance. This increased revenue means increased cost to the manufacturer. Consequently, the manufacturer may become inefficient unless it adjusts its current operating policy. Measuring supply chain performance becomes a difficult and challenging task because of the need to deal with the multiple performance measures related to the supply chain members, and to integrate and coordinate the performance of those members.

As noted in Zhu (2003), although DEA is an effective tool for measuring efficiencies of peer decision making units (DMUs), it cannot be applied directly to the problem of evaluating the efficiency of supply chains. This is because some measures linked to supply chain members cannot be simply classified as "outputs" or "inputs" of the supply chain. In fact, with respect to those measures, conflicts between supply chain members are likely present. For example, the supplier's revenue is an output for the supplier, and it is in the supplier's interest to maximize it; at the same time it is also an input to the manufacturer who wishes to minimize it. Simply minimizing the total supply chain cost or maximizing the total supply chain revenue (profit) does not properly model and resolve the inherent conflicts.

Within the context of DEA, there are a number of methods that have the potential to be used in supply chain efficiency evaluation. Seiford and Zhu (1999) and Chen and Zhu (2004) provide two approaches in modeling efficiency as a two-stage process. Färe and Grosskopf (2000) develop the network DEA approach to model general multi-stage processes with intermediate inputs and outputs (see also chapter 12). Golany, Hackman and Passy (2006) provide an efficiency measurement framework for systems composed of two subsystems arranged in series that simultaneously compute the efficiency of the aggregate system and each subsystem. Zhu (2003), on the other hand, presents a DEA-based supply chain model to both define and measure the efficiency of a supply chain and that of its members, and yield a set of optimal values of the (intermediate) performance measures that establish an efficient supply chain.

We point out that there are number of DEA studies on supply chain efficiency. Yet, all of them tend to focus on a single supply chain member, not the entire supply chain, where at least two members are present. This can be partly due to the lack of DEA models for supply chain or multi-stage structures.

Evaluation of supply chain efficiency, using DEA, has its advantages. In particular, it eliminates the need for unrealistic assumptions inherent in typical supply chain optimization models and probabilistic models; e.g., a typical EOQ model assumes constant and known demand rate and lead-time for delivery. These conventional approaches typically fail, however, to consider the cooperation within the supply chain system.

Recently, Liang et al. (2006) developed two classes of DEA-based models for supply chain efficiency evaluation, using a seller-buyer supply chain as an example. First, the relationship between the buyer and the seller is modeled as a non-cooperative two-stage game, and second, the relationship is assumed to be cooperative. In the non-cooperative two-stage game, they use the concept of a leader-follower structure. In the cooperative game, it is assumed that the members of the supply chain cooperate on the intermediate measures. The resulting DEA models are non-linear and can be solved using a parametric linear programming technique.

The current chapter presents the approaches of Zhu (2003) and Liang et la. (2006).

2. NOTIONS AND STANDARD DEA MODELS

Suppose there are N similar supply chains or N observations on one supply chain. Consider a buyer-seller supply chain as described in Figure 11-1, where for $j = 1, ..., N$, $X_A = (x_{ij}^A, i = 1, ..., I)$ is the input vector of the seller, and $Y_A = (y_{rj}^A, r=1,...., R)$ is the seller's output vector. Y_A is also an input vector of the buyer. The buyer also has an input vector $X_B = (x_{sj}^B, s = 1,...., S)$ and the output vector for the buyer is $Y_B (= y_{tj}^B, t=1, ..., T)$.

We can use the following DEA model to measure the efficiency of the supply chain shown in Figure 11-1 (Charnes, Cooper and Rhodes, 1978; CCR)

Maximize $V_{P0} = \dfrac{\sum_{t=1}^{T} u_t^B y_{t0}^B}{\sum_{i=1}^{I} v_i^A x_{i0}^A + \sum_{s=1}^{S} v_s^B x_{s0}^B}$ (1)

subject to

$\dfrac{\sum_{t=1}^{T} u_t^B y_{tj}^B}{\sum_{i=1}^{I} v_i^A x_{ij}^A + \sum_{s=1}^{S} v_s^B x_{sj}^B} \leq 1 \qquad j = 1, \ldots, N$

$u_t^B, v_i^A, v_s^B \geq 0, t=1, \ldots, T, i = 1, \ldots, I, s = 1, \ldots, S$

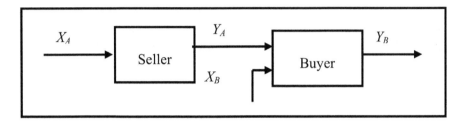

Figure 11-1. Seller-Buyer Supply Chain

Zhu (2003) shows that DEA model (1) fails to correctly characterize the performance of supply chains, because it only considers the inputs and outputs of the supply chain system and ignores measures Y_A associated with supply chain members. Zhu (2003) also shows that if Y_A are treated as both input and output measures in model (1), all supply chains become efficient. Zhu (2003) further shows that an efficient performance indicated by model (1) does not necessarily indicate efficient performance in individual supply chain members. For example, consider the example shown in Table 11-1.

Table 11-1. Efficient supply chains with inefficient supply chain members

DMU	X_A	Y_A	Y_B	Seller Efficiency	Buyer Efficiency	Supply Chain Efficiency
1	1	4	10	1	0.5	1
2	1	2	10	0.5	1	1
3	1	2	5	0.5	0.5	0.5

The above example shows that improvement to the best-practice can be distorted. i.e., the performance improvement of one supply chain member affects the efficiency status of the other, because of the presence of intermediate measures.

Alternatively, we may consider the average efficiency of the buyer and seller as in the following DEA model

$$\text{Maximize } \frac{1}{2}[\frac{\sum_{r=1}^{R}u_r^A y_{r0}^A}{\sum_{i=1}^{I}v_i^A x_{i0}^A} + \frac{\sum_{t=1}^{T}u_t^B y_{t0}^B}{\sum_{r=1}^{R}c_r^A y_{r0}^A + \sum_{s=1}^{S}v_s^B x_{s0}^B}]$$

subject to

$$\frac{\sum_{r=1}^{R}u_r^A y_{rj}^A}{\sum_{i=1}^{I}v_i^A x_{ij}^A} \leq 1 \qquad j = 1, \ldots, N \qquad (2)$$

$$\frac{\sum_{t=1}^{T}u_t^B y_{tj}^B}{\sum_{r=1}^{R}c_r^A y_{rj}^A + \sum_{s=1}^{S}v_s^B x_{sj}^B} \leq 1 \qquad j = 1, \ldots, N$$

$$c_r^A, u_r^A, u_t^B, v_i^A, v_s^B \geq 0$$

$r=1, \ldots, R,\ t=1, \ldots, T,\ i = 1, \ldots, I,\ s = 1, \ldots, S$

Although model (2) considers Y_A, it does not reflect the relationship between the buyer and the seller. Model (2) treats the seller and the buyer as two independent units. This does not reflect an ideal supply chain operation.

3. ZHU (2003) APPROACH[1]

Using the notation in the above section, the Zhu (2003) model for situations when there are only two members in the supply chain can be expressed as

$$\min(\theta^{seller} + \theta^{buyer})$$

subject to
(seller)

$$\sum_{j=1}^{N}\lambda_j x_{ij}^A \leq \theta^{seller} x_{i0}^A \qquad i = 1,\ldots,I$$

$$\sum_{j=1}^{N}\lambda_j y_{rj}^A \geq \tilde{y}_{r0}^A \qquad r = 1,\ldots,R$$

$$\lambda_j \geq 0, \quad j = 1,\ldots,N$$

$$\tilde{y}_{r0}^A \geq 0 \quad r = 1,\ldots,R$$

---

[1] Chapter 8 of Zhu, J., 2003, *Quantitative Models for Performance Evaluation and Benchmarking: Data Envelopment Analysis with Spreadsheets.* Kluwer Academic Publishers, Boston

(buyer)

$$\sum_{j=1}^{N} \beta_j x_{sj}^B \leq \theta^{buyer} x_{s0}^B \qquad i = 1,...,S$$

$$\sum_{j=1}^{N} \beta_j y_{rj}^A \leq \tilde{y}_{r0}^A \qquad r = 1,...,R$$

$$\sum_{j=1}^{N} \beta_j y_{tj}^B \leq y_{t0}^B \qquad t = 1,...,T$$

$$\beta_j \geq 0, \quad j = 1,...,N$$

where $\tilde{y}_{r0}^A$ are decision variables.

Additional constraints can be added into the above model. For example, if y_r^A represents the number of units of a product shipped from the manufacturer to the distributor, and if the capacity of this manufacturer in producing the product is C_r, then we may add $\tilde{y}_{r0}^A \leq C_r$.

4. COOPERATIVE AND NON-COOPERATIVE APPROACHES

In this section, we present several models due to Liang et al. (2006) that directly evaluate the performance of the supply chain as well as its members, while considering the relationship between the buyer and the seller. The modeling processes are based upon the concept of non-cooperative and cooperative games (see, e.g., Simaan and Cruz, 1973; Li, Huang and Ashley, 1995; Huang 2000).

4.1 The Non-cooperative Model

We propose the seller-buyer interaction be viewed as a two-stage non-cooperative game with the seller as the leader and the buyer as the follower. First, we use the CCR model to evaluate the efficiency of the seller as the leader:

Maximize $E_{AA} = \dfrac{\sum_{r=1}^{R} u_r^A y_{r0}^A}{\sum_{i=1}^{I} v_i^A x_{i0}^A}$

subject to

(3)

$$\frac{\sum_{r=1}^{R} u_r^A y_{rj}^A}{\sum_{i=1}^{I} v_i^A x_{ij}^A} \leq 1 \qquad j = 1, \ldots, N$$

$$u_r^A, v_i^A \geq 0 \qquad r=1, \ldots, R,\ i = 1, \ldots, I$$

This model is equivalent to the following standard DEA multiplier model:

Maximize $E_{AA} = \sum_{r=1}^{R} \mu_r^A y_{r0}^A$

subject to

$$\sum_{i=1}^{I} \omega_i^A x_{ij}^A - \sum_{r=1}^{R} \mu_r^A y_{rj}^A \geq 0 \qquad j = 1, \ldots, N \qquad (4)$$

$$\sum_{i=1}^{I} \omega_i^A x_{i0}^A = 1$$

$$\mu_r^A, \omega_i^A \geq 0 \qquad r=1, \ldots, R,\ i = 1, \ldots, I$$

Suppose we have an optimal solution of model (4) $\mu_r^{A*}, \omega_i^{A*}, E_{AA}^*$ ($r=1, \ldots, R$, $i = 1, \ldots, I$), and denote the seller's efficiency as E_{AA}^*. We then use the following model to evaluate the buyer's efficiency:

Maximize $E_{AB} = \dfrac{\sum_{t=1}^{T} u_t^B y_{t0}^B}{D \times \sum_{r=1}^{R} \mu_r^A y_{r0}^A + \sum_{s=1}^{S} v_s^B x_{s0}^B}$

subject to

$$\dfrac{\sum_{t=1}^{T} u_t^B y_{tj}^B}{D \times \sum_{r=1}^{R} \mu_r^A y_{rj}^A + \sum_{s=1}^{S} v_s^B x_{sj}^B} \leq 1 \qquad j = 1, \ldots, N \qquad (5)$$

$$\sum_{r=1}^{R} \mu_r^A y_{r0}^A = E_{AA}^*$$

$$\sum_{i=1}^{I} \omega_i^A x_{ij}^A - \sum_{r=1}^{R} \mu_r^A y_{rj}^A \geq 0 \qquad j = 1, \ldots, N$$

$$\sum_{i=1}^{I} \omega_i^A x_{i0}^A = 1$$

$$\mu_r^A, \omega_i^A, u_t^B, v_s^B, D \geq 0$$

$r=1, \ldots, R,\ t=1, \ldots, T,\ i = 1, \ldots, I,\ s = 1, \ldots, S$

Note that in model (5), we try to determine the buyer's efficiency given that the seller's efficiency remains at E_{AA}^*. Model (5) is equivalent to the following non-linear model:

Maximize $E_{AB} = \sum_{t=1}^{T} \mu_t^B y_{t0}^B$

subject to

$$d \times \sum_{r=1}^{R} \mu_r^A y_{rj}^A + \sum_{s=1}^{S} \omega_s^B x_{sj}^B - \sum_{t=1}^{T} \mu_t^B y_{tj}^B \geq 0 \quad j=1,...,N$$

$$d \times \sum_{r=1}^{R} \mu_r^A y_{r0}^A + \sum_{s=1}^{S} \omega_s^B x_{s0}^B = 1$$

$$\sum_{r=1}^{R} \mu_r^A y_{r0}^A = E_{AA}^*$$

$$\sum_{i=1}^{I} \omega_i^A x_{ij}^A - \sum_{r=1}^{R} \mu_r^A y_{rj}^A \geq 0 \quad j=1,...,N$$

$$\sum_{i=1}^{I} \omega_i^A x_{i0}^A = 1$$

$$\mu_r^A, \omega_i^A, \mu_t^B, \omega_s^B, d \geq 0$$

$r=1, ..., R, t=1, ..., T, i=1, ..., I, s=1, ..., S$

(6)

Note that $d \times \sum_{r=1}^{R} \mu_r^A y_{r0}^A + \sum_{s=1}^{S} \omega_s^B x_{s0}^B = 1$ and $\sum_{r=1}^{R} \mu_r^A y_{r0}^A = E_{AA}^*$. Thus, we have $0 \leq d < 1/\sum_{r=1}^{R} \mu_r^A y_{r0}^A = 1/E_{AA}^*$. i.e, we have the upper and lower bounds on d. Therefore, d can be treated as a parameter, and model (6) can be solved as a parametric linear program.

In computation, we set the initial d value as the upper bound, namely, $d_0 = 1/E_{AA}^*$, and solve the resulting linear program. We then start to decrease d according to $d_t = 1/E_{AA}^* - \varepsilon \times t$ for each step t, where ε is a small positive number[2]. We solve each linear program of model (6) corresponding to d_t and denote the optimal objective value as $E_{BA}^*(d_t)$.

Let $E_{BA}^* = \underset{t}{Max}\, E_{BA}^*(d_t)$. Then we obtain a best heuristic search solution E_{BA}^* to model (6)[3]. This E_{AB}^* represents the buyer's efficiency when the seller is given the pre-emptive priority to achieve its best performance. The efficiency of the supply chain can then be defined as

$$e_{AB} = \frac{1}{2}(E_{AA}^* + E_{AB}^*)$$

---

[2] In the current study, we set $\varepsilon = 0.01$. If we use a smaller ε, the difference only shows in the fourth decimal place in the current study.

[3] The proposed procedure is a global solution approximation using a heuristic technique, as it searches through the entire feasible region of d when d is decreased from its upper bound to lower bound of zero. It is likely that estimation error exists. The smaller the decreased step, the better the heuristic search solution will be.

Similarly, one can develop a procedure for the situation when the buyer is the leader and the seller the follower. For example, in the October 6, 2003 issue of the Business Week, its cover story reports that Walmart dominates its suppliers and not only dictates delivery schedules and inventory levels, but also heavily influences product specifications.

We first evaluate the efficiency of the buyer using the standard CCR ratio model:

$$\text{Maximize } E_{BB} = \frac{\sum_{t=1}^{T} u_t^B y_{t0}^B}{\sum_{r=1}^{R} u_r^A y_{r0}^A + \sum_{s=1}^{S} v_s^B x_{s0}^B}$$

subject to

$$\frac{\sum_{t=1}^{T} u_t^B y_{tj}^B}{\sum_{r=1}^{R} u_r^A y_{rj}^A + \sum_{s=1}^{S} v_s^B x_{sj}^B} \leq 1 \qquad j = 1, \ldots, N \qquad (7)$$

$$u_r^A, u_t^B, v_s^B \geq 0$$

$r=1, \ldots, R,\ t=1, \ldots, T,\ s = 1,\ldots, S$

Model (7) is equivalent to the following standard CCR multiplier model:

$$\text{Maximize } E_{BB} = \sum_{t=1}^{T} \mu_t^B y_{t0}^B$$

subject to

$$\sum_{r=1}^{R} \mu_r^A y_{rj}^A + \sum_{s=1}^{S} \omega_s^B x_{sj}^B - \sum_{t=1}^{T} \mu_t^B y_{tj}^B \geq 0 \qquad j = 1, \ldots, N \qquad (8)$$

$$\sum_{r=1}^{R} \mu_r^A y_{r0}^A + \sum_{s=1}^{S} \omega_s^B x_{s0}^B = 1$$

$$\mu_r^A, \mu_t^B, \omega_s^B \geq 0$$

$r=1, \ldots, R,\ t=1, \ldots, T,\ s = 1,\ldots, S$

Let $\mu_r^{A*}, \mu_t^{B*}, \omega_s^{B*}, E_{BB}^*$ ($r=1, \ldots, R,\ t=1, \ldots, T,\ s = 1,\ldots, S$) be an optimal solution from model (8), where E_{BB}^* represents the buyer's efficiency score. To obtain the seller's efficiency given that the buyer's efficiency is equal to E_{BB}^*, we solve the following model:

$$\text{Maximize } E_{BA} = \frac{U \times \sum_{r=1}^{R} \mu_r^A y_{r0}^A}{\sum_{i=1}^{I} v_i^A x_{i0}^A}$$

subject to

$$\frac{U \times \sum_{r=1}^{R} \mu_r^A y_{rj}^A}{\sum_{i=1}^{I} v_i^A x_{ij}^A} \leq 1 \qquad j = 1, \ldots, N \tag{9}$$

$$\sum_{t=1}^{T} \mu_t^B y_{t0}^B = E_{BB}^*$$

$$\sum_{r=1}^{R} \mu_r^A y_{rj}^A + \sum_{s=1}^{S} \omega_s^B x_{sj}^B - \sum_{t=1}^{T} \mu_t^B y_{tj}^B \geq 0 \qquad j = 1, \ldots, N$$

$$\sum_{r=1}^{R} \mu_r^A y_{r0}^A + \sum_{s=1}^{S} \omega_s^B x_{s0}^B = 1$$

$$v_i^A, \mu_r^A, \mu_t^B, \omega_s^B, U \geq 0$$

$r=1, \ldots, R,\ t=1, \ldots, T,\ i=1, \ldots, I,\ s=1, \ldots, S$

Model (9) is equivalent to the following non-linear program:

Maximize $E_{BA} = u \times \sum_{r=1}^{R} \mu_r^A y_{r0}^A$

subject to

$$\sum_{i=1}^{I} \omega_i^A x_{ij}^A - u \times \sum_{r=1}^{R} \mu_r^A y_{rj}^A \geq 0 \qquad j = 1, \ldots, N$$

$$\sum_{i=1}^{I} \omega_i^A x_{i0}^A = 1 \tag{10}$$

$$\sum_{t=1}^{T} \mu_t^B y_{t0}^B = E_{BB}^*$$

$$\sum_{r=1}^{R} \mu_r^A y_{rj}^A + \sum_{s=1}^{S} \omega_s^B x_{sj}^B - \sum_{t=1}^{T} \mu_t^B y_{tj}^B \geq 0 \qquad j = 1, \ldots, N$$

$$\sum_{r=1}^{R} \mu_r^A y_{r0}^A + \sum_{s=1}^{S} \omega_s^B x_{s0}^B = 1$$

$$\omega_i^A, \mu_r^A, \mu_t^B, \omega_s^B, u \geq 0$$

$r=1, \ldots, R,\ t=1, \ldots, T,\ i=1, \ldots, I,\ s=1, \ldots, S$

This model (10) is similar to model (6) and can be treated as a linear program with u as the parameter. We next show how to select the initial value of this parameter.

We first solve the following model:

Maximize $EF_{BA} = \dfrac{U \times \sum_{r=1}^{R} \mu_r^{A*} y_{r0}^A}{\sum_{i=1}^{I} v_i^A x_{i0}^A}$

subject to

$$\frac{U \times \sum_{r=1}^{R} \mu_r^{A*} y_{rj}^A}{\sum_{i=1}^{I} v_i^A x_{ij}^A} \leq 1 \qquad j = 1, \ldots, N \tag{11}$$

$v_i^A, U \geq 0 \quad i = 1, \ldots, I$

where μ_r^{A*} ($r=1, \ldots, R$) is an optimal solution from model (8).

Model (11) is equivalent to the following linear program:

Maximize $EF_{BA} = u \times \sum_{r=1}^{R} \mu_r^{A*} y_{r0}^A$

subject to

$$\sum_{i=1}^{I} \omega_i^A x_{ij}^A - u \times \sum_{r=1}^{R} \mu_r^{A*} y_{rj}^A \geq 0 \qquad j=1, \ldots, N \qquad (12)$$

$$\sum_{i=1}^{I} \omega_i^A x_{i0}^A = 1$$

$$\omega_i^A, u \geq 0 \quad i=1, \ldots, I$$

Let $\omega_i^{A*}, u^*, EF_{BA}^*$ ($i = 1, \ldots, I$) be an optimal solution from model (12). Note that the optimal value to model (12), EF_{BA}^*, may not be the maximum value for the seller because of possible multiple optima in model (8). We have $u \times \sum_{r=1}^{R} \mu_r^A y_{r0}^A \geq EF_{BA}^*$. Further, based upon $\sum_{r=1}^{R} \mu_r^A y_{r0}^A + \sum_{s=1}^{S} \omega_s^B x_{s0}^B = 1$, we have $\sum_{r=1}^{R} \mu_r^A y_{r0}^A \leq 1$. Therefore, $u \geq EF_{BA}^*$. We then utilize EF_{BA}^* as the lower bound for the parameter u when solving for seller's efficiency using model (10). However, this lower bound can be converted into an upper bound as follows.

Let $u \times \mu_r^A = c_r^A$, $g = 1/u$, then model (10) is equivalent to the following model:

Maximize $E_{BA} = \sum_{r=1}^{R} c_r^A y_{r0}^A$

subject to

$$\sum_{i=1}^{I} \omega_i^A x_{ij}^A - \sum_{r=1}^{R} c_r^A y_{rj}^A \geq 0 \qquad j=1, \ldots, N$$

$$\sum_{i=1}^{I} \omega_i^A x_{i0}^A = 1$$

$$\sum_{t=1}^{T} \mu_t^B y_{t0}^B = E_{BB}^* \qquad (13)$$

$$\frac{1}{g}\sum_{r=1}^{R} c_r^A y_{rj}^A + \sum_{s=1}^{S} \omega_s^B x_{sj}^B - \sum_{t=1}^{T} \mu_t^B y_{tj}^B \geq 0 \quad j=1, \ldots, N$$

$$\frac{1}{g}\sum_{r=1}^{R} c_r^A y_{r0}^A + \sum_{s=1}^{S} \omega_s^B x_{s0}^B = 1$$

$$\omega_i^A, c_r^A, \mu_t^B, \omega_s^B, g \geq 0,$$

$r=1, \ldots, R, t=1, \ldots, T, i = 1, \ldots, I, s = 1,\ldots, S$

where $0 \leq g \leq 1/EF_{BA}^*$ can be treated as a parameter.

We solve model (13) for the seller's efficiency. The computational procedure is similar to the one used in model (6). Denote the heuristic search solution to (13) as E_{BA}^*. Then the efficiency of the supply chain can be defined as

$$e_{BA} = \frac{1}{2}(E_{BA}^* + E_{BB}^*)$$

Table 11-2. Numerical Example

DMU	x_1^A	x_2^A	y_1^A	y_2^A	x^B	y^B
1	8	50	20%	10	8	100
2	10	18	10%	15	10	70
3	15	30	10%	20	8	95
4	8	25	20%	25	10	80
5	10	40	15%	20	15	85

Table 11-3. Leader-follower Results

DMU	Model (1)	Model (2)	Model (4) Seller	Model (6) Buyer	e_{AB}	Model (8) Buyer	Model (13) Seller	e_{BA}
1	1	1	1	1	1	1	1	1
2	1	0.854	0.833	0.56	0.697	0.875	0.766	0.821
3	1	0.833	0.667	0.95	0.808	1	0.546	0.773
4	1	0.827	1	0.653	0.827	0.653	1	0.827
5	0.794	0.698	0.64	0.453	0.547	0.756	0.621	0.688

We now illustrate the above DEA procedures with five supply chain operations (DMUs) given in Table 11-2. The seller has two inputs, x_1^A (labor) and x_2^A (cost) and two outputs, y_1^A (buyer's fill rate) and y_2^A (number of product shipped). The buyer has another input x^B (labor) and one output: y^B. Note that the buyer's fill rate is actually a cost measure to the buyer, because the fill rate is associated with inventory holding cost and the amount of products required from the seller. The buyer's fill rate implies benefit to the seller, since more products are needed from the seller (meaning more revenue to the seller) if the buyer wishes to maintain a higher fill rate. Therefore, the buyer's fill rate is treated as an output from the seller and an input to the buyer. From a buyer's point of view, the buyer always tries to meet the needs of its customer while maintaining a fill rate as low as possible, because unnecessary high fill rates incur additional costs to the buyer.

Table 11-3 reports the efficiency scores obtained from the supply chain efficiency models. It can be seen from models (1) that the supply chain is rated as efficient while its two members are inefficient (e.g., DMUs 2). This is because the intermediate measures are ignored in model (1). This also indicates that supply chain efficiency cannot be measure by the conventional DEA approach.

When the seller is treated as the leader, two seller operations in DMUs 1 and 4 are efficient with only one efficient buyer operation in DMU1. This indicates that only DMU1 is the efficient supply chain.

When the buyer is treated as the leader, model (8) shows that three buyer operations are inefficient and model (13) shows that only two seller operations are efficient. This also implies that only DMU1 is efficient.

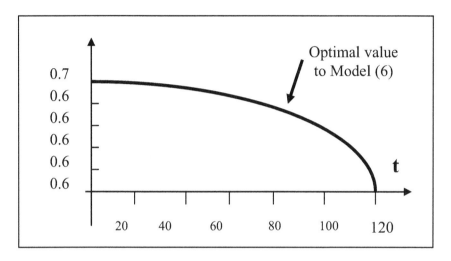

Figure 11-2. Solving Non-Cooperative Model for DMU2

Figure 11-2 shows how the best heuristic search is obtained when solving model (6) for DMU2. We set $d_t = 1/E_{AA}^* - 0.01 \times t$, where $E_{AA}^* = 0.833$ and $t = 0, \ldots, 120$. Note that when $t = 120$, the parameter $d = 0$, the lower bound, and the optimal value to model (6) is 0.650. Therefore, we have completed the search over the entire feasible region of d and the best solution is obtained at $t = 0$, that is $E_{AB}^* = 0.697$.

4.2 The Cooperative Model

In game theory, when the buyer-seller relation was treated as leader-follower, the buyer does not have control over the seller, and the seller determines the optimal strategy (optimal weights for the intermediate measures). Recent studies however have demonstrated that many retailers (buyers) have increased their bargaining power relative to the manufactures' (sellers) bargaining power (Porter, 1974; Li, Huang and Ashley, 1996). The shift of power from manufacturers to retailers is one of the most significant phenomena in manufacturing and retailing. Walmart is an extreme case where the manufacturer becomes a "follower". Therefore, it is in the best interest of the supply chain to encourage cooperation. This section considers the case where both the seller and buyer have the same degree of power to influence the supply chain system. Our new DEA model seeks to maximize both the seller's and buyer's efficiency, subject to a condition that the weights on the intermediate measures must be equal:

$$\text{Maximize } \frac{1}{2}[\frac{\sum_{r=1}^{R} c_r y_{r0}^A}{\sum_{i=1}^{I} v_i^A x_{i0}^A} + \frac{\sum_{t=1}^{T} u_t^B y_{t0}^B}{\sum_{r=1}^{R} c_r y_{r0}^A + \sum_{s=1}^{S} v_s^B x_{s0}^B}]$$

subject to

$$\frac{\sum_{r=1}^{R} c_r y_{rj}^A}{\sum_{i=1}^{I} v_i^A x_{ij}^A} \leq 1 \qquad j=1,\ldots,N \qquad (14)$$

$$\frac{\sum_{t=1}^{T} u_t^B y_{tj}^B}{\sum_{r=1}^{R} c_r y_{rj}^A + \sum_{s=1}^{S} v_s^B x_{sj}^B} \leq 1 \qquad j=1,\ldots,N$$

$$c_r, u_t^B, v_i^A, v_s^B \geq 0$$
$r=1,\ldots,R,\ t=1,\ldots,T,\ i=1,\ldots,I,\ s=1,\ldots,S$

We call model (14) the cooperative efficiency evaluation model, because it maximizes the joint efficiency of the buyer and seller, and forces the two players to agree on a common set of weights on the intermediate measures[4].

We apply the following Charnes-Cooper transformation to model (14):

$$t_1 = \frac{1}{\sum_{i=1}^{I} v_i^A x_{i0}^A} \qquad t_2 = \frac{1}{\sum_{r=1}^{R} c_r^A y_{r0}^A + \sum_{s=1}^{S} v_s^B x_{s0}^B}$$

$$\omega_i^A = t_1 v_i^A \qquad c_r^A = t_1 c_r \qquad \mu_t^B = t_2 u_t^B$$

$$\omega_s^B = t_2 v_s^B \qquad c_r^B = t_2 c_r$$

$r=1,\ldots,R,\ t=1,\ldots,T,\ i=1,\ldots,I,\ s=1,\ldots,S$

Note that in the above transformation, $c_r^A = t_1 c_r$ and $c_r^B = t_2 c_r$ imply a linear relationship between c_r^A and c_r^B. Therefore, we can assume $c_r^B = k \times c_r^A$, $k \geq 0$. Then model (14) can be changed into:

---

[4] In cooperative game theory, the joint profit of seller and buyer is maximized.

Maximize $\frac{1}{2}[\sum_{r=1}^{R} c_r^A y_{r0}^A + \sum_{t=1}^{T} \mu_t^B y_{t0}^B]$

subject to

$$\sum_{i=1}^{I} \omega_i^A x_{ij}^A - \sum_{r=1}^{R} c_r^A y_{rj}^A \geq 0 \qquad j = 1, \ldots, N$$

$$\sum_{r=1}^{R} c_r^B y_{rj}^A + \sum_{s=1}^{S} \omega_s^B x_{sj}^B - \sum_{t=1}^{T} \mu_t^B y_{tj}^B \geq 0 \qquad j = 1, \ldots, N \qquad (15)$$

$$\sum_{i=1}^{I} \omega_i^A x_{i0}^A = 1$$

$$\sum_{r=1}^{R} c_r^B y_{r0}^A + \sum_{s=1}^{S} \omega_s^B x_{s0}^B = 1$$

$$c_r^B = k \times c_r^A$$

$$c_r^A, c_r^B, \mu_t^B, \omega_i^A, \omega_s^B, k \geq 0$$

$r=1, \ldots, R, t=1, \ldots, T, i = 1, \ldots, I, s = 1, \ldots, S$

Model (15) is a non-linear programming problem, and can be converted into the following model:

Maximize $\frac{1}{2}[\sum_{r=1}^{R} c_r^A y_{r0}^A + \sum_{t=1}^{T} \mu_t^B y_{t0}^B]$

subject to

$$\sum_{i=1}^{I} \omega_i^A x_{ij}^A - \sum_{r=1}^{R} c_r^A y_{rj}^A \geq 0 \qquad j = 1, \ldots, N$$

$$\sum_{r=1}^{R} k \times c_r^A y_{rj}^A + \sum_{s=1}^{S} \omega_s^B x_{sj}^B - \sum_{t=1}^{T} \mu_t^B y_{tj}^B \geq 0 \qquad j = 1, \ldots, N \qquad (16)$$

$$\sum_{i=1}^{I} \omega_i^A x_{i0}^A = 1$$

$$\sum_{r=1}^{R} k \times c_r^A y_{r0}^A + \sum_{s=1}^{S} \omega_s^B x_{s0}^B = 1$$

$$c_r^A, \mu_t^B, \omega_i^A, \omega_s^B, k \geq 0$$

$r=1, \ldots, R, t=1, \ldots, T, i = 1, \ldots, I, s = 1, \ldots, S$

Note that

$\sum_{r=1}^{R} k \times c_r^A y_{r0}^A + \sum_{s=1}^{S} \omega_s^B x_{s0}^B = 1$, $\sum_{r=1}^{R} c_r^A y_{r0}^A \leq 1$, $\sum_{s=1}^{S} \omega_s^B x_{s0}^B > 0$ in model (15). We have $k = (1 - \sum_{s=1}^{S} \omega_s^B x_{s0}^B) / \sum_{r=1}^{R} c_r^A y_{r0}^A < 1 / \sum_{r=1}^{R} c_r^A y_{r0}^A$.
Note also that the optimal $\sum_{r=1}^{R} c_r^{A*} y_{r0}^A$ in model (15) will not be less than E_{BA}^* in model (10). Thus, we have $0 \leq k < 1/E_{BA}^*$. That is, model (15) can be treated as a parametric linear program, and we can obtain a heuristic search solution using the procedure developed for models (6) and (13).

At the optima, let $\theta_A^* = \sum_{r=1}^{R} c_r^{A*} y_{r0}^A$ and $\theta_B^* = \sum_{t=1}^{T} \mu_t^{B*} y_{t0}^B$ represent the efficiency scores for the seller and buyer respectively. The following two remarks show that in general, the supply chain efficiency under the assumption of cooperation will not be less than the efficiency under the assumption of non-cooperation.

Remark 1: If we set $\sum_{r=1}^{R} c_r^A y_{r0}^A = E_{AA}^*$ as a constraint in model (15), then the feasible region of model (15) is the same as that of model (6). Therefore, $V_P^* = e_{AB}$.

Remark 2: If we set $\sum_{t=1}^{T} \mu_t^B y_{t0}^B = E_{BB}^*$ as a constraint in model (15), then the feasible region of model (15) is the same as that of model (13). Therefore, $V_P^* = e_{BA}$.

We consider again the numerical example in Table 11-2. Table 11-4 reports the results from model (15), where columns 2 and 3 report the efficiency scores for the seller and buyer respectively, and the last column reports the optimal value to model (15), the supply chain efficiency.

Table 11-4. Cooperative structure results

DMU	θ_A^*	θ_B^*	Supply chain
1	1	1	1
2	0.766	0.875	0.821
3	0.667	0.95	0.808
4	1	0.653	0.827
5	0.621	0.756	0.688

Table 11-5. Comparison of Non-cooperative and Cooperative results

DMU	e_{AB}	e_{BA}	Model (15)
1	1	1	1
2	0.697	0.821	0.821
3	0.808	0.773	0.808
4	0.827	0.827	0.827
5	0.547	0.688	0.688

Table 11-5 compares the efficiency scores for the cooperative and non-cooperative assumptions. In this numerical example, for all DMUs, one of the two leader-follower models achieves efficiency under the cooperative assumption. This indicates that no better solution can be found to yield a higher efficiency in the cooperative assumption. However, in other examples, the supply chain is likely to show a better performance when assuming cooperative operation.

4.3 The Cooperative Model

Although the above discussion is based upon two supply chain members, the models can be extended to include multiple supply chain members. Suppose there are N supply chains and each has P supply chain members as described in Figure 11-3, where $P \geq 3$.

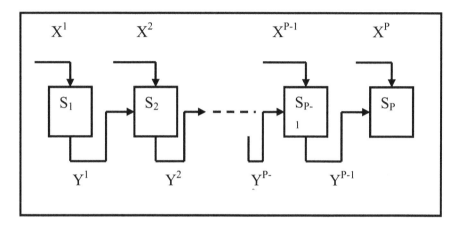

Figure 11-3. Multi-member supply chain

In the j-th supply chain, S_{dj} means the d-th member, X^{dj} means inputs of member d, Y^{dj} are outputs of member d and also inputs of member $d+1$. Here, $X^{dj}=(x_i^{dj}, i = 1, ..., I_d)$, $Y^{dj}=(y_s^{dj}, s = 1, ..., S_d)$, $d = 1, ..., P, j = 1, ..., N$. Suppose the weight of x_i^{dj} is v_i^d and the weight of y_s^{dj} is u_s^d, then we have the following model:

Maximize
$$\frac{1}{P}[\frac{\sum_{s=1}^{S_d} u_s^d y_s^{d0}}{\sum_{i=1}^{I_d} v_i^d x_i^{d0}}(d=1) + \sum_{d=2}^{P} \frac{\sum_{s=1}^{S_d} u_s^d y_s^{d0}}{\sum_{i=1}^{I_d} v_i^d x_i^{d0} + \sum_{s=1}^{S_{d-1}} u_s^{d-1} y_s^{(d-1)0}}]$$

subject to
$$\frac{\sum_{s=1}^{S_d} u_s^d y_s^{dj}}{\sum_{i=1}^{I_d} v_i^d x_i^{dj}} \leq 1 \qquad d=1, \qquad j = 1, ..., N \qquad (17)$$

$$\frac{\sum_{s=1}^{S_d} u_s^d y_s^{dj}}{\sum_{i=1}^{I_d} v_i^d x_i^{dj} + \sum_{s=1}^{S_{d-1}} u_s^{d-1} y_s^{(d-1)j}} \leq 1 \qquad d = 2, ..., P, j = 1, ..., N$$

$v_i^d, u_s^d \geq 0$

$i = 1, \ldots, I_d, s = 1, \ldots, S_d, d = 1, \ldots, P$

By using the following Charnes-Cooper transformation:

$$t_1 = \frac{1}{\sum_{i=1}^{I_d} v_i^d x_i^{d0}} \quad t_d = \frac{1}{\sum_{i=1}^{I_d} v_i^d x_i^{d0} + \sum_{s=1}^{S_{d-1}} u_s^{d-1} y_s^{(d-1)0}} \quad (d = 2, \ldots, P)$$

$\omega_i^d = t_d \times v_i^d \ (d = 1, \ldots, P)$

$\mu_s^d = t_d \times u_s^d \ (d = 1, \ldots, P)$

$c_s^{d-1} = t_d \times u_s^{d-1} \ (d = 2, \ldots, P)$

model (17) can be transformed into the following model:

Maximize $\frac{1}{P}[\sum_{s=1}^{S_d} \mu_s^d y_s^{d0} (d=1) + \sum_{d=2}^{P} \sum_{s=1}^{S_d} \mu_s^d y_s^{d0}]$

subject to

$\sum_{i=1}^{I_d} \omega_i^d x_i^{dj} - \sum_{s=1}^{S_d} \mu_s^d y_s^{dj} \geq 0 \qquad d=1, \qquad j=1, \ldots, N$

$\sum_{i=1}^{I_d} \omega_i^d x_i^{dj} + \sum_{s=1}^{S_{d-1}} c_s^{d-1} y_s^{(d-1)j} - \sum_{s=1}^{S_d} \mu_s^d y_s^{dj} \geq 0$ (18)

$\qquad\qquad\qquad\qquad\qquad\qquad\qquad d = 2, \ldots, P, j = 1, \ldots, N$

$\sum_{i=1}^{I_d} \omega_i^d x_i^{d0} = 1, \qquad\qquad\qquad\qquad d=1$

$\sum_{i=1}^{I_d} \omega_i^d x_i^{d0} + \sum_{s=1}^{S_{d-1}} c_s^{d-1} y_s^{(d-1)0} = 1 \qquad d = 2, \ldots, P$

for $\forall s, \mu_s^d = h_d \times c_s^d \qquad\qquad\qquad d = 2, \ldots, P$

$\omega_i^d, \mu_s^d, c_s^d, h_d \geq 0 \quad i = 1, \ldots, I_d, s = 1, \ldots, S_d, d = 1, \ldots, P$

5. CONCLUSIONS

The current chapter presents several DEA models for evaluating the performance of a supply chain and its members. The non-cooperative model is modeled as a leader-follower structure, where in our case, the leader is first evaluated by the regular DEA model, and then the follower is evaluated using the leader-optimized weights on the intermediate measures. The cooperative model tries to maximize the joint efficiency of the seller and buyer, and imposes weights on the intermediate measures that are the same when applied to the measures as outputs of the supplier as when applied to those measures as inputs of the buyer. Although the models are nonlinear programming problems, they can be solved as parametric linear programming problems, and a best solution can be found using a heuristic technique.

We point out that although the current chapter uses the concept of a cooperative game, it does not try to examine whether the members of a specific supply chain are behaving in a cooperative or non-cooperative manner. This is a topic for further research. Specifically, the current chapter is not an empirical one and we, therefore, do not pursue these opportunities. Instead, a simple numerical example has been used to demonstrate the theoretical contributions of the current chapter. Other useful theoretical developments include the idea that one echelon can use knowledge about another echelons (supplier or customer), to improve it own performance or the mutual performance of the members. This is consistent with the theory by games as it applies, for example, to the 'bullwhip effect' such as in the earlier-referenced beer game.

REFERENCES

1. Charnes A., W.W. Cooper, and E. Rhodes, 1978, Measuring the efficiency of decision making units, *European Journal of Operational Research*, 2(6), 428-444.
2. Chen, Y. and J. Zhu, 2004, Measuring information technology's indirect impact on firm performance, *Information Technology & Management Journal*, 5(1-2), 9-22.
3. Färe, R. and S. Grosskopf, 2000, Network DEA. *Socio-Economic Planning Sciences*, 34, 35–49.
4. Golany, B., S.T. Hackman and U. Passy, 2006, An efficiency measurement framework for multi-stage production systems, *Annals of Operations Research*, 145(1), 51-68.
5. Huang, Z., 2000, Franchising cooperation through chance cross-constrained games, *Naval Research Logistics*, 47, 669-685.
6. Lee, H.L. and C. Billington, 1992, Managing supply chain inventory: pitfalls and opportunities, *Sloan Management Review* 33(3), 65-73.
7. Li, S.X., Z. Huang and A. Ashley, 1995, Seller-buyer system co-operation in a monopolistic market, *Journal of the Operational Research Society*, 46, 1456-1470.
8. Li, S.X., Z. Huang and A. Ashley, 1996, Improving buyer-seller system cooperation through inventory control, *International Journal of Production Economics*, 43, 37-46.

9. Liang, L. F. Yang, W. D. Cook and J. Zhu, 2006, DEA models for supply chain efficiency evaluation, *Annals of Operations Research*, 145(1), 35-49.
10. Porter, M.E., 1974, Consumer behavior, retailer power and market performance in consumer goods industries, *Review of Econom. Statist.* LVI, 419-436.
11. Seiford, L.M. and J. Zhu, 1999, Profitability and marketability of the top 55 US commercial banks, *Management Science*, 45(9), 1270-1288.
12. Simaan, M. and J.B. Cruz, 1973, On the Stackelberg strategy in nonzero-sum games, *Journal of Optimization Theory and Applications* 11, 533-555.
13. Weber, C.A., and A. Desai, 1996, Determinants of paths to vendor market efficiency using parallel coordinates representation: a negotiation tool for manufacturers, *European Journal of Operational Research* 90, 142–155.
14. Zhu, J., 2003, *Quantitative Models for Performance Evaluation and Benchmarking: Data Envelopment Analysis with Spreadsheets*. Kluwer Academic Publishers, Boston.

This chapter is based upon Liang, L. F. Yang, W. D. Cook and J. Zhu, 2006, DEA models for supply chain efficiency evaluation, *Annals of Operations Research*, Vol. 145, No. 1, 35-49.

Chapter 12

NETWORK DEA

Rolf Färe[1], Shawna Grosskopf[2] and Gerald Whittaker[3]

[1] *Department of Economics and Department of Agricultural Economics, Oregon State University, Corvallis, OR 97331 USA*

[2] *Department of Economics, Oregon State University, Corvallis, OR 97331 USA, shawna.grosskopf@orst.edu*

[3] *National Forage Seed Production Research Center, Agricultural Research Sevice, USDA, Corvallis, OR 97331 USA*

Abstract: This chapter describes network DEA models, where a network consists of sub-technologies. A DEA model typically describes a technology to a level of abstraction necessary for the analyst's purpose, but leaves out a description of the sub-technologies that make up the internal functions of the technology. These sub-technologies are usually treated as a "black box", i.e., there is no information about what happens inside them. The specification of the sub-technologies enables the explicit examination of input allocation and intermediate products that together form the production process. The combination of sub-technologies into networks provides a method of analyzing problems that the traditional DEA models cannot address. We apply network DEA methods to three examples; a static production technology with intermediate products, a dynamic production technology, and technology adoption (or embodied technological change). The data and GAMS code for two examples of network DEA models are listed in appendices.

Key words: Data Envelopment Analysis (DEA), Network, Intermediate Products, Dynamic Production, Technology Adoption.

1. INTRODUCTION

Many production systems (technologies) may be conceptualized as the joint, interacting action of a finite number of production sub-technologies called Activities. *(Shephard and Färe, 1975 p. 43)*

In this chapter we will apply the idea of sub-technologies to enter into the "black box" of DEA. We will discuss three "black boxes": *i)* a static production technology *ii)* a dynamic production technology and *iii)* technology adoption, *i.e.* embodied technical change.

Production of electricity by a coal-fired plant can be seen as a static network model. One sub-technology produces electricity and sulfur dioxide (SO_2), where electricity is a final output and sulfur dioxide is a by-product which may then be an input into a second technology, pollution abatement. We can think of the optimization problem of this power plant that uses the sub-technologies of electricity production and pollution abatement as a network model. This facilitates the explicit analysis of allocation across sub-technologies of both inputs (e.g., labor) and intermediate products (SO_2) that also serve as inputs.

The discrete (Ramsey, 1928) model can be formulated as a dynamic network model. Using potato production as an example application, a network is constructed where one year of potato production may be either consumed as a final output or saved to become an input for the next growing season. The inter-temporal dependence can be used to study optimal savings.

Our third network model, technology adoption, can be formulated on the basis of two machines, each with different technologies. These sub-technologies are modeled as a network where inputs are allocated among the machines, and hence determine which technology will be adopted. This is more generally known in economics as embodied technical change.

In the chapter, we will discuss the theoretical content of each of the three "black boxes" described above, and present simple empirical examples in a DEA or Activity Analysis framework. We apply the network DEA methods to three examples; a static production technology with intermediate products, a dynamic production technology, and technology adoption (or embodied technological change). The data and GAMS code for two examples of network DEA models are listed in appendices.

2. STATIC NETWORK MODEL

Here we introduce the static network model. It consists of a finite set of sub-technologies (or activities) that are connected to form a network. This network model enables us to study the processes that usually remain hidden within the "black box" of DEA. The static model is particularly useful for analyzing the allocation of intermediate products. This model also provides the basic structure for the dynamic model and the technology adoption model discussed in the following sections. We end the section by analyzing property rights specified using a network model.

We restrict our presentation to three sub-technologies P^1, P^2, and P^3. These three sub-technologies are connected by the directed network shown in Figure 12-1.

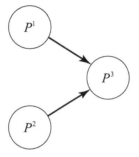

Figure 12-1. Representation of sub-technologies

To complete a network of these three sub-technologies, we add a distribution process and a sink, or collection of final outputs[1]. Inputs are denoted by $x = (x_1, \ldots, x_N) \in \Re_+^N$, the network exogenous vector, i.e., total availability is attached to the three sub-technologies, $_0^i x, i = 1, 2, 3\ldots$ where 0 denotes source and i denotes use (Figure 12-2). For example, $_0^1 x$ is the input vector from source 0 used in activity 1. The total amounts used in the three activities cannot exceed the total amount available, i.e.,

$$x \geq \sum_{i=1}^{3} {}_0^i x.$$

In Figure 12-2 activity P^1 uses $_0^1 x$ as an exogenous input and produces

$$_1^3 y + _1^4 y$$

---

[1] This model is adopted from Färe and Grosskopf (1996).

as outputs. The $_1^3y$ is an input into activity 3, while $_1^4y$ is the final output from P^1. Activity P^3 uses $_0^3x$ as an exogenous input and $_1^3y$, $_2^3y$ as intermediate inputs. The final product from activity P^3 is output vector $_3^4y$. The total network output is the sum of the final output of the three activities,

$$_1^4y + {}_2^4y + {}_3^4y \ .$$

Note that even if an activity does not produce one of the listed outputs, that output is set at zero and the network structure remains the same. It is also noteworthy that some outputs may be both final and intermediate outputs, e.g. spare parts.

Based on the description above, a generic network model takes the form

$$P(x) = \left\{ \left({}_1^4y + {}_2^4y + {}_3^4y \right) : \right.$$
$$\left({}_1^4y + {}_1^3y \right) \in P^1 \left({}_0^1x \right)$$
$$\left({}_2^4y + {}_2^3y \right) \in P^2 \left({}_0^2x \right) \quad (12.1)$$
$$_3^4y \in P^3 \left({}_0^3x, {}_1^3y + {}_2^3y \right)$$
$$\left. {}_0^1x + {}_0^2x + {}_0^3x \leq x \right\} \ ,$$

where $P^i(\bullet)\ i = 1, 2, 3$ are output sets, i.e., $P(x) = \{y : x \text{ can produce } y\}$. Thus the network model $P(x)$ is formed by the individual sub-technologies.

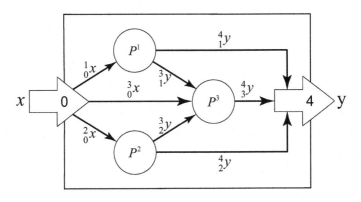

Figure 12-2. A network technology

A variety of estimations can be accomplished using $P(x)$ as the reference technology, including

i) profit maximization: let $p \in \Re_+^M$ be output prices and $w \in \Re_+^N$ be input prices. Then profit maximization is

$$\max py - wx \quad s.t. \quad y \in P(x)$$

ii) revenue maximization:

$$\max py \quad s.t. \quad y \in P(x)$$

iii) directional output distance function estimation: let $g \in \Re_+^M, g \neq 0$ be a directional vector. Then the directional output distance function is estimated as

$$\max \beta \quad s.t. \quad (y + \beta g) \in P(x)$$

iv) output distance function estimation:

$$\min \lambda \quad s.t. \quad y/\lambda \in P(x).$$

The DEA network model is obtained by formulating a DEA or Activity Analysis model for each activity.

Recall that if there are $k = 1,\ldots,K$ observations of inputs and outputs (x^k, y^k), then a DEA output set is

$$P(x) = \{ \ y : \sum_{k=1}^{k} z_k y_{km} \geq y_m, \quad m = 1,\ldots,M$$

$$\sum_{k=1}^{k} z_k y_{kn} \leq x_n, \quad n = 1,\ldots,N \quad (12.2)$$

$$z_k \geq 0, \quad k = 1,\ldots,K \ \}.$$

The z_k are the intensity variables, here non-negative, making $P(x)$ homogeneous of degree one, i.e., it satisfies constant returns to scale (CRS), $P(\lambda x) = \lambda P(x), \lambda > 0$. By restricting the intensity variables as

$$\sum_{n=1}^{K} z_k \leq 1,$$

the model will exhibit non-increasing returns to scale (NIRS), $P(\lambda x) \subseteq \lambda P(x)$. If the sum is constrained to equal one, variable returns to scale (VRS) are obtained.

The following restrictions on the data are imposed [2]

$$
\begin{aligned}
&(i) & \sum_{k=1}^{K} y_{km} &> 0 & m &= 1,\ldots,M \\
&(ii) & \sum_{m=1}^{M} y_{km} &> 0 & k &= 1,\ldots,K \\
&(iii) & \sum_{k=1}^{K} x_{kn} &> 0 & n &= 1,\ldots,N \\
&(iv) & \sum_{n=1}^{N} x_{kn} &> 0 & k &= 1,\ldots,K
\end{aligned}
\qquad (12.3)
$$

Output condition (i) states that each output is produced by some k (DMU), and (ii) requires that each activity produce some output. The input conditions (iii) and (iv) say that each input is used by some k and that each k uses at least one input. When the DEA model in (12.2) satisfies the conditions in (12.3), one can prove that it has the following properties[3],

A.1 $P(0) = \{0\}$; no free lunch.

A.2 $P(x)$ is bounded for each $x \in \Re_+^N$; scarcity.

A.3 $x \geq x^0 \Rightarrow P(x) \supseteq P(x^0)$; free disposability of inputs.

A.4 $y \leq y^0, y^0 \in P(x) \Rightarrow y \in P(x)$; free disposability of outputs

A.5 $P(x)$ is a closed set for each $x \in \Re_+^N$.

A.6 $P(\lambda x) = \lambda P(x)$; constant returns to scale.

If each of the sub-technologies satisfies A.1 - A.6 one can show that the network technology $P(x)$ also satisfies these conditions[4]. Clearly the network model allows for a number of sub-technologies, and is not limited to those described above.

Next we give a numerical example how the output distance function (iv) can be estimated in the network model (12.1). The data for the three sub-technologies P^1, P^2, and P^3 are given in Tables 12-1, 12-2 and 12-3.

The sub-technology P^1 consists of three observations. These observations of the sub-technology are referred to as sub-DMU's in this chapter. As usual, these willl form the technology P^1.

---

[2] These conditions are due to Kemeny, Morgenstern and Thompson Kenmeny, J. G., Morgenstern, O., and Thompson, G. L. (1956). "A generalization of the von Neumann model of an expanding economy." *Econometrica*, 24, 115-135.

[3] See Shephard (1970)

[4] Färe and Grosskopf (1996)

Table 12-1. Data for sub-technology P^1.

Sub-DMU	Input 1 $_0^1x_1$	Input 2 $_0^1x_2$	Output $_1^4y + _1^3y$
1	1	2	4
2	2	1	3
3	2	2	5

Observations for the second example sub-technology are set out in Table 12-2.

Table 12-2. Data for sub-technology P^2.

Sub-DMU	Input 1 $_0^2x_1$	Input 2 $_0^2x_2$	Output $_2^3y + _2^4y$
1	2	3	4
2	3	2	3
3	3	3	5

Table 12-3. Data for sub-technology P^3

Sub-DMU	Input 1 $_0^3x_1$	Input 2 $_0^3x_2$	Intermediate Input $_1^3y$	Intermediate Input $_2^3y$	Output $_3^4y$
1	3	2	2	2	2
2	4	2	1	2	3
3	4	3	3	2	2

Table 12-4. Data for the network technology

Sub-DMU	Input 1 x_1	Input 2 x_2	Output 1 $_1^4y$	Output 2 $_2^4y$	Output 3 $_3^4y$
1	6	7	2	2	2
2	9	5	2	1	3
3	9	8	2	3	4

As shown in Figure 12-2, the sub-technology P^3 uses outputs from the sub-technologies P^1 and P^2 as inputs.

In addition to the three sub-technologies we have the network technology data (Figure 12-4).

To understand how this table relates the data of the three sub-technologies, consider DMU 1. Input 1 for DMU 1 is the sum $({}_0^1x_1 + {}_0^2x_1 + {}_0^3x_1) = (1+2+3) = 6$. Output 1 for DMU 1 is $y_1 = (({}_1^4y + {}_1^3y) - {}_1^3y) = (4-2) = 2$, and so on for the other elements of the network.

Next let us specify the DEA model that solves for the output distance function for DMU 1. This problem is

$$(1/\lambda_1) = \max \gamma_1$$

s.t.

Sub-technology 1

$$\gamma_1 \bullet 2 + {}_1^3y \leq z_1^1 \bullet 4 + z_2^1 \bullet 3 + z_3^1 \bullet 5$$
$$z_1^1 \bullet 1 + z_2^1 \bullet 2 + z_3^1 \bullet 2 \leq {}_0^1x_1$$
$$z_1^1 \bullet 2 + z_2^1 \bullet 1 + z_3^1 \bullet 2 \leq {}_0^1x_2$$
$$z_1^1 \geq 0, z_2^1 \geq 0, z_3^1 \geq 0$$

Sub-technology 2

$$\gamma_1 \bullet 2 + {}_2^3y \leq z_1^2 \bullet 4 + z_2^2 \bullet 3 + z_3^2 \bullet 5$$
$$z_1^2 \bullet 2 + z_2^2 \bullet 3 + z_3^2 \bullet 3 \leq {}_0^2x_1$$
$$z_1^2 \bullet 3 + z_2^2 \bullet 2 + z_3^2 \bullet 3 \leq {}_0^2x_2$$
$$z_1^2 \geq 0, z_2^2 \geq 0, z_3^2 \geq 0$$

Sub-technology 3

$$\gamma_1 \bullet 4 \leq z_1^3 \bullet 2 + z_2^3 \bullet 3 + z_3^3 \bullet 4$$
$$z_1^3 \bullet 3 + z_2^3 \bullet 4 + z_3^3 \bullet 4 \leq {}_0^3x_1$$
$$z_1^3 \bullet 2 + z_2^3 \bullet 2 + z_3^3 \bullet 3 \leq {}_0^3x_2$$
$$z_1^3 \bullet 2 + z_2^3 \bullet 1 + z_3^3 \bullet 3 \leq {}_1^3y$$
$$z_1^3 \bullet 2 + z_2^3 \bullet 2 + z_3^3 \bullet 2 \leq {}_2^3y$$

Total Input

$$ {}_0^1x_1 + {}_0^2x_1 + {}_0^3x_1 \leq 6$$
$$ {}_0^1x_2 + {}_0^2x_2 + {}_0^3x_2 \leq 7$$

In Appendix I, we provide GAMS programs, including the data sets and results, that show how this type of problem can be solved.

A DEA network model less general than that presented here was used by Lewis and Sexton (2004) to study major league baseball. There they "demonstrate the advantages of the Network DEA Model over the standard DEA Model. Specifically, the Network DEA Model can detect inefficiencies that the standard DEA Model misses" by allowing examination of the individual stages of the production process. Färe and Whittaker (1996) use the Network DEA model to analyze the efficiency of cattle producers with permits for grazing on public land. The network in that study had sub-technologies for grain and cattle production, and for different time periods.

Before we provide a simple empirical example, let us look at how this model was used by Färe, Grosskopf and Lee (2004) to analyze property rights. We assume there are two firms modeled as P^1 and P^2. The upstream firm of P^1 produces two outputs using input x^1. One output is a marketed commodity y^1 (paper, for example) and the other an undesirable externality $_1^2u$ (waste water, for example). The downstream firm P^2 produces a marketed output y^2 using x^2 and $_1^2u$ as inputs. This makes the downstream firm's production dependent on the upstream firm. This model is illustrated in Figure 12-3.

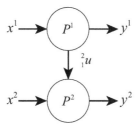

Figure 12-3. The Network model with externalities

Some additional assumptions need to be introduced to complete the model. First we assume that y^1 and $_1^2u$ are nulljoint, i.e.,

$$\text{if } (y^2) \in P^2(x^2, {}_1^2u) \text{ and } {}_1^2u = 0 \text{ then } y^2 = 0.$$

In words this condition requires that if no externality is produced, then no marketed output can be produced. Or, conversely, to produce y^1, some externality $_1^2u$ (also called a bad) also is produced. The externality $_1^2u$ is assumed to be a pollutant that affects the downstream firm; i.e.,

$${}^2_1u' \geq {}^2_1u \Rightarrow P^2\left(x^2, {}^2_1u'\right) \subseteq P^2\left(x^2, {}^2_1u\right).$$

To evaluate the effects of assigning the property rights and the effect of internalizing the externality we can compute the profit under different arrangements. Assume that the upstream firm has the property right, then this firm maximizes its profit thus [5];

$$\max_{(x^1, y^1, {}^2_1u)} p^1 y^1 - w^1 x^1 \quad \text{s.t.} \quad (p^1, {}^2_1u) \in P^1(x^1) \tag{12.4}$$

Equation (12.4) gives us a solution vector including the output of pollutant ${}^2_1u^*$. The downstream firm must use ${}^2_1u^*$ as an input, and so solves the problem

$$\max_{(x^2, y^2)} p^2 y^2 - w^2 x^2 \quad \text{s.t.} \quad y^2 \in P^2(x^2, {}^2_1u^*) \tag{12.5}$$

If instead the downstream firm is assigned the property right, then it solves the problem

$$\max_{(x^2, y^2, {}^2_1u)} p^2 y^2 - w^2 x^2 \quad \text{s.t.} \quad y^2 \in P^2(x^2, {}^2_1u). \tag{12.6}$$

Now ${}^2_1u$ may take the value ${}^2_1u^{**}$, which the upstream has to take as its parameter in solving

$$\max_{(x^1, y^1)} p^1 y^1 - w^1 x^1 \quad \text{s.t.} \quad (p^1, {}^2_1u^{**}) \in P^1(x^1). \tag{12.7}$$

The solutions of (12.4)-(12.7) may be compared to the case of internalizing the cost of disposing of the pollutant. To accomplish this we solve

$$\max_{(x^1, y^1, x^2, y^2, {}^2_1u)} p^1 y^1 - w^1 x^1 + p^2 y^2 - w^2 x^2$$
$$\text{s.t.} \quad (p^1, {}^2_1u) \in P^1(x^1) \tag{12.8}$$
$$y^2 \in P^2(x^2, {}^2_1u).$$

---

[5] We cannot use the CRS model here, since it only yields zero profit. Thus the NIRS or the VRS model should be used.

By comparing the five problems (12.4)-(12.8), one may evaluate the assignment of property rights versus internalization of the externality[6]. Next we use sub-technologies to specify a dynamic model.

3. DYNAMIC NETWORK MODEL

Let us start by formulating a dynamic model as a network. Assume there are three time periods $t-1$, t, $t+1$, and that each has its production technology P^τ, $\tau = t-1, t, t+1$. A dynamic model has the property that a decision in one time period impacts on later time periods. For example, if I save now, then my possible consumption may increase later. Therefore we introduce intermediate products, i.e., those products that are held over between time periods, $_\tau^{\tau+1}y \in \Re_+^M$. If $\tau = 1$, then $_t^{t+1}y$ is the intermediate vector of outputs produced at time t and entering the production process at $t+1$, i.e., an input at P^{t+1}. Figure 12-4 illustrates this setup.

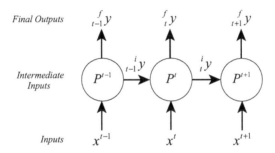

Figure 12-4. Dynamic sub-technologies

Each of the production sub-technologies P^τ uses exogenous inputs x^τ to produce the final output $_{\tau-1}^\tau y$ and intermediate inputs $_{\tau-1}^\tau y$. To complete the network model (see Figure 12-5) we add initial conditions (distribution process in the static model) and transversality conditions (sink in the static model).

---

[6] Of course, this simple model can be extended by the addition of more variables, firms, and sub-technologies.

The initial condition is given by ${}^i\bar{y}$ and may be thought of as the stocks of "capital" initially available. The transversality condition could include the number of periods, say $t+1=T$, the state of the system at T, ${}_{t+1}^i\bar{y}$, and the final output vector from the last period, ${}_t^f\bar{y}$. The chosen conditions are specific to the research problem to be analyzed.

Thus far we have not introduced discounting. However, if outputs are given in value terms, a discount factor of $\delta^\tau, 0 \leq \delta^\tau \leq 1$, may be introduced to account for the lesser of future income compared to present income. For example, $\delta^t {}_t^f y$ are the discounted values of the final output in period t.

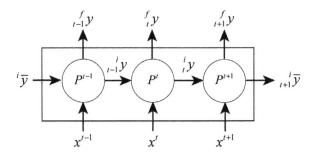

Figure 12-5. A dynamic network model

The dynamic network DEA model consists of the interaction of a finite number of static models. Therefore let us start by studying the sub-technology P^t from above. This technology uses inputs x^t and intermediate inputs ${}_{t-1}^i y$ to produce outputs ${}_t^f y + {}_t^i y$, where ${}_t^f y$ is the final output and ${}_t^i y$ is the intermediate output that is used as an input in the next period. Thus we may write

$$P^t\left(x^t, {}_{t-1}^i y\right) = \{\left({}_t^f y + {}_t^i y\right):$$

$$\left({}_t^f y_m + {}_t^i y_m\right) \leq \sum_{k=1}^{K^t} z_k^t \left({}_t^f y_{km} + {}_t^i y_{km}\right) \quad m=1,\ldots,M,$$

$$\sum_{k=1}^{K^t} z_k^t {}_{t-1}^i y_{km} \leq {}_{t-1}^i y_{k'm}, \quad m=1,\ldots,M, \quad (12.9)$$

$$\sum_{k=1}^{K^t} z_k^t x_{km}^t \leq x_{k'n}, \quad n=1,\ldots,N,$$

$$z_k^t \geq 0, \quad k=1,\ldots,K^t\},$$

where ${}_t^f y_{km}, {}_t^i y_{km}, {}_{t-1}^i y_{km}$, and x_{kn} are observed inputs and outputs. We allow the number of observations to differ between periods, hence the notation K^t.

The output vector $\left({}_{t}^{f}y + {}_{t}^{i}y\right)$ is specified so that one can include M^{t} for all t, i.e., $M = M^{t-1} + M^{t} + \ldots$ by including appropriate zeroes.

Recall from the static model, that if the sub-technologies have the properties A.1-A.6, then the whole network model also has those properties. This observation also applies to the dynamic model below:

$$P^{t}\left(x^{t-1}, x^{t}, x^{t+1}, {}^{i}\overline{y}\right) = \left\{\left({}_{t-1}^{f}y, {}_{t}^{f}y, ({}_{t+1}^{f}y + {}_{t+1}^{i}y)\right):\right.$$

$$\left({}_{t-1}^{f}y_{m} + {}_{t-1}^{i}y_{m}\right) \leq \sum_{k=1}^{K^{t-1}} z_{k}^{t-1}\left({}_{t-1}^{f}y_{km} + {}_{t-1}^{i}y_{km}\right) \quad m = 1,\ldots,M,$$

$$\sum_{k=1}^{K^{t-1}} z_{k}^{t-1}\, {}^{i}y_{km} \leq {}^{i}\overline{y}_{km}, \quad m = 1,\ldots,M, \quad (12.10)$$

$$\sum_{k=1}^{K^{t-1}} z_{k}^{t-1} x_{kn}^{t-1} \leq x_{n}^{t-1}, \quad n = 1,\ldots,N,$$

$$z_{k}^{t-1} \geq 0, \quad k = 1,\ldots,K^{t-1},$$

$$\left({}_{t}^{f}y_{m} + {}_{t}^{i}y_{m}\right) \leq \sum_{k=1}^{K^{t}} z_{k}^{t}\left({}_{t}^{f}y_{km} + {}_{t}^{i}y_{km}\right) \quad m = 1,\ldots,M,$$

$$\sum_{k=1}^{K^{t}} z_{k}^{t}\, {}_{t-1}^{i}y_{km} \leq {}_{t-1}^{i}\overline{y}_{m}, \quad m = 1,\ldots,M,$$

$$\sum_{k=1}^{K^{t}} z_{k}^{t} x_{kn}^{t} \leq x_{n}^{t}, \quad n = 1,\ldots,N,$$

$$z_{k}^{t} \geq 0, \quad k = 1,\ldots,K^{t},$$

$$\left({}_{t+1}^{f}y_{m} + {}_{t+1}^{i}y_{m}\right) \leq \sum_{k=1}^{K^{t+1}} z_{k}^{t+1}\left({}_{t+1}^{f}y_{km} + {}_{t+1}^{i}y_{km}\right) \quad m = 1,\ldots,M,$$

$$\sum_{k=1}^{K^{t+1}} z_{k}^{t+1}\, {}_{t}^{i}y_{km} \leq {}_{t}^{i}\overline{y}_{km}, \quad m = 1,\ldots,M,$$

$$\sum_{k=1}^{K^{t+1}} z_{k}^{t+1} x_{kn}^{t+1} \leq x_{n}^{t+1}, \quad n = 1,\ldots,N,$$

$$\left. z_{k}^{t+1} \geq 0, \quad k = 1,\ldots,K^{t+1}\right\},$$

Note that each sub-technology has its own intensity vector z^{τ}, $\tau = t-1, t, t+1$, and that the interaction between time periods comes through the intermediate outputs.

Färe and Grosskopf (1997) use this model to study the inefficiency of APEC countries due to dynamic misallocation of resources. They used the sum of Shephard (1970) sub-technology distance functions as their optimization criterion. Nemota and Gota (2003) applied the dynamic network model to study Japanese electricity production over time. They used cost minimization for the optimization criterion. Jaenicke (2000) applied the dynamic model in the analysis of the yield effects of crop rotation. Nemota and Gota (1999) applied the dual linear programming problem formulation to the cost minimization and derived the fundamental equation (Hamilton-Jacobi-Bellman) of dynamic programming.

4. TECHNOLOGY ADOPTION

We now turn to our simplest, yet quite powerful network model, the technology adoption model. Here we give two applications, one to "embodied technical change" and the other to permit allocation. Although the applications appear different, the basic approach is the same; allocation of one or more resources among two or more technologies. Figure 12-6 illustrates a simple version of the problem.

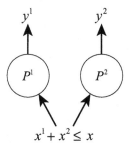

Figure 12-6. Distribution of resources among technologies

The two technologies P^1 and P^2 use inputs x^1 and x^2 respectively to produce outputs y^1 and y^2. The total amount of inputs that can be allocated among the technologies is x. In this illustration, to analyze permit allocation, for example, we would define one of the components of x as a permit for pollution emission.

Let us start with the case of technology adoption, where one of the technologies is vintage. The original vintage models due to Johansen (1959) and Solow (1960) were based on continuous time. Here we follow Färe and Grosskopf (1996) and discuss the discrete time version. We distinguish between two forms of inputs, durable and instantaneous. The durable inputs are vintage specific, and they form sub-technologies, P^1 and P^2 above. The instantaneous inputs are allocated between the vintage sub-technologies, x^1 and x^2 above.

The adoption of technology is generated by the reallocation of instantaneous inputs away from "old" vintages to new ones. Hence making the old obsolete and replaced by the newer vintages.

Assume there are $v = 1,\ldots,V$ vintages and K^v observations of outputs $y^{k,v} \in \Re_+^M$, durable inputs $X^{k,v} \in \Re_+^L$ and instantaneous inputs $x^{k,v} \in \Re_+^N$[7]. The vintage technology v at time t, $t \geq v$ has the following DEA formulation:

$$P^v\left(X^v(t), x^v(t)\right) = \{y^v(t) \in \Re_+^M :$$

$$y_m^v(t) \leq \sum_{k=1}^{K^v} z_k^v y_{km}^v, \qquad m = 1,\ldots,M,$$

$$X_l^v(t) \geq \sum_{k=1}^{K^v} z_k^v X_{kl}^v, \qquad l = 1,\ldots,L, \qquad (12.11)$$

$$x_n^v(t) \geq \sum_{k=1}^{K^v} z_k^l x_{kn}^v, \qquad n = 1,\ldots,N,$$

$$z_k^v \geq 0, \qquad k = 1,\ldots,K^V \}.$$

This vintage model has outputs, inputs both durable and instantaneous, strong disposability, and satisfies constant returns to scale (CRS). Other conditions may be substituted in the usual manner.

The technology is unchanged over time, y_{km}^v, X_{kl}^v and x_{kn}^v are independent of time. However, the durable inputs age or degrade over time and may lose their capabilities. We model this observation with an aging function $A_l(t-v)$ so that

$$X_l^V(t) = A_l(t-v) X_l^V, \qquad l = 1,\ldots,L.$$

Thus, as t increases, the effectiveness of $X_l^V(t)$ may decline as $A_l(t-v)$ decreases. Characteristics of aging are specific to each application, and we leave the properties of $A_l(t-v)$ to the researcher to address.

If there are V vintages, the multi-vintage model at time t is:

---

[7] This part is based on Färe and Grosskopf (1996).

$$P\left(X^1(t),\ldots,X^V(t),x(t)\right) = \left\{\sum_{v=1}^{V} y^v(t):\right.$$

$$y_m^v(t) \leq \sum_{k=1}^{K^v} z_k^v y_{km}^v, \qquad m=1,\ldots,M,$$

$$X_l^v(t) \geq \sum_{k=1}^{K^v} z_k^v X_{kl}^v, \qquad l=1,\ldots,L, \qquad (12.12)$$

$$x_n^v(t) \geq \sum_{k=1}^{K^v} z_k^v x_{kn}^v, \qquad n=1,\ldots,N,$$

$$z_k^v \geq 0, \qquad k=1,\ldots,K^V, v=1,\ldots,V$$

$$\left.\sum_{v=1}^{V} x_n^v(t) \leq x_n(t), \qquad n=1,\ldots,N\right\}.$$

Equation (12.12) may also be written as

$$P\left(X^1(t),\ldots,X^V(t),x(t)\right) = \left\{\sum_{v=1}^{V} P^v\left(X^v(t),x^v(t)\right):\right.$$
$$\left.\sum_{v=1}^{V} x^v(t) \leq x(t)\right\}. \qquad (12.13)$$

Equation (12.13) states that the multi-vintage model is the sum of the single-vintage models, with the instantaneous input allocated over all vintages. This model can be used in combination with any of the objective functions discussed in section 2.

Some prior information is needed to specify the permit allocation model. The technologies in question are assumed to produce desirable and undesirable outputs, as in our network model with externalities above. Therefore we assume that the output vector of "goods", $y \in \Re_+^M$, is nulljoint with the "bads" output vector $u \in \Re_+^3$, i.e.

$$\text{if } (y,u) \in P(x) \text{ and } u=0, \text{ then } y=0,$$

where $P(x)$ is the output set and $x \in \Re_+^N$ is an input vector. In addition we assume that y and u are weakly disposable, i.e.

$$\text{if } (y,u) \in P(x) \text{ and } 0 \leq \theta \leq 1, \text{ then } (\theta y, \theta u) \in P(x).$$

A typical output set of a polluting technology may then be illustrated for one good and one bad as in Figure 12-7.

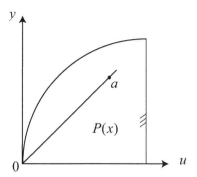

Figure 12-7. Polluting technology

The output set defined above is such that if $u = 0$, then for (y,u) to belong to it, y must also equal zero. That is y is nulljoint with u. In addition, for any element, say $a \in P(x)$, its proportional contraction $(\theta a,\ 0 \leq \theta \leq 1)$ also belongs to $P(x)$. Thus the technology has weakly disposable outputs (y,u).

To regulate the bad outputs u, we introduce a restriction on the bads, namely,

$$B(\bar{u}) = \{(y,u) : y \geq 0 \text{ and } u \leq \bar{u}\}.$$

The addition of these regulatory constraints to the polluting technology in Figure 12-7 is illustrated in Figure 12-8.

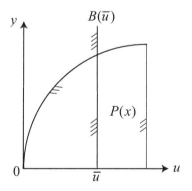

Figure 12-8. Regulated polluting technology

This diagram shows that the regulated technology is constrained by both the technology $P(x)$ and the regulation $B(\bar{u})$. Thus, for feasibility $(y,u) \in P(x) \cap B(\bar{u})$.

To study permit allocation, we specify a number of firms $f = 1,\ldots,F$, and a total amount of pollutant that society will allow, u^*. Then our model should allocate u^* optimally among the F firms so that the sum does not exceed u^*,

$$\sum_{f=1}^{F} u^f \leq \overline{u}, \quad \overline{u}, u^f \in \Re_+^J.$$

Here we follow Brännlund, Färe and Grosskopf (Brännlund et al., 1995) and assume that each firm is a profit maximizer. The first question we address is the calculation of the optimal allocation of u^* under profit maximization, $(u^{*1},\ldots,u^{*F})$. To find this allocation we solve

$$\Pi = \max_{(y,u,x)} \sum_{f=1}^{F} \left(\sum_{m=1}^{M} p_m y_m^f - \sum_{n=1}^{N} w_n x_n^f \right)$$

$$\text{s.t.} \quad y_m^f \leq \sum_{k=1}^{K^f} z_k^f y_{km}^f, \quad m=1,\ldots,M,$$

$$u_j^f = \sum_{k=1}^{K^f} z_k^f u_{kj}^f, \quad j=1,\ldots,J,$$

$$x_n^f = \sum_{k=1}^{K^f} z_k^f x_{kn}^f, \quad n=1,\ldots,N, \quad (12.14)$$

$$\sum_{k=1}^{K^f} z_k^f \leq 1, \quad z_k^f \geq 0, \quad k=1,\ldots,K^f$$

$$\sum_{f=1}^{F} u_j^f \leq \overline{u}_j, \quad j=1,\ldots,J,$$

$$f=1,\ldots,F.$$

Equation (12.14) consists of F sub-problems, one for each firm. These problems are connected by the pollution allocation expressions

$$\sum_{k=1}^{K^f} u_j^f = \overline{u}_j, \quad j=1,\ldots,J.$$

Each firm is here assumed to satisfy non-increasing returns to scale

$$\sum_{k=1}^{K^f} z_n^f \leq 1.$$

Thus, we have imposed the condition of no negative profits. This condition can be relaxed by imposing variable returns to scale, if required by the application.

The solution to this problem yields the total profit Π^*, the firm's profit Π^{*f} and the optimal allocation of undesirables u_j^{*f}, $f=1,\ldots,F$ and $j=1,\ldots,J$. Thus we may compare u_j^{*f} to the permit amount allocated to each firm, denoted by $\bar{u}_j^f$. This comparison measures if there is an opportunity for permit trading. We compute $\bar{u}_j^f$ as

$$\Pi^{*f}\left(\bar{u}^f\right) = \max \sum_{m=1}^{M} p_m y_m^f - \sum_{n=1}^{N} w_n x_n^f$$

$$\text{s.t.} \quad y_m^f \leq \sum_{k=1}^{K^f} z_k^f y_{km}^f, \quad m = 1,\ldots,M,$$

$$\bar{u}_j^f = \sum_{k=1}^{K^f} z_k^f u_{kj}^f, \quad j = 1,\ldots,J,$$

$$x_n^f = \sum_{k=1}^{K^f} z_k^f x_{kn}^f, \quad n = 1,\ldots,N, \quad (12.15)$$

$$\sum_{f=1}^{F} z_k^f \leq 1,$$

$$z_k^f \geq 0, \quad k = 1,\ldots,K^f.$$

If

$$\sum_{f=1}^{F} \Pi^{*f}\left(\bar{u}^f\right) < \sum_{f=1}^{F} \Pi^{*f} = \Pi^*$$

then there is a possibility for permit trading, although the trading cost must also considered. If firm f has

$$\Pi^{*f}\left(\bar{u}^f\right) < \Pi^{*f}$$

then there is a possibility of adjustment from $\bar{u}^f$ to u^{*f}.

5. EPILOG

We end this chapter by referring the reader to some other relevant literature. The classic book, *Linear Programming & Economic Analysis*, Dorfman, Samuelson and Solow (1958) (often referred to as "DOSSO") provides a basis for further study for someone new to this area of study. The book covers both static and dynamic linear models, including the Leontief system.

For the German speaker, May's book, *Dynamische Produktionstheorie auf Basis der Aktivitätsanalys* (1992), connects the models discussed here to production models developed in Europe.

We also refer the reader to the work of J.K. Sengupta. A search for "J.K. Sengupta dynamic models" using the Google search engine shows examples of the large number of dynamic non-parametric models he has analyzed. See also the paper "Nonparametric approaches to dynamic efficiency" (Sengupta,

1992). Last, we would like to direct attention to an interesting paper by Silva and Stefanou, "Dynamic Efficiency Measurement: Theory and Application" (in press).

REFERENCES

1. Brännlund, R., Färe, R., and Grosskopf, S. (1995). "Environmental Regulation and Profitability: An Application to Swedish Pulp and Paper Mills." *Environmental and Resource Economics*, 6(1), 23-36.
2. Dorfman, R., Samuelson, P., and Solow, R. (1958). *Linear Programming & Economic Analysis*, McGraw-Hill, New York.
3. Färe, R., and Grosskopf, S. (1996). *Intertemporal production frontiers: With dynamic DEA*, Kluwer Academic, In collaboration with R. Brännlund et al. Boston; London and Dordrecht.
4. Färe, R., and Grosskopf, S. (1997). "Efficiency and Productivity in Rich and Poor Countries." Dynamics, economic growth, and international trade, B. S. Jensen and K. Wong, eds., University of Michigan Press, Studies in International Economics. Ann Arbor, 243-63.
5. Färe, R., Grosskopf, S., and Lee, W. F. (2004). "Property rights and profitability." New Directions: Efficiency and Productivity, R. Färe and S. Grosskopf, eds., Kluwer Academic Publishers, Boston, 65-77.
6. Färe, R., and Whittaker, G. (1996). "Dynamic measurement of efficiency: an application to western public grazing." Intertemporal production frontiers: With dynamic DEA, R. Fare and S. Grosskopf, eds., Kluwer Academic, Boston; London and Dordrecht, 168-186.
7. Jaenicke, E. C. (2000). "Testing for Intermediate Outputs in Dynamic DEA Models: Accounting for Soil Capital in Rotational Crop Production and Productivity Measures." *Journal of Productivity Analysis*, 14(3), 247-266.
8. Johansen, L. (1959). "Substitution versus fixed production coefficient in the theory of economic growth: a synthesis." *Econometrica*, 27(2), 157-176.
9. Kemeny, J. G., Morgenstern, O., and Thompson, G. L. (1956). "A generalization of the von Neumann model of an expanding economy." *Econometrica*, 24, 115-135.
10. Lewis, H. F., and Sexton, T. R. (2004). "Network DEA: Efficiency analysis of organizations with complex internal structure." *Computers and Operations Research*, 31(9), 1365.
11. May, E. (1992). *Dynamische Produktionstheorie auf Basis der Aktivitätsanalyse*, Physica-Verlag, Heidelberg.
12. Nemota, J., and Gota, M. (1999). "Dynamic data envelopment analysis: modeling intertemporal behavior of a firm in the presence of productive inefficiencies." *Economic Letters*, 64, 51-56.

13. Nemota, J., and Gota, M. (2003). "Measuring dynamic efficiency in production: An application of data envelopment analysis to Japanese electric utilities." *Journal of Productivity Analysis*, 19, 191-210.
14. Ramsey, R. (1928). "A Mathematical Theory of Saving." *Economic Journal*, 38, 543-559.
15. Sengupta, J. K. (1992). "Non-parametric approaches to dynamic efficiency: a non-parametric application of cointegration fo production frontiers." *Applied Economics*, 24, 153-159.
16. Shephard, R. W. (1970). *Theory of Cost and Production Functions*, Princeton University Press, Princeton.
17. Shephard, R. W. and Färe, R (1975). "A dynamic theory of production correspondences," ORC 75-13, OR Center, U.C. Berkeley.
18. Silva, E., and Stefanou, S. E. (in press). "Dynamic efficiency measurement: theory and applications." *American Journal of Agricultural Economics*.
19. Solow, R. M. (1960). "Investment and technical progress." Mathematical methods in the social sciences, K. J. Arrow, S. Karlin, and P. Suppes, eds., Stanford University Press, Stanford, CA.

Appendix I

This appendix contains a program listing and output using the General Algebraic Modeling System (GAMS) for the network shown in Figure 12-2. The example data are called from separate file, which is listed first. The variables are named to follow the nomenclature in Figure 12-2, e.g., $_0^1 x$ is X10 in the program. The page numbers and equation names refer to Färe and Grosskopf (1996).

Listing of fig2.dat (example data):

TABLE D(FM,R)	x0	x10	x20	x30	y31	y32	y43	y41	y42	y0
FM1	10	3	5	2	4	2	3	2	4	9
FM2	7	2	1	3	1	1	2	1	1	4
FM3	11	4	2	4	2	2	2	2	2	6

Listing of GAMS program:

```
* NETWORK MODEL FOR FIGURE 12-2
*
*

$TITLE NETWORK FIGURE 12-2
$OFFSYMXREF
$OFFSYMLIST
```

SETS

FM DMU / FM1 * FM3 /

R /
  X0, X10, X20, X30, Y31, Y32, Y43, Y41, Y42, Y0
/

OUTPUTS(R) / Y31, Y32, Y41, Y42, Y43 /

INTERMED(R) / Y31, Y32 /

INPUTS(R) / X10, X20, X30, Y31, Y32 /

FSS(FM);

ALIAS (FM,OBS) ;
ALIAS (FM,J) ;

OPTION RESLIM=450000;
OPTION INTEGER1=2;
OPTION INTEGER2=2;
OPTION INTEGER3=2;
OPTION SOLPRINT = OFF;
OPTION LIMROW = 23;
OPTION LIMCOL = 23;
OPTION SYSOUT = OFF;

PARAMETER REPORT(*,*);
PARAMETER PPERACRE;
SCALAR TOTFERT
SCALAR TOTPROFIT
SCALAR RUNOFF

$INCLUDE FIG2.DAT

* PRICES: P = 2 W = 1

VARIABLES
ZE1(FM)
ZE2(FM)
ZE3(FM)
TH1
TH2
TH3

```
  THALL
  YV(OUTPUTS)     OUTPUT SOLUTION VECTOR
  XV(INPUTS)      INPUT SOLUTION VECTOR
  ;

POSITIVE VARIABLES ZE1,ZE2,ZE3,XV,YV,LAMBDA;

EQUATIONS
 EQA
 EQB
 EQC
 EQD
 EQE
 EQF
 EQG
 EQH
 EQI
 EQJ
 EQK
 EQL
 OBJ1
 OBJ2
 OBJ3
 OBJALL
 ;


*----------------------------------------
* P. 22, (A) - (E)  P(3)
*----------------------------------------
 EQA .. SUM(FM,(D(FM,'Y43'))*ZE3(FM)) =G= YV('Y43');
 EQB .. SUM(FM,(D(FM,'X30'))*ZE3(FM)) =G= XV('X30');
 EQC .. SUM(FM,(D(FM,'Y31'))*ZE3(FM)) =G= YV('Y31');
 EQD .. SUM(FM,(D(FM,'Y32'))*ZE3(FM)) =G= YV('Y32');
 EQE .. SUM(FM,ZE3(FM)) =L= 1;


*----------------------------------------
* P. 22, (F) - (H)  P(1)
*----------------------------------------
 EQF .. SUM(FM,(D(FM,'Y31')+D(FM,'Y41'))*ZE1(FM)) =G= YV('Y31') + YV('Y41');
 EQG .. SUM(FM,(D(FM,'X10'))*ZE1(FM)) =L= XV('X10');
 EQH .. SUM(FM,ZE1(FM)) =L= 1;


*----------------------------------------
* P. 22, (I) - (L)  P(2)
*----------------------------------------
 EQI .. SUM(FM,(D(FM,'Y32')+D(FM,'Y42'))*ZE2(FM)) =G= YV('Y32') + YV('Y42');
```

EQJ .. SUM(FM,(D(FM,'X20'))*ZE2(FM)) =L= XV('X20');
EQK .. SUM(FM,ZE2(FM)) =L= 1;
EQL .. XV('X10') + XV('X20') + XV('X30') =E= SUM(FSS,D(FSS,'X0')) ;

*--
* OBJ PROFIT, HOLD OUTPUT FIXED, VARY INPUT
*--

OBJ1 .. TH1 =E= (YV('Y31') + YV('Y41'))*2 - (XV('X10'))*1.0;
OBJ2 .. TH2 =E= (YV('Y32') + YV('Y42'))*2 - (XV('X20'))*1.0;
OBJ3 .. TH3 =E= (YV('Y43'))*2 - (XV('X30') - YV('Y31') - YV('Y32'))*1.0;
OBJALL .. THALL =E= (YV('Y41') + YV('Y42') + YV('Y43'))*2 - (XV('X10') + XV('X20') + XV('X30'))*1.0;

MODEL SUBP1 /
  EQF
  EQG
  EQH
  OBJ1
   /;

MODEL SUBP2 /
  EQI
  EQJ
  EQK
  EQL
  OBJ2
   /;

MODEL SUBP3 /
  EQA
  EQB
  EQC
  EQD
  EQE
  OBJ3
   /;

MODEL OVERALL /
  EQA
  EQB
  EQC
  EQD
  EQE
  EQF

```
      EQG
      EQH
      EQI
      EQJ
      EQK
      EQL
      OBJALL
         /;

* SET OUTPUT FILE & FORMAT
FILE RES1 /SUBP1.RES / ;
FILE RES2 /SUBP2.RES / ;
FILE RES3 /SUBP3.RES / ;
FILE RESALL /SUBPALL.RES / ;

PUT RES1;
* COLUMN LABELS
   PUT @10,"TH1"
       @20,"XV('X10')"
       @30,"XV('X20')"
       @40,"XV('X30')"
       @50,"YV('Y31')"
       @60,"YV('Y41')"
       @70,"YV('Y42')"
       @80,"YV('Y43')"
       @90,"MODELSTAT"
   /;

PUT RES2;
* COLUMN LABELS
   PUT @10,"TH1"
       @20,"XV('X10')"
       @30,"XV('X20')"
       @40,"XV('X30')"
       @50,"YV('Y31')"
       @60,"YV('Y41')"
       @70,"YV('Y42')"
       @80,"YV('Y43')"
       @90,"MODELSTAT"
   /;

PUT RES3;
* COLUMN LABELS
   PUT @10,"TH1"
       @20,"XV('X10')"
       @30,"XV('X20')"
```

```
        @40,"XV('X30')"
        @50,"YV('Y31')"
        @60,"YV('Y41')"
        @70,"YV('Y42')"
        @80,"YV('Y43')"
        @90,"MODELSTAT"
    /;

PUT RESALL;
* COLUMN LABELS
   PUT @10,"TH1"
       @20,"XV('X10')"
       @30,"XV('X20')"
       @40,"XV('X30')"
       @50,"YV('Y31')"
       @60,"YV('Y41')"
       @70,"YV('Y42')"
       @80,"YV('Y43')"
       @90,"MODELSTAT"
    /;

LOOP (OBS,
    FSS(FM) = NO; FSS(OBS) = YES;

************************************************
    SOLVE SUBP1 USING LP MAXIMIZING TH1 ;
   PUT RES1;
       PUT @1, OBS.TL;
       PUT @8,TH1.L:10:4;
       PUT @18,XV.L('X10'):10:4;
       PUT @28,XV.L('X20'):10:4;
       PUT @38,XV.L('X30'):10:4;
       PUT @48,YV.L('Y31'):10:4;
       PUT @58,YV.L('Y41'):10:4;
       PUT @68,YV.L('Y42'):10:4;
       PUT @78,YV.L('Y43'):10:4;
       PUT @88,SUBP1.MODELSTAT:10:4
    /;


************************************************
    SOLVE SUBP2 USING LP MAXIMIZING TH2 ;
   PUT RES2;
       PUT @1, OBS.TL;
       PUT @8,TH1.L:10:4;
       PUT @18,XV.L('X10'):10:4;
```

```
        PUT @28,XV.L('X20'):10:4;
        PUT @38,XV.L('X30'):10:4;
        PUT @48,YV.L('Y31'):10:4;
        PUT @58,YV.L('Y41'):10:4;
        PUT @68,YV.L('Y42'):10:4;
        PUT @78,YV.L('Y43'):10:4;
        PUT @88,SUBP2.MODELSTAT:10:4
    /;

*********************************************************
    SOLVE SUBP3 USING LP MAXIMIZING TH3 ;
    PUT RES3;
        PUT @1, OBS.TL;
        PUT @8,TH1.L:10:4;
        PUT @18,XV.L('X10'):10:4;
        PUT @28,XV.L('X20'):10:4;
        PUT @38,XV.L('X30'):10:4;
        PUT @48,YV.L('Y31'):10:4;
        PUT @58,YV.L('Y41'):10:4;
        PUT @68,YV.L('Y42'):10:4;
        PUT @78,YV.L('Y43'):10:4;
        PUT @88,SUBP3.MODELSTAT:10:4
    /;
*************************************************************
    SOLVE OVERALL USING LP MAXIMIZING THALL ;

    PUT RESALL;
        PUT @1, OBS.TL;
        PUT @8,TH1.L:10:4;
        PUT @18,XV.L('X10'):10:4;
        PUT @28,XV.L('X20'):10:4;
        PUT @38,XV.L('X30'):10:4;
        PUT @48,YV.L('Y31'):10:4;
        PUT @58,YV.L('Y41'):10:4;
        PUT @68,YV.L('Y42'):10:4;
        PUT @78,YV.L('Y43'):10:4;
        PUT @88,OVERALL.MODELSTAT:10:4
    /;
);
```

Output:

subp1.res

	TH1	XV('X10')	XV('X20')	XV('X30')	YV('Y31')	YV('Y41')	YV('Y42')	YV('Y43')	MODELST.
FM1	9	3	0	0	6	0	0	0	1
FM2	9	3	5	2	0	6	6	3	1
FM3	9	3	4	0	0	6	5.3333	3	1

subp2.res

	TH1	XV('X10')	XV('X20')	XV('X30')	YV('Y31')	YV('Y41')	YV('Y42')	YV('Y43')	MODELST.
FM1	9	5	5	0	6	0	0	0	1
FM2	9	2	5	0	0	6	6	3	1
FM3	9	6	5	0	0	6	6	3	1

subp3.res

	TH1	XV('X10')	XV('X20')	XV('X30')	YV('Y31')	YV('Y41')	YV('Y42')	YV('Y43')	MODELST.
FM1	9	5	5	0	4	0	0	3	1
FM2	9	2	5	0	4	6	6	3	1
FM3	9	6	5	0	4	6	6	3	1

subpall.res

	TH1	XV('X10')	XV('X20')	XV('X30')	YV('Y31')	YV('Y41')	YV('Y42')	YV('Y43')	MODELST.
FM1	9	3	5	2	0	6	6	3	1
FM2	9	3	4	0	0	6	5.3333	3	1
FM3	9	3	6	2	0	6	6	3	1

Appendix II

This appendix contains a program listing and output using the General Algebraic Modeling System (GAMS) for the network shown in Figure 12-4. The example data are called from separate file, which is listed first. The variables are named to follow the nomenclature in Figure 12-4, i.e., $_0^1 x$ is X10 in the program.

Example data, FIG4.DAT:

	TABLE D(FM,R)							
	X1	X2	X3	Y1F	Y2F	Y3F	Y1I	Y2I
FM1	3	4	4	2	5	5	1	1
FM2	4	5	6	3	6	6	1	2
FM3	5	6	7	4	7	6	2	2

Listing of GAMS program:

```
* NETWORK MODEL FOR FIGURE 12-4
*
*

$TITLE NETWORK FIGURE 12-4
$OFFSYMXREF
$OFFSYMLIST
```

SETS

  FM DMU / FM1 * FM3 /

  R /
    X1, X2, X3, Y1F, Y2F, Y3F, Y1I, Y2I
    /

  OUTPUTS(R) / Y1F, Y2F, Y3F, Y1I, Y2I /

  INTERMED(R) / Y1I, Y2I /

  INPUTS(R) / X1, X2, X3, Y1I, Y2I /

  FSS(FM);

  ALIAS (FM,OBS) ;
  ALIAS (FM,J) ;

  OPTION RESLIM=450000;
  OPTION INTEGER1=2;
  OPTION INTEGER2=2;
  OPTION INTEGER3=2;
  OPTION SOLPRINT = OFF;
  OPTION LIMROW = 23;
  OPTION LIMCOL = 23;
  OPTION SYSOUT = OFF;

   PARAMETER REPORT(*,*);
   PARAMETER PPERACRE;
   SCALAR TOTFERT
   SCALAR TOTPROFIT
   SCALAR RUNOFF

  $INCLUDE FIG4.DAT

  * PRICES: P = 2 W = 1

   VARIABLES
   ZE1(FM)
   ZE2(FM)
   ZE3(FM)
   TH1
   TH2
   TH3

THALL
YV(OUTPUTS)   OUTPUT SOLUTION VECTOR
XV(INPUTS)   INPUT SOLUTION VECTOR
;

POSITIVE VARIABLES ZE1,ZE2,ZE3,XV,YV,LAMBDA;

EQUATIONS
EQA
EQB
EQC
EQD
EQE
EQF
EQG
EQH
OBJ1
;

*--
* INPUTS
*--
EQA .. SUM(FM,(D(FM,'X1'))*ZE1(FM)) =L= SUM(FSS,D(FSS,'X1'));
EQB .. SUM(FM,(D(FM,'X2'))*ZE2(FM)) =L= SUM(FSS,D(FSS,'X2'));
EQC .. SUM(FM,(D(FM,'X3'))*ZE3(FM)) =L= SUM(FSS,D(FSS,'X3'));

*--
* INTERMEDIATE OUTPUTS
*--
EQD .. SUM(FM,(D(FM,'Y1I'))*ZE2(FM)) =L= YV('Y1I');
EQE .. SUM(FM,(D(FM,'Y2I'))*ZE3(FM)) =L= YV('Y2I');

*--
* OUTPUTS
*--
EQF .. YV('Y1F')+YV('Y1I') =L= SUM(FM,(D(FM,'Y1F')+D(FM,'Y1I'))*ZE1(FM));
EQG .. YV('Y2F')+YV('Y2I') =L= SUM(FM,(D(FM,'Y2F')+D(FM,'Y2I'))*ZE2(FM));
EQH .. YV('Y3F') =L= SUM(FM,D(FM,'Y3F')*ZE3(FM));

*--
* OBJ PROFIT, HOLD OUTPUT FIXED, VARY INPUT
*--
* PRICE OF ALL OUTPUTS = 1
OBJ1 .. TH1 =E= YV('Y1F') + YV('Y2F') + YV('Y3F');

MODEL FIG4 /

```
        EQA
        EQB
        EQC
        EQD
        EQE
        EQF
        EQG
        EQH
        OBJ1
          /;

* SET OUTPUT FILE & FORMAT

*FILE RES /SUB_TECH_P3.RES / ;
FILE RES1 /FIG4.RES / ;

PUT RES1;

* COLUMN LABELS
  PUT @12,"TH1"
      @30,"YV('Y1I')"
      @40,"YV('Y2I')"
*     @48,"YV('X3F')"
      @50,"YV('Y1F')"
      @60,"YV('Y2F')"
      @70,"YV('Y3F')"
      @80,"MODELSTAT"
  /;

  LOOP (OBS,
    FSS(FM) = NO; FSS(OBS) = YES;
*     OPTION LP=MPSWRITE;
*     OPTION LP = CONVERT;
*     UPPER.OPTFILE = 1;
***********************************************
    SOLVE FIG4 USING LP MAXIMIZING TH1 ;
  PUT RES1;
    PUT @1, OBS.TL;
    PUT @8,TH1.L:10:4;
    PUT @28,YV.L('Y1I'):10:4;
    PUT @38,YV.L('Y2I'):10:4;
    PUT @48,YV.L('Y1F'):10:4;
    PUT @58,YV.L('Y2F'):10:4;
    PUT @68,YV.L('Y3F'):10:4;
    PUT @78,FIG4.MODELSTAT :7:2
  /;
```

);

FIG4.RES
	TH1	YV('Y1I')	YV('Y2I')	YV('Y1F')	YV('Y2F')	YV('Y3F')	MODELST.
FM1	13.2	0.8	1	2.8	5.4	5	1
FM2	17.8	1	1.5	3.8	6.5	7.5	1
FM3	21.4	1.2	1.75	4.8	7.85	8.75	1

Chapter 13

CONTEXT-DEPENDENT DATA ENVELOPMENT ANALYSIS AND ITS USE

Hiroshi Morita[1] and Joe Zhu[2]
[1]*Department of Information and Physical Sciences, Osaka University, Suita, 565-0871, Japan, morita@ist.osaka-u.ac.jp*

[2]*Department of Management, Worcester Polytechnic Institute, Worcester, MA 01609, jzhu@wpi.edu*

Abstract: Data envelopment analysis (DEA) is a methodology for identifying the efficient frontier of decision making units (DMUs). Context-dependent DEA refers to a DEA approach where a set of DMUs is evaluated against a particular evaluation context. Each evaluation context represents an efficient frontier composed by DMUs in a specific performance level. The context-dependent DEA measures the attractiveness and the progress when DMUs exhibiting poorer and better performance are chosen as the evaluation context, respectively. This chapter also presents a slack-based context-dependent DEA approach. In DEA, nonzero input and output slacks are very likely to present after the radial efficiency score improvement. The slack-based context-dependent DEA allows us to fully evaluate the inefficiency in a DMU's performance.

Key words: Data Envelopment Analysis (DEA), Attractiveness, Progress, Value judgment, Slack-based measure, Context-dependent, Benchmarking

1. INTRODUCTION

Data envelopment analysis (DEA) uses linear programming technique to evaluating the relative efficiency of decision making units (DMUs) with multiple outputs and multiple inputs. Adding or deleting an inefficient DMU does not alter the efficient frontier and the efficiencies of the existing DMUs. The inefficiency scores change only if the efficient frontier is altered. The performance of DMUs depends only on the identified efficient frontier characterized by the DMUs with a unity efficiency score.

If the performance of inefficient DMUs deteriorates or improves, the efficient DMUs still may have a unity efficiency score. Although the performance of inefficient DMUs depends on the efficient DMUs, efficient DMUs are only characterized by a unity efficiency score. The performance of efficient DMUs is not influenced by the presence of inefficient DMUs, once the DEA frontier is identified.

However, evaluation is often influenced by the context. A DMU's performance will appear more attractive against a background of less attractive alternatives and less attractive when compared to more attractive alternatives. Researchers of the consumer choice theory point out that consumer choice is often influenced by the context. e.g., a circle appears large when surrounded by small circles and small when surrounded by larger ones. Similarly, a product may appear attractive against a background of less attractive alternatives and unattractive when compared to more attractive alternatives (Tversky and Simonson, 1993).

Considering this influence within the framework of DEA, one could ask "what is the relative attractiveness of a particular DMU when compared to others?" As in Tversky and Simonson (1993), one agrees that the relative attractiveness of DMU x compared to DMU y depends on the presence or absence of a third option, say DMU z (or a group of DMUs). Relative attractiveness depends on the evaluation context constructed from alternative options (or DMUs).

In fact, a set of DMUs can be divided into different levels of efficient frontiers. If we remove the (original) efficient frontier, then the remaining (inefficient) DMUs will form a new second-level efficient frontier. If we remove this new second-level efficient frontier, a third-level efficient frontier is formed, and so on, until no DMU is left. Each such efficient frontier provides an evaluation context for measuring the relative attractiveness. e.g., the second-level efficient frontier serves as the evaluation context for measuring the relative attractiveness of the DMUs located on the first-level (original) efficient frontier. On the other hand, we can measure the performance of DMUs on the third-level efficient frontier with respect to the first or second level efficient frontier.

The context-dependent DEA (Seiford and Zhu, 1999a, Zhu, 2003 and Seiford and Zhu, 2003) is introduced to measure the relative attractiveness of a particular DMU when compared to others. Relative attractiveness depends on the evaluation context constructed from a set of different DMUs.

The context-dependent DEA is a significant extension to the original DEA approach. The original DEA approach evaluates each DMU against a set of efficient DMUs and cannot identify which efficient DMU is a better option with respect to the inefficient DMU. This is because all efficient DMUs have an efficiency score of one. Although one can use the super-efficiency DEA model (Andersen and Petersen, 1993) and Seiford and Zhu, 1999b) to rank the performance of efficient DMUs, the evaluation context changes in each evaluation for each efficient, and the efficient DMUs are not evaluated against the same reference set.

In the context-dependent DEA, the evaluation contexts are obtained by partitioning a set of DMUs into several levels of efficient frontiers. Each efficient frontier provides an evaluation context for measuring the relative attractiveness and progress. When DMUs in a specific level are viewed as having equal performance, the attractiveness measure allows us to differentiate the "equal performance" based upon the same specific evaluation context. A combined use of attractiveness and progress measures can further characterize the performance of DMUs.

The original context-dependent DEA model is developed by using radial efficiency measure, where slack values are not taken into account. If DMU with efficiency score of one has non-zero slack value, it would be categorized into the same efficiency level together with efficient DMUs. The slack-based measure (Tone, 2002a)) is introduced to evaluate the efficiency based on the slack values. When we apply the slack-based measure to the context-depend DEA, we can have a more appropriate stratification of the DMU performance levels. In the current chapter, we present the slack-based context-dependent DEA developed by Morita, Hirokawa and Zhu (2005).

The rest of this chapter is organized as follows. The next section introduces the context-dependent DEA with a numerical example. We then introduce the slack-based context-dependent DEA. An application and conclusions then follow.

2. CONTEXT-DEPENDENT DATA ENVELOPMENT ANALYSIS

2.1 Stratification DEA Model

Assume that there are n DMUs which produce s outputs by using m inputs. We define the set of all DMUs as J^1 and the set of efficient DMUs in J^1 as E^1. Then the sequences of J^l and E^l are defined interactively as $J^{l+1} = J^l - E^l$. The set of E^l can be found as the DMUs with optimal value ϕ_k^l of 1 to the following linear programming problem:

$$\begin{aligned}
\underset{\lambda,\theta}{\text{minimize}} \quad & \theta_k^l = \theta \\
\text{subject to} \quad & \sum_{j \in J^l} \lambda_j x_{ij} \leq \theta x_{ik}, i = 1,...,m \\
& \sum_{j \in J^l} \lambda_j y_{rj} \geq y_{rk}, r = 1,...,s \\
& \lambda_j \geq 0, j \in J^l
\end{aligned}$$

(1)

where x_{ij} and y_{rj} are i-th input and r-th output of DMU j. When $l = 1$, model (1) becomes the original input-oriented CCR model (Charnes, Cooper and Rhodes, 1978) and E^1 consists of all the radially efficient DMUs. A radially efficient DMU may have non-zero input/output slack values. The DMUs in set E^1 define the first-level efficient frontier. When $l = 2$, model (1) gives the second-level efficient frontier after the exclusion of the first-level efficient DMUs. In this manner, we identify several levels of efficient frontiers. We call E^l the l-th level efficient frontier. The following algorithm accomplishes the identification of these efficient frontiers by model (1).

Step 1: Set $l = 1$. Evaluate the entire set of DMUs, J^1, by model (1) to obtain the first-level efficient DMUs, set E^1 (the first-level efficient frontier).

Step 2: Let $J^{l+1} = J^l - E^l$ to exclude the efficient DMUs from future DEA runs. If $J^{l+1} = \varnothing$ then stop.

Step 3: Evaluate the new subset of "inefficient" DMUs, J^{l+1}, by model (1) to obtain a new set of efficient DMUs E^{l+1} (the new efficient frontier).

Step 4: Let $l = l + 1$. Go to step 2.

Stopping rule: $J^{l+1} = \varnothing$, the algorithm stops.

Model (1) yields a stratification of the whole set of DMUs, which partitions into different subgroups of efficiency levels characterized by E^l. It is easy to show that these sets of DMUs have the following properties:

(i) $J^1 = \bigcup E^l$ and $E^l \cap E^{l'} = \phi$ for $l \neq l'$;

(ii) The DMUs in $E^{l'}$ are dominated by the DMUs in E^l if $l' > l$;

(iii) Each DMU in set E^l is efficient with respect to the DMUs in set $J^{l'}$ for all $l' > l$.

Figure 13-1 plots the three levels of efficient frontiers of 10 DMUs with two inputs and one single output of one indicated in Table 13-1.

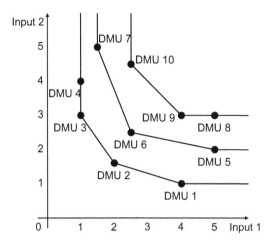

Figure 13-1. Efficient Frontiers in Different Levels

Table 13-1. Numerical Example

DMU	1	2	3	4	5	6	7	8	9	10
Input 1	4	2	1	1	5	2.5	1.5	5	4	2.5
Input 2	1	1.5	3	4	2	2.5	5	3	3	4.5

2.2 Attractiveness and Progress

Based upon the evaluation context E^l, the context-dependent DEA measures the relative attractiveness of DMUs. Consider a specific DMU q in $E^{l'}$. The following model is used to characterize the attractiveness with respect to levels exhibiting poorer performance in $E^{l'}$ for $l' > l$.

$$\underset{\lambda,\theta}{\text{minimize}} \quad \theta_q^{l'} = \theta$$

$$\text{subject to} \quad \sum_{j \in J^{l'}} \lambda_j x_{ij} \leq \theta x_{iq}, i = 1,\ldots,m$$

$$\sum_{j \in J^{l'}} \lambda_j y_{rj} \geq y_{rq}, r = 1,\ldots,s \quad (2)$$

$$\lambda_j \geq 0, j \in J^{l'}$$

It is easy to show that $\theta_q^{l'} > 1$ for $l' > l$, and $\theta_q^{l_1} > \theta_q^{l_2}$ for $l_1 > l_2$. Then $\theta_q^{l'}$ is called the input-oriented d-degree attractiveness of DMU q from a specific level $E^{l'}$, where $d = l' - l$.

In model (2), each efficient frontier represents an evaluation context for evaluating the relative attractiveness of DMUs in $E^{l'}$. Note that the bigger the value of $\theta_q^{l'} > 1$, the more attractive DMU q is, because DMU q makes itself more distinctive from the evaluation context $E^{l'}$. We are able to rank the DMUs in $E^{l'}$ based upon their attractiveness scores and identify the best one.

To obtain the progress measure for a specific DMU q in $E^{l'}$, we use the following context-dependent DEA, which is used to characterize the progress with respect to levels exhibiting better performance in $E^{l'}$ for $l' < l$.

$$\begin{aligned}
\underset{\lambda,\varphi}{\text{minimize}} \quad & \varphi_q^{l'} = \varphi \\
\text{subject to} \quad & \sum_{j \in J^{l'}} \lambda_j x_{ij} \leq \varphi x_{iq}, i = 1,\ldots,m \\
& \sum_{j \in J^{l'}} \lambda_j y_{rj} \geq y_{rq}, r = 1,\ldots,s \\
& \lambda_j \geq 0, j \in J^{l'}
\end{aligned} \quad (3)$$

We have that $\varphi_q^{l'} < 1$ for $l' < l$, and $\varphi_q^{l_1} < \varphi_q^{l_2}$ for $l_1 > l_2$. Then $\varphi_q^{l'}$ is called the input-oriented g-degree progress of DMU q from a specific level $E^{l'}$, where $g = l - l'$.

2.3 Output oriented context-dependent DEA model

Here we provide the output-oriented context-dependent DEA model. Consider the following linear programming problem for DMU q in specific level $E^{l'}$ based upon the evaluation context E^{l} for $l' > l$.

$$\begin{aligned}
\underset{\lambda,H}{\text{maximize}} \quad & H_q^{l'} = H \\
\text{subject to} \quad & \sum_{j \in J^{l'}} \lambda_j x_{ij} \leq x_{iq}, i = 1,\ldots,m \\
& \sum_{j \in J^{l'}} \lambda_j y_{rj} \geq H y_{rq}, r = 1,\ldots,s \\
& \lambda_j \geq 0, j \in J^{l'}
\end{aligned} \quad (4)$$

This problem is used to characterize the attractiveness with respect to levels exhibiting poorer performance in $E^{l'}$. Note that dividing each side of the constraint of (4) by H gives

$$\sum_{j \in J^{l'}} \tilde{\lambda}_j x_{ij} \leq \frac{1}{H} x_{iq}$$

$$\sum_{j \in J^{l'}} \tilde{\lambda}_j y_{rj} \geq y_{rq}$$

$$\tilde{\lambda}_j = \frac{\lambda_j}{H} \geq 0, j \in J^{l'}$$

Therefore, (4) is equivalent to (2), and we have that $H_q^{l'} < 1$ for $l' > l$ and $H_q^{l'} = 1/\theta_q^{l'}$. Then $H_q^{l'}$ is called the output-oriented d-degree attractiveness of DMU q from a specific level $E^{l'}$, where $d = l' - l$. The smaller the value of $H_q^{l'}$ is, the more attractive DMU q is. Model (4) determines the relative attractiveness score for DMU q when inputs are fixed at their current levels.

To obtain the progress measure for DMU q in $E^{l'}$, we develop the following linear programming problem, which is used to characterize the progress with respect to levels exhibiting better performance in $E^{l'}$ for $l' < l$.

$$\begin{aligned}
\underset{\lambda, G}{\text{maximize}} \quad & G_q^{l'} = G \\
\text{subject to} \quad & \sum_{j \in J^{l'}} \lambda_j x_{ij} \leq x_{iq}, i = 1, \ldots, m \\
& \sum_{j \in J^{l'}} \lambda_j y_{rj} \geq G y_{rq}, r = 1, \ldots, s \\
& \lambda_j \geq 0, j \in J^{l'}
\end{aligned} \quad (5)$$

We have that $G_q^{l'} > 1$ for $l' < l$ and $G_q^{l'} = 1/\varphi_q^{l'}$. Then $G_q^{l'}$ is called the output-oriented g-degree progress of DMU q from a specific level $E^{l'}$, where $g = l - l'$.

2.4 Context-dependent DEA with Value Judgment

Both attractiveness and progress are measured radially with respect to different *levels* of efficient frontiers. The measurement does not require *a priori* information on the importance of the attributes (input/output) that features the performance of DMUs. However different attributes play different roles in the evaluation of a DMU's overall performance. Therefore, we introduce value judgment into the context-dependent DEA.

In order to incorporate such *a priori* information into our measures of attractiveness and progress, we first specify a set of weights related to the m

inputs, $v_i, i = 1,...,m$ such that $\sum_{i=1}^{m} v_i = 1$. Based upon Zhu (1996), we develop the following linear programming problem for DMU q in E^l.

$$\begin{aligned}
\underset{\lambda_j, \Theta_q^i}{\text{Maximize}} \quad & \Theta_q^{l'*} = \sum_{r=1}^{s} v_i \Theta_{iq} \\
\text{subject to} \quad & \sum_{j \in E^{l'}} \lambda_j x_{ij} \leq \Theta_{iq} x_{iq}, i = 1,...,m \\
& \sum_{j \in E^{l'}} \lambda_j y_{rj} \geq y_{rq}, r = 1,...,s \\
& \Theta_{iq} \leq 1, i = 1,...,m \\
& \lambda_j \geq 0, j \in E^{l'}
\end{aligned} \quad (6)$$

$\Theta_q^{l'*}$ is called the input-oriented value judgment d-degree attractiveness of DMU q from a specific level $E^{l'}$, where $d = l'-l$. Obviously, $\Theta_q^{l'*} > 1$. The larger the $\Theta_q^{l'*}$ is, the more attractive the DMU q appears under the weights $v_i, i = 1,...,m$. We now can rank DMUs in the same level by their attractiveness scores with value judgment which are incorporated with the preferences over outputs.

If one wishes to prioritize the options (DMUs) with higher values of the i_o-th input, then one can increase the value of the corresponding weight v_{i_o}. These user-specified weights reflect the relative degree of desirability of the corresponding outputs. For example, if one prefers a printer with faster printing speed to one with higher print quality, then one may specify a larger weight for the speed. The constraints of $\Theta_{iq} \leq 1, i = 1,...,m$ ensure that in an attempt to make itself as distinctive as possible, DMU q is not allowed to decrease some of its outputs to achieve higher levels of other preferred outputs.

Note that $\Theta_q^{l'*}$ is an overall attractiveness of DMU q in terms of inputs while keeping the outputs at their current levels. On the other hand, each individual optimal value of $\Theta_{iq}, i = 1,...,m$ measures the attractiveness of DMU q in terms of each input dimension. Θ_{iq}^* is called the input-oriented value judgment input-specific attractiveness measure for DMU q.

With the input-specific attractiveness measures, one can further identify which inputs play important roles in distinguishing a DMU's performance. On the other hand, if $\Theta_{i_o q}^* = 1$, then other DMUs in $E^{l'}$ or their combinations can also produce the amount of the i_o-th input of DMU q, i.e., DMU q does not exhibit better performance with respect to this specific output dimension. Therefore, DMU q should improve its performance on the i_o-th input to distinguish itself in the future.

Similar to the development in the previous section, we can define the output-oriented value judgment progress measure:

$$\begin{aligned}
\text{Maximize}_{\lambda_j, \Phi_{iq}} \quad & \Phi_q^{l'*} = \sum_{r=1}^{s} v_l \Phi_{iq} \\
\text{subject to} \quad & \sum_{j \in E^{l'}} \lambda_j x_{ij} \leq \Phi_{iq} x_{iq}, i = 1, \ldots, m \\
& \sum_{j \in E^{l'}} \lambda_j y_{rj} \geq y_{rq}, r = 1, \ldots, s \\
& \Phi_{iq} \geq 1, i = 1, \ldots, m \\
& \lambda_j \geq 0, j \in E^{l'}
\end{aligned} \quad (7)$$

The optimal value $\Phi_q^{l'*}$ is called the input-oriented value judgment g-degree progress DMU q from a specific level $E^{l'}$, where $g = l - l'$. The larger the $\Phi_q^{l'*}$ is, the greater the amount of progress is expected for DMU q. Here the user-specified weights reflect the relative degree of desirability of improvement on the individual output levels. Let $\Phi_{iq}^*, i = 1, \ldots, m$ represent the optimal value of (7) for a specific level l'. By Zhu (1996), we know that $\sum_{j \in E^{l'}} \lambda_j^* x_{ij} = \Phi_{iq}^* x_{iq}$ holds at optimality for each $i = 1, \ldots, m$ s. Consider the following linear programming problem:

$$\begin{aligned}
\text{Maximize} \quad & \sum_{i=1}^{m} s_i^- \\
\text{subject to} \quad & \sum_{j \in E^{l'}} \lambda_j x_{ij} = \Phi_{iq}^* x_{iq}, i = 1, \ldots, m \\
& \sum_{j \in E^{l'}} \lambda_j y_{rj} - s_r^+ = y_{rq}, r = 1, \ldots, s \\
& s_r^+ \geq 0, r = 1, \ldots, s \\
& \lambda_j \geq 0, j \in E^{l'}
\end{aligned} \quad (8)$$

The following point

$$\begin{cases} \hat{x}_{iq} = \Phi_{iq}^* x_{iq}, i = 1, \ldots, m \\ \hat{y}_{rq} = y_{rq} + s_r^{+*}, r = 1, \ldots, s \end{cases}$$

is called a *preferred global efficient target* for DMU q in level E^l for $l' = l - 1$; otherwise, if $l' < l - 1$, it represents a *preferred local efficient target*, where Φ_{iq}^* is the optimal value in (7), and s_r^{+*} represent the optimal values in (8).

3. SLACK-BASED CONTEXT-DEPENDENT DEA

To improve the performance of inefficient DMU, the target of improvement should be given among the efficient DMUs. The reference set suggests the target of improvement for the inefficient DMUs. Actually, when $l = 1$, model (1) gives the reference set of DMUs from the efficient DMUs for inefficient DMUs. It may be a final goal of improvement, however, for some inefficient DMUs, this goal may be quite different from the current performance and difficult to achieve the improvement. Therefore, it is not appropriate to set a benchmark target for improvement from the efficient DMUs directly. Step-by-step improvement is a useful way to improve the performance, and the benchmark target at each step is provided based on the evaluation context at each level of efficient frontier.

The above context-dependent DEA is developed by using a radial efficiency measure, which ignores possible non-zero slack values. A slack-based measure (SBM) of efficiency is introduced to evaluate the efficiency together with the slack value (Tone, 2002b). The following index ρ

$$\rho = \frac{1 - \frac{1}{m}\sum_{i=1}^{m}\frac{s_i^-}{x_{io}}}{1 + \frac{1}{s}\sum_{r=1}^{s}\frac{s_r^+}{y_{ro}}} \tag{9}$$

is defined in terms of the amount of slack, and has the value between 0 and 1. The SBM efficiency score is obtained from the following linear program,

$$\text{Minimize} \quad \rho = \frac{1 - \frac{1}{m}\sum_{i=1}^{m}\frac{s_i^-}{x_{io}}}{1 + \frac{1}{s}\sum_{r=1}^{s}\frac{s_r^+}{y_{ro}}}$$

$$\text{subject to} \quad \sum_{j=1}^{n}\lambda_j x_{ij} + s_i^- = x_{ik} \tag{10}$$

$$\sum_{j=1}^{n}\lambda_j y_{rj} - s_r^+ = y_{rk}$$

$$\lambda \geq 0, s^- \geq 0, s^+ \geq 0$$

The SBM efficiency score is less than CCR efficiency score, and CCR inefficient DMU never become SBM efficient. The SBM efficiency score is normalized between 0 and 1, and we have that if, and only if, $\rho^* = 1$, then it is efficient, because $\rho^* = 1$ implies that all slacks are zero and the DMU locates on the efficient frontier. The slack-based score is units invariant and has monotonic property.

Base upon (4), we propose a stratification procedure in the same manner to the original context-dependent DEA as $J^{l+1} = J^l - E^l$, where the set of efficient DMUs E^l is determined from the slack-based efficiency score. That is, the set of E^l can be found as the DMUs with optimal value ρ_k^l of 1 to the following programming problem:

$$\text{Minimize} \quad \rho_k^l = \frac{1 - \frac{1}{m}\sum_{i=1}^{m} \frac{s_i^-}{x_{ik}}}{1 + \frac{1}{s}\sum_{r=1}^{s} \frac{s_r^+}{y_{rk}}}$$

$$\text{subject to} \quad \sum_{j \in J^l} \lambda_j x_{ij} + s_i^- = x_{ik} \tag{11}$$

$$\sum_{j \in J^l} \lambda_j y_{rj} - s_r^+ = y_{rk}$$

$$\lambda \geq 0, s^- \geq 0, s^+ \geq 0$$

The intensity vector λ in (11) shows the reference set $R_k^{SBM}(l)$ of DMU k in the efficiency level l_k based on the context E^l, where $l < l_k$.

$$R_k^{SBM}(l) = \{j \in J^l | \lambda_j > 0 \text{ in (11)}\} \tag{12}$$

For context-dependent DEA by radial efficiency measure, the reference set $R_k^{CCR}(l)$ of DMU k in the efficiency level l_k based on the context E^l is similarly given by

$$R_k^{CCR}(l) = \{j \in J^l | \lambda_j > 0 \text{ in (2)}\} \tag{13}$$

Note that the reference set based on the context E^1 is the same to the reference set of slack-based DEA model. The DMUs in the reference set can be used as benchmark targets for inefficient DMU. The context-dependent DEA provides several benchmark targets by setting evaluation context. To reduce the inefficiency, it is effective to improve the performance step by step. So the benchmark targets should be provided according to the efficiency level that the DMU exists. Therefore, the sequence of reference sets $R_k(l), R_k(l-1), ..., R_k(1)$ is used as the step by step benchmark targets.

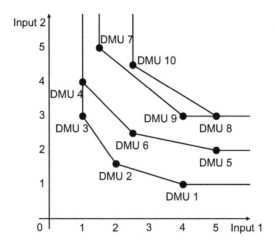

Figure 13-2. Stratification by Slack-based Measure

Now, the attractiveness based on the evaluation context E^l is measured with respect to the DMUs in the subset J^l. For example, the attractiveness for DMU k based on the evaluation context E^l is obtained from the following programming problem.

$$\text{maximize} \quad \delta = \frac{\frac{1}{m}\sum_{i=1}^{m}\frac{\bar{x}_i}{x_{ik}}}{\frac{1}{s}\sum_{r=1}^{s}\frac{\bar{y}_r}{y_{rk}}}$$

$$\text{subject to} \quad \bar{x} \geq \sum_{j \in J^l} \lambda_j x_j \tag{14}$$

$$\bar{y} \leq \sum_{j \in J^l} \lambda_j y_j$$

$$\bar{x} \geq x_k, \bar{y} \leq y_k, \lambda \geq 0$$

Note that this fractional programming problem can be transformed into a linear programming problem using the Charnes-Cooper transformation as shown in Charnes, Cooper and Rhodes (1978).

The slack-based attractiveness and slack-based super-efficiency score (Tone, 2002a) share some similarity. We consider the subset $\bar{P}_{-k}$ of the production possibility set excluding DMU k.

$$\bar{P}_{-k} = \left\{(\bar{x},\bar{y}) \middle| \bar{x} \geq \sum_{j \neq k}\lambda_j x_j, \bar{y} \leq \sum_{j \neq k}\lambda_j y_j, \lambda \geq 0\right\} \tag{15}$$

$$\cap \{(\bar{x},\bar{y}) | \bar{x} \geq x_k, \bar{y} \leq y_k\}$$

The L_1 distance between (x_k, y_k) and $\overline{P}_{-k}$ is used as the index of slack-based super-efficiency, which is obtained from the following programming problem.

$$\text{Maximize} \quad \frac{\frac{1}{m}\sum_{i=1}^{m}\frac{\overline{x}_i}{x_{ik}}}{\frac{1}{s}\sum_{r=1}^{s}\frac{\overline{y}_r}{y_{rk}}}$$

$$\text{subject to} \quad \overline{x}_i \geq \sum_{j \neq k} \lambda_j x_{ij}, i = 1, ..., m \quad (16)$$

$$\overline{y}_r \leq \sum_{j \neq k} \lambda_j y_{rj}, r = 1, ..., s$$

$$\overline{x}_i \geq x_{ik}, \overline{y}_r \leq y_{rk}, \lambda \geq 0$$

Note that in slack-based super-efficiency, the evaluation context (reference set) changes for each efficient. Our slack-based context-dependent DEA uses the same evaluation context.

4. APPLICATION

A company in Osaka, Japan has 14 sales branches, and each sales branch does business independently in its covered district. When a sales branch having poor performance tried to improve its performance by benchmarking the best sales branches, it was very difficult to adopt their performance, because there was a huge gap between the best practices and the underperforming sales branches. This indicates that it is necessary to provide an attainable benchmark target via a step-wise improvement.

This company measures the performance of sales branches in terms of four indices, namely Management, Mobility, Planning and Presentation. Table 13-2 shows the data of 14 sales branches. The purpose of the analysis is to find the good performer, and to provide an appropriate benchmark target for poor performers.

Each sales branch is viewed as a DMU and the four indices are used as four outputs with one input of unity. By calculating (1) for $l = 1$, we obtain $\theta_k^1 = 1$ for k = H, J and L. The first-level efficient frontier is E^1 = {H, J, L} (the original DEA frontier). Next, we exclude the DMUs in set E^1 from J^1 and obtain J^2 = {A, B, C, D, E, F, G, I, K, M, N}. We have $\theta_k^2 = 1$ for k = A, F, G, I and N. Therefore, the efficient frontier of J^2 is E^2 = {A, F, G, I, N} (second level efficient frontier). By repeating this process, we finally obtain E^5 = {C} (the fifth-level efficient frontier) and $L = 5$. Next, by applying the slack-based context dependent DEA, we have the efficiency scores and the stratification of sales branches. Table 13-3 reports the radial and slack-based efficiency scores for each sales branch.

Table 13-2. Index Values for 14 Branches

Branch	Management	Mobility	Planning	Presentation
A	14.1	25.4	4.9	4.4
B	13.2	20.5	6.4	4.7
C	11.5	16.5	4.0	4.4
D	15.6	23.6	6.4	7.4
E	15.2	18.8	6.4	0.8
F	15.2	22.4	9.0	6.0
G	14.7	23.9	9.1	8.3
H	18.0	29.0	11.3	10.2
I	16.4	23.5	6.3	7.9
J	18.1	26.4	13.0	10.0
K	10.4	20.0	4.8	8.6
L	15.1	25.6	8.8	10.2
M	12.9	16.9	7.9	7.4
N	12.8	20.4	8.8	8.7

Table 13-3. Efficiency Scores and Levels

Branch	Radial measure		Slack-based measure	
	θ	Level	ρ	Level
A	0.758	2	0.352	3
B	0.465	4	0.423	4
C	0.402	5	0.162	5
D	0.728	3	0.558	3
E	0.681	4	0.101	4
F	0.681	2	0.626	2
G	0.804	2	0.698	2
H	1.000	1	1.000	1
I	0.813	2	0.568	2
J	1.000	1	1.000	1
K	0.835	3	0.247	4
L	1.000	1	0.868	2
M	0.711	3	0.375	4
N	0.845	2	0.543	3

Table 13-4 and Table 13-5 show the stratification of efficient frontier by radial measure and slack-based measure, respectively, which indicate five levels for both measures. C has the least efficiency and form the last level of efficient frontier. Note that the efficiency level does not necessarily follow the order of the efficiency scores. For example, D on level 3 has a larger efficiency score than F on level 2 by radial measure.

Table 13-4. Levels by CRS Radial Measure

Levels	Frontier branches	CRS efficiency range
Level 1	H, J, L	1
Level 2	A, F, G, I, N	0.681 – 0.845
Level 3	D, K, M	0.711 – 0.835
Level 4	B, E	0.465 – 0.681
Level 5	C	0.402

Table 13-5. Levels by Slack-based Measure

Levels	Frontier branches	Slack-based efficiency range
Level 1	H, J	1
Level 2	F, G, I, L	0.568 – 0.868
Level 3	A, D, N	0.352 – 0.558
Level 4	B, E, K, M	0.101 – 0.423
Level 5	C	0.162

The difference between radial and slack-based measures is found in the sales branches with non-zero slack value. For example, L has a radial efficiency score of one, however, it is not efficient because of the non-zero slacks in indices of management, mobility and planning. The slack-based efficiency score reflects the amount of non-zero slack, and the efficiency score is less than one. L is categorized into level 2 by slack-based context-dependent DEA.

Table 13-6. Reference Sets as Benchmark Targets by Radial Measure

Branch	Level	$R^{CCR}(1)$	$R^{CCR}(2)$	$R^{CCR}(3)$	$R^{CCR}(4)$
A	2	H			
B	4	H, J	G, I, A	D, M	
C	5	J	I, A	D	B
D	3	J, H	I, G, A		
E	4	J	I, F	D, M	
F	2	J, H			
G	2	H			
H	1				
I	2	J			
J	1				
K	3	H	N		
L	1				
M	3	H	G, N		
N	2	H			

The reference sets and their diagram for benchmark targets $R^{CCR}(l)$ by radial measure are shown in Table 13-6 and Table 13-3. Table 13-6 tells us that, for example, the benchmark targets of C are found as J on level 1, I and A on level 2, D on level 3 and B on level 4. J may not be an appropriate benchmark target for C, because J on level 1 is far from C on level 5. At the first step, C should benchmark B on level 4. Note that although L is in level 1 by the radial measure, it is not benchmarked by other branches because of the non-zero slacks. However, the performance of L is excellent and should be used as a benchmark for inefficient branches. Table 13-7 and Table 13-5 show the reference sets and their benchmark targets $R^{SBM}(l)$ by slack-based measure. Actually, by using the slack-based measure, L is benchmarked from branches on level 3 based on the second level evaluation context.

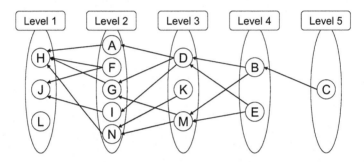

Figure 13-3. Diagram of Benchmark Targets by Radial Measure

Table 13-7. Reference Sets as Benchmark Targets by Slack-based Measure

Branch	Level	$R^{SBM}(1)$	$R^{SBM}(2)$	$R^{SBM}(3)$	$R^{SBM}(4)$
A	3	J	L		
B	4	J	L	D, N	
C	5	J	L	N	M
D	3	J	L, I		
E	4	H	L, I	D, N	
F	2	J			
G	2	J			
H	1				
I	2	J			
J	1				
K	4	J	L	N	
L	2	H			
M	4	J	L	N	
N	3	J	L		

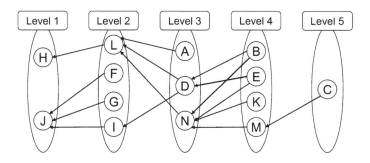

Figure 13-4. Diagram of Benchmark Targets by Slack-based Measure

Table 13-8. Attractiveness Scores by Radial Measure for the First and Second Levels Branches

Level	Branch	Evaluation context	
		E^2	E^3
1	H	1.242	1.523
	J	1.429	1.646
	L	1.190	1.319
2	F	1	1.200
	G	1	1.237
	I	1	1.062
	A	1	1.076
	N	1	1.162

Table 13-9. Attractiveness Scores by Slack-based Measure for the First and Second Levels Branches

Level	Branch	Evaluation context	
		E^2	E^3
1	H	1.143	1.313
	J	1.149	1.322
2	L	1	1.144
	F	1	1.061
	G	1	1.078
	I	1	1.029

We now turn to the attractiveness scores for the branches in the first and second levels. Table 13-8 and Table 13-9 report the attractiveness scores by radial measure and slack-based measure, respectively. On both of the evaluation contexts E^2 and E^3, J is the best branch because it has the largest attractiveness score.

Consider the sales branches on the first and second levels based on the evaluation context E^3. We can rank them in the order of J, H, L, G, F and I by slack-based measure. L is ranked in the third position by radial as well as slack-based context-dependent DEA, which indicate that the performance of L is good. It is reasonable that L is used as a benchmark by less efficient branches, which can be realized by the slack-based measure, not by the radial measure.

5. CONCLUDING REMARKS

This chapter presented a relative new DEA approach--context-dependent DEA. Interested read should refer to Seiford and Zhu (1999a), Seiford and Zhu (2003), Zhu (2003) and Morita, Hirokawa Zhu (2005) for detailed applications. Chen, Morita and Zhu (2005) also provided an illustrative application to measuring the performance of Tokyo public libraries. This chapter also show that non-zero slacks can be incorporated into the context-dependent DEA.

REFERENCES

1. Andersen, P. and N. C. Petersen, (1993), "A procedure for ranking efficient units in data envelopment analysis", Management Sciences, vol. 39, no. 10, 1261-1264.
2. Charnes, A., W. W. Cooper and E. Rhodes, (1978), "Measuring the efficiency of decision making units", European Journal of Operational Research, vol. 2, 429-444.
3. Chen, Y., H. Morita and J. Zhu, (2005), "Context-dependent DEA with an application to Tokyo public libraries", International Journal of Information Technology and Decision Making, Vol. 4, No. 3, 385-394.
4. Morita, H., K. Hirokawa and J. Zhu, (2005), "A slack-based measure of efficiency in context-dependent data envelopment analysis", Omega, vol. 33, 357-362.
5. Seiford, L.M. and J. Zhu, (1999a), "Profitability and marketability of the top 55 US commercial banks", Management Science, Vol. 45, No. 9, 1270-1288.
6. Seiford, L.M. and J. Zhu, (1999b), "Infeasibility of super-efficiency DEA models", INFOR, vol. 37, no. 2, 174-187.
7. Seiford, L.M. and J. Zhu, (2003), "Context-dependent data envelopment analysis: measuring attractiveness and progress", Omega, vol. 31, no. 5, 397-408.

8. Tone, K. (2002a), "A slack-based measure of super-efficiency in data envelopment analysis", European Journal of Operational Research, vol. 143, no. 1, 32-41.
9. Tone, K. (2002b), "A slack-based measure of efficiency in data envelopment analysis", European Journal of Operational Research, vol. 143, no. 3, 498-509.
10. Tversky, A. and I. Simonson, (1993), "Context-dependent preferences", Management Sciences, vol. 39, 1179-1189.
11. Zhu, J. (1996), Data envelopment analysis with preference structure. Journal of Operational Research Society, Vol. 47, No.1, 136-150
12. Zhu, J. (2003), "Quantitative models for performance evaluation and benchmarking –Data envelopment analysis with spreadsheets and DEA Excel solver", Kluwer, Boston.

Part of the material in this chapter is based upon OMEGA, vol. 31, no. 5, L. M. Seiford and J. Zhu, Context-dependent data envelopment analysis: measuring attractiveness and progress, pp. 397-408 (2003), and OMEGA, vol. 33, H. Morita, K. Hirokawa and J. Zhu, A slack-based measure of efficiency in context-dependent data envelopment analysis, pp. 357-362 (2005), with permission from Elsevier Science.

Chapter 14

FLEXIBLE MEASURES–CLASSIFYING INPUTS AND OUTPUTS

Wade D. Cook[1] and Joe Zhu[2]
[1]*Schulich School of Business, York University, Toronto, Ontario, Canada, M3J 1P3, wcook@shulich.yorku.ca*

[2]*Department of Management, Worcester Polytechnic Institute, Worcester, MA 01609, jzhu@wpi.edu*

Abstract: In standard data envelopment analysis (DEA), it is assumed that the input versus output status of each of the chosen analysis measures is known. In some situations, however, certain measures can play either input or output roles. Consider using the number of interns on staff in a study of hospital efficiency. Such a factor clearly constitutes an output measure for a hospital, being one form of training provided by the organization, but at the same time is an important component of the hospital's total staff complement, hence is an input. This chapter presents DEA models to accommodate such flexible measures. Both an individual DMU model and an aggregate model are suggested as methodologies for deriving the most appropriate designations for flexible measures.

Key words: Data Envelopment Analysis (DEA), Flexible, Inputs, Outputs

1. INTRODUCTION

In data envelopment analysis (DEA) (Charnes, Cooper and Rhodes 1978; CCR), efficiency of peer decision making units (DMUs) is defined as the ratio of weighted outputs to weighted inputs. It is assumed that the input versus output status of each performance measure related to the DMUs is known priori to the application of DEA model. For example, in a study of power plant efficiency by Cook et. al. (1998, 2005), one of the outputs is a function of what is termed 'outages'. This measure is designed to represent the percentage of time that the plant is available to be in operation. On the other hand, this variable can legitimately be considered as an environmental input that represents the ability of the plant to perform as it should.

Similar arguments to those above can be made regarding the evaluation of research productivity by universities, such as described in Beasley (1990, 1995). There, "research income" is treated as both an output and input. A related problem setting is that in which research - granting agencies (e.g., NSERC in Canada and NSF in the USA) wish to allocate funds to those researchers and universities such as to have the greatest impact. In this environment, graduate students can play the role of either an input (a resource available to the faculty member, effecting his/her productivity), or as an output (trained personnel, hence a benefit resulting from research funding). Medical interns have a similar interpretation in the evaluation of hospital efficiency. In a very different environment, Cook et al (1992), in evaluating robotics installations, use the measure "uptime" as an output that represents the percentage of production time available. At the same time, one might make the argument that this measure is, as well, an input that clearly influences the overall operation of the technology.

In many problem situations such as those described, the input versus output status of certain measures can be deemed as *flexible*. Cook and Zhu (2007) develop DEA models for classifying a measure into an input or output. Cook, Green and Zhu (2006) presents a methodology for dealing with performance evaluation settings where factors can simultaneously play both input and output roles.

This chapter presents the approach in Cook and Zhu (2007) in an effort to facilitate the derivation of the input/output status of variables when flexibility is an issue. The approach is illustrated with an example.

2. IDENTIFYING THE INPUT OUTPUT STATUS OF FLEXIBLE MEASURES

Suppose we have n peer DMUs, $\{DMU_j: j = 1, 2, ..., n\}$, and each DMU_j produces multiple outputs y_{rj}, ($r = 1, 2, ..., s$), by utilizing multiple inputs x_{ij}, ($i = 1, 2, ..., m$). When a DMU_o is under evaluation by the CCR model, we have (Charnes, Cooper and Rhodes, 1978)

$$\max \frac{\sum_{r=1}^{s}\mu_r y_{ro}}{\sum_{i=1}^{m}v_i x_{io}}$$

$$\text{s.t.} \quad \frac{\sum_{r=1}^{s}\mu_r y_{rj}}{\sum_{i=1}^{m}v_i x_{ij}} \leq 1, \quad j=1,2,\ldots,n \qquad (1)$$

$$\mu_r, v_i \geq 0, \quad \forall r, i$$

In Cook and Zhu (2007), it is assumed that there are L "flexible measures", whose input/output status we wish to determine. The values assumed by these measures are denoted as w_{lj} for DMU_j ($l=1,..,L$). For each measure l, binary variables $d_l \in \{0,1\}$ are introduced, where $d_l = 1$ designates that factor l is an output, and $d_l = 0$ designates it as an input.

Let γ_l be the weight for each measure l. The following programming problem is established to let DMU_o to choose the best status (input or output) for each measure l'[1].

$$\max \frac{\sum_{r=1}^{s}\mu_r y_{ro} + \sum_{l=1}^{L}d_l \gamma_l w_{lo}}{\sum_{i=1}^{m}v_i x_{io} + \sum_{l=1}^{L}(1-d_l)\gamma_l w_{lo}}$$

$$\text{s.t.} \quad \frac{\sum_{r=1}^{s}\mu_r y_{rj} + \sum_{l=1}^{L}d_l \gamma_l w_{lj}}{\sum_{i=1}^{m}v_i x_{ij} + \sum_{l=1}^{L}(1-d_l)\gamma_l w_{lj}} \leq 1 \quad j=1,2,\ldots,n \qquad (2)$$

$$d_l \in \{0,1\}, \forall l \quad \mu_r, v_i, \gamma_l \geq 0, \quad \forall r, i, l$$

For any *fixed* value d_l (whether 0 or 1), model (2) becomes the following linear program

---

[1] Note that since either $d_l = 0$ or $d_l = 1$, then measure l will end up either as an output or an input.

$$\max \sum_{r=1}^{s} \mu_r y_{ro} + \sum_{l=1}^{L} d_l \gamma_l w_{lo}$$

$$\text{s.t.} \quad \sum_{r=1}^{s} u_r y_{rj} + \sum_{l=1}^{L} d_l \gamma_l w_{lj} - \sum_{i=1}^{s} v_i x_{ij} - \sum_{l=1}^{L} (1-d_l) \gamma_l w_{lj} \leq 0 \quad j=1,\ldots,n; \quad (3)$$

$$\sum_{i=1}^{m} v_i x_{io} + \sum_{l=1}^{L} (1-d_l) \gamma_l w_{lo} = 1$$

$$d_l \in \{0,1\}, \quad \mu_r, v_i, \gamma_l \geq 0, \quad \forall r,i,l$$

Model (3) is clearly nonlinear. It can, however, be linearized by letting $\delta_l = d_l \gamma_l$, and, for each l, imposing the constraints

$$\delta_l \leq M d_l$$
$$\delta_l \leq \gamma_l \quad (4)$$
$$\gamma_l \leq \delta_l + M(1-d_l),$$

where M is a large positive number. These linear integer restrictions capture the nonlinear expression $d_l \gamma_l = \delta_l$, without actually having to directly specify it in the optimization model. Note that if $d_l = 1$ then $\gamma_l = \delta_l$, and if $d_l = 0$, then $\delta_l = 0$. We, therefore, have the following mixed integer linear program

$$\max \sum_{r=1}^{s} \mu_r y_{ro} + \sum_{l=1}^{L} \delta_l w_{lo}$$

$$\text{s.t.} \quad \sum_{r=1}^{s} \mu_r y_{rj} + 2 \sum_{l=1}^{L} \delta_l w_{lj} - \sum_{i=1}^{s} v_i x_{ij} - \sum_{l=1}^{L} \gamma_l w_{lj} \leq 0 \quad j=1,\ldots,n;$$

$$\sum_{i=1}^{m} v_i x_{io} + \sum_{l=1}^{L} \gamma_l w_{lo} - \sum_{l=1}^{L} \delta_l w_{lo} = 1$$

$$\delta_l \leq M d_l \quad (5)$$
$$\delta_l \leq \gamma_l$$
$$\gamma_l \leq \delta_l + M(1-d_l)$$
$$d_l \in \{0,1\}, \quad \delta_l, \gamma_l \geq 0, \quad \mu_r, v_i \geq 0, \quad \forall r,i$$

We point out that when only one flexible measure w is present, model (3) becomes

$$\max \sum_{r=1}^{s} \mu_r y_{ro} + d\gamma w_o \qquad (6)$$

s.t. $\sum_{r=1}^{s} \mu_r y_{rj} + d\gamma w_j - \sum_{i=1}^{s} v_i x_{ij} - (1-d)\gamma w_j \le 0 \qquad j=1,\ldots,n;$

$\sum_{i=1}^{m} v_i x_{io} + (1-d)\gamma w_o = 1$

$\gamma \ge 0,\ \mu_r, v_i \ge 0,\ \forall r, i\ \ d \in \{0,1\}.$

Replacing $d\gamma = \delta$ by a single constraint set as given by (4) for L=1, model (6) reduces to the integer linear program

$$\max \sum_{r=1}^{s} \mu_r y_{ro} + \delta w_o$$

s.t. $\sum_{r=1}^{s} \mu_r y_{rj} + 2\delta w_j - \sum_{i=1}^{s} v_i x_{ij} - \gamma w_j \le 0 \qquad j=1,\ldots,n;$

$\sum_{i=1}^{m} v_i x_{io} + \gamma w_o - \delta w_o = 1 \qquad (7)$

$\delta \le Md$

$\delta \le \gamma$

$\gamma \le \delta + M(1-d)$

$d \in \{0,1\},\ \gamma, \delta \ge 0,\ \mu_r, v_i \ge 0, \qquad \forall r, i$

If $d = 1$, then $\gamma = \delta$ and the factor w becomes an output. Model (7) then becomes

$$\max \sum_{r=1}^{s} \mu_r y_{ro} + \delta w_o$$

s.t. $\sum_{r=1}^{s} \mu_r y_{rj} + \delta w_j - \sum_{i=1}^{s} v_i x_{ij} \le 0 \qquad j=1,\ldots,n;$

$\sum_{i=1}^{m} v_i x_{io} = 1$

$\delta \ge 0,\ \mu_r, v_i \ge 0, \qquad \forall r, i$

If $d = 0$, then $\delta = 0$ and the factor w becomes an input, and model (7) becomes

$$\max \sum_{r=1}^{s} \mu_r y_{ro}$$

$$\text{s.t.} \sum_{r=1}^{s} \mu_r y_{rj} - \sum_{i=1}^{s} v_i x_{ij} - \gamma w_j \leq 0 \quad j=1,\ldots,n;$$

$$\sum_{i=1}^{m} v_i x_{io} + \gamma w_o = 1$$

$$\gamma \geq 0, \ \mu_r, v_i \geq 0, \quad \forall r, i$$

In the above development, we solve model (5), and obtain a set of optimal d_l^* for each DMU. One criterion for deciding the input versus output status of the measure, would be to base it on the majority choice among the DMUs. An alternative approach would be to use the designation (input or output), that renders the *aggregate efficiency* of the collection of DMUs as large as possible. Such a model would be helpful if ties are encountered using model (5)

Cook and Zhu (2007), therefore, assume that the optimal input/output will be the one created by the model which optimizes the aggregate or average ratio of outputs to inputs, namely

$$\max \frac{\sum_{r=1}^{s} \mu_r (\sum_{j=1}^{n} y_{rj}) + \sum_{l=1}^{L} d_l \gamma_l \sum_{j=1}^{n} w_{lj}}{\sum_{i=1}^{m} v_i (\sum_{j=1}^{n} x_{ij}) + \sum_{l=1}^{L} (1-d_l) \gamma_l \sum_{j=1}^{n} w_{lj}}$$

$$\text{s.t.} \quad \frac{\sum_{r=1}^{s} \mu_r y_{rj} + \sum_{l=1}^{L} d_l \gamma_l w_{lj}}{\sum_{i=1}^{m} v_i x_{ij} + \sum_{l=1}^{L} (1-d_l) \gamma_l w_{lj}} \leq 1 \quad j=1,2,\ldots,n \quad (8)$$

$$d_l \in \{0,1\}, \ \mu_r, v_i, \gamma_l \geq 0, \quad \forall r, i, l.$$

For notational convenience, let $\tilde{y}_r = \sum_{j=1}^{n} y_{rj}$, $\tilde{x}_i = \sum_{j=1}^{n} x_{ij}$, and $\tilde{w}_l = \sum_{j=1}^{n} w_{lj}$. Using (4), the aggregate problem is then equivalent to the following integer linear programming problem

$$\max \sum_{r=1}^{s} \mu_r \tilde{y}_r + \sum_{l=1}^{L} \delta_l \tilde{w}_l$$

s.t. $\sum_{r=1}^{s} \mu_r y_{rj} + 2\sum_{l=1}^{L} \delta_l w_{lj} - \sum_{i=1}^{s} v_i x_{ij} - \sum_{l=1}^{L} \gamma_l w_{lj} \leq 0, \quad j=1,\ldots,n$

$$\sum_{i=1}^{m} v_i \tilde{x}_i + \sum_{l=1}^{L} \gamma_l \tilde{w}_l - \sum_{l=1}^{L} \delta_l \tilde{w}_l = 1 \quad (9)$$

$\delta_l \leq M d_l$

$\delta_l \leq \gamma_l$

$\gamma_l \leq \delta_l + M(1 - d_l)$

$d_l \in \{0,1\}, \quad \mu_r, v_i, \gamma_l \geq 0, \quad \forall r, i, l$

3. APPLICATION

In this section, we apply the models to the data set in Beasley (1990). Table 14-1 reports the data set which consists of two inputs, General Expenditure and Equipment Expenditure and three outputs, consisting of three types of students. The "flexible measure" here is the Research Income.

Table 14-2 reports the results from model (7), where the second column shows the optimal d and the third column, the optimal value to model (7). In this case, 20 out of the 50 universities treat the research income measure as an output, i.e., the majority of 30 treat it as an input.

When we apply our aggregate model (9), the optimal $d = 1$, with the optimal objective function value being 0.69329. This indicates that from an aggregate efficiency perspective, research income is treated as an output. Part of the explanation for the different results between the two approaches may be that for the 30 cases where the input was the preferred status using model (7), that preference may not have been as strong as was the preference of output versus input in the other 20 cases. Hence, the majority concept may be flawed by failing to take account of the *difference* in efficiency for each DMU, when the flexible measure is used as an output versus an input.

Table 14-1. University data

DMU	General expenditure	Equipment expenditure	UG Students	PG Teaching	PG Research	Research Income
University1	528	64	145	0	26	254
University2	2605	301	381	16	54	1485
University3	304	23	44	3	3	45
University4	1620	485	287	0	48	940
University5	490	90	91	8	22	106
University6	2675	767	352	4	166	2967
University7	422	0	70	12	19	298
University8	986	126	203	0	32	776
University9	523	32	60	0	17	39
University10	585	87	80	17	27	353
University11	931	161	191	0	20	293
University12	1060	91	139	0	37	781
University13	500	109	104	0	19	215
University14	714	77	132	0	24	269
University15	923	121	135	10	31	392
University16	1267	128	169	0	31	546
University17	891	116	125	0	24	925
University18	1395	571	176	14	27	764
University19	990	83	28	36	57	615
University20	3512	267	511	23	153	3182
University21	1451	226	198	0	53	791
University22	1018	81	161	5	29	741
University23	1115	450	148	4	32	347
University24	2055	112	207	1	47	2945
University25	440	74	115	0	9	453
University26	3897	841	353	28	65	2331
University27	836	81	129	0	37	695
University28	1007	50	174	7	23	98
University29	1188	170	253	0	38	879
University30	4630	628	544	0	217	4838
University31	977	77	94	26	26	490
University32	829	61	128	17	25	291
University33	898	39	190	1	18	327
University34	901	131	168	9	50	956
University35	924	119	119	37	48	512
University36	1251	62	193	13	43	563
University37	1011	235	217	0	36	714
University38	732	94	151	3	23	297
University39	444	46	49	2	19	277
University40	308	28	57	0	7	154
University41	483	40	117	0	23	531
University42	515	68	79	7	23	305
University43	593	82	101	1	9	85
University44	570	26	71	20	11	130
University45	1317	123	293	1	39	1043
University46	2013	149	403	2	51	1523
University47	992	89	161	1	30	743
University48	1038	82	151	13	47	513
University49	206	1	16	0	6	72
University50	1193	95	240	0	32	485

Table 14-2. Results from model (7)

DMU	D	Efficiency	DMU	D	Efficiency
University1	1	1	University26	0	1
University2	0	1	University27	1	0.855471
University3	0	0.837244	University28	0	1
University4	1	0.685697	University29	1	0.824968
University5	0	1	University30	0	1
University6	0	1	University31	1	0.775853
University7	1	1	University32	0	0.896402
University8	1	0.811941	University33	1	1
University9	0	1	University34	0	1
University10	1	0.906595	University35	1	1
University11	0	0.890126	University36	0	0.8369
University12	1	0.709313	University37	1	0.830789
University13	0	0.803249	University38	0	0.833414
University14	0	0.767744	University39	0	0.791219
University15	0	0.704214	University40	1	0.741404
University16	0	0.54274	University41	1	1
University17	1	0.819451	University42	0	0.847172
University18	1	0.627824	University43	0	0.920638
University19	1	1	University44	0	1
University20	0	1	University45	0	1
University21	0	0.699625	University46	0	1
University22	1	0.716738	University47	1	0.688445
University23	0	0.617112	University48	0	0.938878
University24	0	1	University49	0	1
University25	1	1	University50	0	0.841683

4. CONCLUSIONS

Conventional DEA analyses require that each variable or measure be assigned an explicit designation specifying whether it is an input or output. In various settings, however, it remains that there are variables whose status is flexible. In cases where a resource can as well represents a tangible product of the organization (medical interns, graduate students, research funding, etc.), this flexibility is present.

The current chapter presents the Cook and Zhu' (2007) modification to the standard CCR model that permits the inclusion of flexible measures in the DEA analysis. The approach assigns an optimal designation, whether input or output to each such variable. Two models are given for accomplishing this, namely an individual DMU model, and one that optimizes the aggregate efficiency of the collection of DMUs.

REFERENCES

1. Beasley J., 1990, "Comparing university departments", Omega, 8(2), 171-183.
2. Beasley, J., 1995, "Determining teaching and research efficiencies", Journal of the Operational Research Society, 46, 441-452.
3. Charnes A., Cooper W., and Rhodes E., 1978, "Measuring the efficiency of decision making units", European Journal of Operational Research, 2(6), 428-444.
4. Cook, W.D. Green, R.H. and Zhu, Joe, 2006, "Dual-role factors in DEA", IIE Transactions, 38(2), 105-115.
5. Cook, W. D. and Green, R.H., 2005, "Evaluating power plant efficiency: a hierarchical model", Computers and Operations Research, 32, 813-823.
6. Cook, W. D., Chai, D., Doyle, J., and Green, R.H., 1998, "Hierarchies and groups in DEA", Journal of Productivity Analysis, 10, 177-198.
7. Cook, W. D, Johnston, D.J., and McCutcheon, D., 1992, "Impletation of robotics: identifying efficient implementers", OMEGA, 10, 221-232.
8. Cook, W. D. and J. Zhu, 2005, "Building performance standards into DEA structures", IIE Transactions, 37(3), 267-275.
9. Cook, W.D. and Zhu, J. 2007, "Classifying inputs and outputs in data envelopment analysis", forthcoming in European Journal of Operational Research.

Part of the material in this chapter is based upon Cook, W.D. and Zhu, Joe, Classifying inputs and outputs in data envelopment analysis, European Journal of Operational Research, Vol. 180, Issue 2 (2007), 692-699 with permission from Elsevier Science.

Chapter 15

INTEGER DEA MODELS
How DEA models can handle integer inputs and outputs

Sebastián Lozano and Gabriel Villa
Department of Industrial Management, University of Seville, slozano@us.es, gvilla@esi.us.es

Abstract: Conventional Data Envelopment Analysis (DEA) models consider that inputs and outputs are continuous (i.e. real-valued) amounts. However, there are many applications in which one or more inputs and/or outputs are necessarily integer quantities. Commonly, in these situations, the non-integer targets are rounded off. However, rounding off may easily lead to an infeasible target (i.e. out of the Production Possibility Set) or to a dominated operation point. In this chapter, a general framework to handle integer inputs and outputs is presented and a number of integer DEA models reviewed. We illustrate the working of the proposed approach with on a problem from the literature.

Key words: Data Envelopment Analysis (DEA), Performance, Efficiency, Integer data

1. INTRODUCTION

Conventional Data Envelopment Analysis (DEA) models consider that inputs and outputs are continuous (i.e. real-valued) amounts. However, there are many applications in which one or more inputs and/or outputs are necessarily integer quantities. Situations in which integer values appear naturally are those involving resources, services or outcomes which cannot be split, for example: number of workers, number of machines, number of customers, number of deliveries, number of transactions, number of stockouts, number of open accounts, number of iterations, etc. When the amounts are integer but large it is reasonable to consider them as continuous since the overall effect of rounding off the solution ex post may be small. That is not however the case with small integer values, which are usually associated to high cost or value items. In that case, one unit more or less makes a significant difference and therefore ad hoc rounding off the solution is not advisable.

Although the described situation can happen often, the subject of integer DEA models that guarantee this type of integrality constraints has not attracted too much attention. To our knowledge, only Lozano and Villa (2006) have dealt with this research topic. One reason why this issue has been previously ignored by researchers may be the attachment of the DEA community to Linear Programming (LP) models. It is true that most DEA models are LP and that LP DEA models can be efficiently solved even for large data sets (Dulá and López 2002). However, some DEA models also involve binary or integer variables (e.g. Lozano and Villa 2005) and they can still be efficiently solved.

One of the first steps in DEA, once inputs and outputs have been identified and the corresponding data gathered, is selecting the appropriate technology, i.e. determining the Production Possibility Set (PPS). The most common choices are the non-convex Free Disposal Hull (FDH) technology (Tulkens 1993) and the convex Constant and Variable Returns to Scale (CRS and VRS respectively) technologies. When some of the inputs and outputs are integer, the assumption of FDH technology carries no problem since FDH DEA models always project the units whose performance is to be assessed (i.e. the Decision Making Units, DMU) onto one of the existing DMU. Therefore FDH targets are always feasible in terms of the required integrality constraints.

On the contrary, in the case of the CRS and VRS technologies, the PPS assumes as feasible operation points that are convex combination of existing DMU but that do not necessarily respect any integrality constraint of inputs or outputs. The approach usually taken is to use a DEA model that ignores the integrality constraints and try to impose them a posteriori via some rounding off. Not only that can be tricky but also, as shown in Lozano and Villa (2006), arbitrarily rounding up or down inputs or outputs may easily lead to an infeasible (i.e. out of the PPS) target or to an inferior (i.e. dominated) operation point.

Although a rigorous ex post rounding procedure can be conceived, it would eventually require solving a Mixed Integer Linear program (MILP) in order to guarantee the integrality of some inputs and/or outputs (e.g. Lozano et al 2004). Instead of taking that longer route, experience has taught us that it is better to modify the formulation of DEA models to explicitly impose the required integrality constraints. This inevitably leads to MILP models but surprisingly they are generally solved easily.

The remainder of this chapter is organized as follows. In section 2 basic concepts and notation are introduced and two radial integer DEA models formulated. In section 3 an illustrative example is presented. In section 4 additional, integer DEA models are discussed. The last section summarizes and concludes.

2. INTEGER RADIAL DEA MODELS

Let

n	number of operating units
m	number of inputs
p	number of outputs
j=1,2,...,n	index for operating units
i=1, 2,...,m	index for inputs
k=1, 2,...,p	index for outputs
x_{ij}	amount of input i consumed by unit j
y_{kj}	quantity of output k produced by unit j
I'⊆{1,2,...,m}	subset of inputs that are integer
O'⊆{1,2,...,p}	subset of outputs that are integer

Of course, since we deal with the case that some data are integer, at least one of the subsets I' and O' is non-empty. As mentioned in the introduction, existing FDH DEA models can handle integer data without problem. Only when the underlying technology is convex it is necessary to modify the conventional DEA models in order to be able to guarantee targets that are integer in the required dimensions.

The first step is the definition of an appropriate integer technology. This can be done modifying the conventional convex technology with the addition of integrality constraints. Thus, given a convex technology T (CRS or VRS, for example), the corresponding integer technology is simply

$$T' = \{(\bar{x}, \bar{y}) \in T : x_i \text{ integer } \forall i \in I' \quad y_k \text{ integer } \forall k \in O'\} \quad (1)$$

where $\bar{x} = (x_1, x_2, ..., x_m)$ is a generic inputs vector and $\bar{y} = (y_1, y_2, ..., y_p)$ a generic outputs vector.

It is clear that T'⊂T. This inclusion property however does not hold for their respective efficient subsets, namely $(T')^{eff} \not\subset (T)^{eff}$ where

$$(T)^{eff} = \{(\bar{x}, \bar{y}) \in T : (\hat{x}, \hat{y}) \in T \cap \hat{x} \leq \bar{x} \cap \hat{y} \geq \bar{y} \Rightarrow (\hat{x}, \hat{y}) = (\bar{x}, \bar{y})\} \quad (2)$$

$$(T')^{eff} = \{(\bar{x}, \bar{y}) \in T' : (\hat{x}, \hat{y}) \in T' \cap \hat{x} \leq \bar{x} \cap \hat{y} \geq \bar{y} \Rightarrow (\hat{x}, \hat{y}) = (\bar{x}, \bar{y})\} \quad (3)$$

This can be shown by means of a counterexample. Consider the case of four DMU each one with a single integer input and a single integer output. The data are shown in Table 15-1 and graphically in Figure 15-1. Let us assume VRS.

Table 15-1. Input and output data for counterexample

DMU	Input	Output
A	2	1
B	5	2
C	6	7
D	7	3

The convex DEA efficient frontier $(T)^{eff}$ is the segment that links points (2,1) and (6,7). The efficient integer set is $(T')^{eff}=\{(2,1), (3,2), (4,4), (5,5)\}$, which includes points (3,2) and (5,5) that do not belong to $(T)^{eff}$.

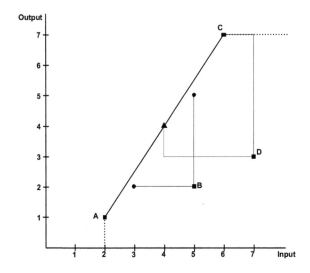

Figure 15-1. Differences between convex and integer efficient frontiers

This simple example also illustrates the following general property
$$\{(\bar{x},\bar{y}) \in (T)^{eff} : x_i \text{ integer } \forall i \in I' \quad y_k \text{ integer } \forall k \in O'\} \subseteq (T')^{eff} \quad (4)$$
i.e., any integer member of the convex efficient subset $(T)^{eff}$ is also a member of the integer efficient subset $(T')^{eff}$. As a corollary, since existing DMU are integer, those existing DMU that are efficient are also integer efficient. Thus, in the example above, the efficient integer subset $(T')^{eff}$ includes the three points (2,1), (4,4) and (6,7) which are the only integer points along the convex efficient frontier $(T)^{eff}$. and two of which correspond to the existing DMU that are efficient.

Summarizing, although the efficient integer frontier is close to the convex efficient frontier and there will always be some points at which they coincide, they are not identical in general.

Once the appropriate integer technology has been defined, the corresponding DEA model can be formulated. For simplicity, we will assume VRS. In what follows basic input and output-oriented radial formulations, analogous to the well-known BCC models (Banker et al 1984), are presented. The adaptation of other DEA models will be proposed in section 4.

Let

J	index of unit to be assessed
θ	radial reduction of inputs
s_i^-	reduction of input i
s_k^+	increase in output k
$(\lambda_1, \lambda_2, ..., \lambda_n)$	coefficients of convex linear combination
ε	non-Archimedean constant

Input-oriented integer radial VRS DEA model

$$\text{Min } \theta - \varepsilon \cdot \left(\sum_i s_i^- + \sum_k s_k^+ \right)$$

s.t.

$$\sum_j \lambda_j x_{ij} = x_i \quad \forall i \notin I'$$

$$\sum_j \lambda_j x_{ij} \leq x_i \quad \forall i \in I'$$

$$x_i = \theta \cdot x_{iJ} - s_i^- \quad \forall i$$

$$\sum_j \lambda_j y_{kj} = y_k \quad \forall k \notin O' \tag{5}$$

$$\sum_j \lambda_j y_{kj} \geq y_k \quad \forall k \in O'$$

$$y_k = y_{kJ} + s_k^+ \quad \forall k$$

$$\sum_j \lambda_j = 1$$

$$\lambda_j \geq 0 \ \forall j \quad s_i^-, x_i \geq 0 \ \forall i \quad s_k^+, y_k \geq 0 \ \forall k \quad \theta \text{ free}$$

$$x_i \text{ integer } \forall i \in I' \quad y_k \text{ integer } \forall k \in O'$$

Note that this model is slightly different to the CRS model presented in Lozano and Villa (2006) and the reason is that when VRS is assumed the integer DEA target may not lie within the convex hull of the original DMU although, according to the definition of the corresponding technology, it may be dominated by one such point. That is why the constraints that specify the relations between the convex linear combination and the target values are different for the continuous versus the integer inputs (an equality in the first case versus a $\leq$ inequality in the second) as well as for the continuous versus the integer outputs (an equality in the first case versus a $\geq$ inequality in the second). In the CRS model that distinction is not necessary and the integer DEA target is always equal to the linear combination of the existing DMU.

Model (5) searches among the integer solutions that dominate DMU J the one that leads to the largest radial input contraction. In order to avoid a weakly efficient target, slacks are added in the objective function so as to exhaust any possible additional input reduction and output increase without relaxing the integrality constraints.

Note that the number of integer variables of this MILP is equal to $|I'|+|O'|$ and therefore it is rather small. The values of variables x_i and y_k in the optimal solution to (5) define the integer efficient target sought. Since the only difference between this model and the conventional BCC-I model lies on the integrality constraints, the radial efficiency score θ^* obtained in (5) is always greater than or equal to the technical efficiency computed using BCC-I.

Also, using similar arguments as those in Lozano and Villa (2006), it is easy to prove that DMU J is integer efficient, i.e. belongs to $(T')^{eff}$, if and only if in the optimal solution to this model

$$\theta^*=1 \cap \left(s_i^-\right)^* = 0 \ \forall i \cap \left(s_k^+\right)^* = 0 \ \forall k \tag{6}$$

Another easy way to characterize existing integer efficient units is through the fact that they are projected onto themselves, whether using input orientation or output orientation.

Table 15-2 compares the results of the input-oriented projection using the proposed integer DEA approach versus convex DEA for the example of Table 15-1. Note that DMU A and C are projected onto themselves in both approaches indicating that they are efficient according to both models. For the other two DMU, the targets are different for the two approaches. The corresponding efficiency scores are also different, with a smaller value of the efficiency score θ^* (i.e. the maximal radial input reduction) for the convex DEA approach than for integer DEA. The integer DEA targets are also shown in Figure 15-1 where it can be graphically seen how the integer projection of DMU C does not lie on the convex DEA efficient frontier.

Table 15-2. Input-oriented convex and integer DEA results for the example

DMU	Convex DEA			Integer DEA		
	θ^*	Target		θ^*	Target	
		Input	Output		Input	Output
A	1.00	2	1	1.00	2	1
B	0.53	2.67	2	0.60	3	2
C	1.00	6	7	1.00	6	7
D	0.48	3.33	3	0.57	4	4

Let

γ radial expansion of inputs

<u>Output-oriented integer radial VRS DEA model</u>

Max $\gamma + \varepsilon \cdot \left(\sum_i s_i^- + \sum_k s_k^+ \right)$

s.t.

$\sum_j \lambda_j x_{ij} = x_i \quad \forall i \notin I'$

$\sum_j \lambda_j x_{ij} \leq x_i \quad \forall i \in I'$

$x_i = x_{iJ} - s_i^- \quad \forall i$

$\sum_j \lambda_j y_{kj} = y_k \quad \forall k \notin O'$ (7)

$\sum_j \lambda_j y_{kj} \geq y_k \quad \forall k \in O'$

$y_k = \gamma \cdot y_{kJ} + s_k^+ \quad \forall k$

$\sum_j \lambda_j = 1$

$\lambda_j \geq 0 \ \forall j \quad s_i^-, x_i \geq 0 \ \forall i \quad s_k^+, y_k \geq 0 \ \forall k \quad \gamma \text{ free}$

x_i integer $\forall i \in I' \quad y_k$ integer $\forall k \in O'$

Analogous to the input-oriented case, the radial efficiency score γ^* obtained in (7) is always lower than or equal to the technical efficiency computed using BCC-O and DMU J belongs to $(T')^{eff}$ if and only if in the optimal solution to this model

$$\gamma^* = 1 \cap \left(s_i^-\right)^* = 0 \ \forall i \cap \left(s_k^+\right)^* = 0 \ \forall k \tag{8}$$

Table 15-3 compares the results of the output-oriented projection for the example of Table 15-1. Again, as expected, DMU A and C are projected onto themselves. Both models project DMU D onto the same point, namely DMU C, giving thus the same value for the radial efficiency score γ^*. On the contrary, DMU C is projected onto a different point by each method giving a higher value of γ^* for the integer DEA approach than for the convex DEA approach.

Table 15-3. Output-oriented convex and integer DEA results for the example

DMU	Convex DEA			Integer DEA		
	γ^*	Target		γ^*	Target	
		Input	Output		Input	Output
A	1.00	2	1	1.00	2	1
B	2.75	5	5.5	2.50	5	5
C	1.00	6	7	1.00	6	7
D	2.33	6	7	2.33	6	7

3. ILLUSTRATION OF INTEGER RADIAL DEA MODEL

In this section we will show the usefulness of the proposed integer DEA approach by considering its application to a problem that the authors have studied in previous research (Lozano and Villa 2002). The application consists in the measurement of the performance of nations at Olympic Games, where essential outputs are the number (integer) of gold, silver and bronze medals. VRS and an output orientation are assumed. The approach considers two (non-discretionary) inputs, namely Gross National Product and Population. Three outputs were considered: Number of Gold (NG), Silver (NS) and Bronze (NB) medals won. Appropriate weight restrictions are required to ensure that gold medals are not given valued less than silver medals or silver medals valued less than bronze medals.

Although the three outputs are integer-valued, in Lozano and Villa (2002) a conventional VRS DEA model that did not take into account these integrality constraints was used. In this section we compare the results of the proposed integer DEA model to the convex DEA approach. Table 15-4 shows, for the Olympic Games of Athens 2004, the results of the convex and integer DEA approaches. Note that computing times are not shown and that is because, although one approach is LP and the other MILP, the computing times are negligible (less than 1 second) in both cases. The rows of the DMU that are found efficient by each approach are indicated in bold. Note that the integer DEA radial efficiency score γ^* is lower than or equal to the convex DEA radial efficiency score γ^* and that the number of integer efficient DMU is larger than the number of convex efficient DMU.

Table 15-4. Convex DEA versus integer DEA for Athens 2004 Olympic Games

DMU	Convex DEA				Integer DEA			
	γ^*	Target			γ^*	Target		
		NG	NS	NB		NG	NS	NB
Argentina	6.70	13.39	11.89	16.87	6.50	13	11	16
Australia	1.00	17.00	16.00	16.00	1.00	17	16	16
Austria	2.62	7.33	6.33	6.93	2.50	7	6	7
Azerbaijan	2.17	2.76	2.55	1.20	1.75	2	2	2
Bahamas	1.00	1.00	0.00	1.00	1.00	1	0	1
Belarus	1.25	5.11	4.22	4.89	1.22	5	4	5
Belgium	9.44	9.44	6.04	6.90	9.00	9	8	8
Brazil	6.80	27.20	26.80	37.26	6.75	27	27	38
Bulgaria	1.84	5.29	4.28	5.31	1.74	5	4	5
Cameroon	3.70	3.70	3.21	2.67	3.00	3	3	3
Canada	4.41	17.95	17.05	18.09	4.25	17	17	18
Chile	5.05	10.10	8.23	12.15	5.00	10	8	12
China	1.00	32.00	17.00	14.00	1.00	32	17	14
Chinese Taipei	7.18	14.79	13.50	15.10	6.83	14	13	15
Colombia	81.14	11.61	9.90	14.91	76.00	11	9	14
Croatia	2.64	4.06	2.73	4.70	2.40	4	2	4
Cuba	1.00	9.00	7.00	11.00	1.00	9	7	11
Czech Republic	4.04	8.39	6.68	9.70	3.93	8	7	9
Denmark	2.41	5.14	4.14	4.89	2.36	5	4	5
Dominican Rep.	5.73	5.73	4.67	5.86	5.00	5	5	6
DPR Korea	2.95	4.14	3.53	3.36	2.75	4	3	3
Egypt	9.61	12.31	10.67	15.96	9.22	12	10	15
Eritrea	1.00	0.00	0.00	1.00	1.00	0	0	1
Estonia	2.43	1.67	0.62	1.77	1.75	1	1	1
Ethiopia	1.04	2.66	2.47	1.04	1.00	2	3	2
Finland	7.00	5.00	4.00	4.75	6.50	5	3	4
France	1.84	20.22	19.55	23.09	1.82	20	19	23
Georgia	1.00	2.00	2.00	0.00	1.00	2	2	0
Germany	1.49	22.07	21.58	27.16	1.48	22	21	27
Great Britain	2.16	20.19	19.51	23.01	2.14	20	19	23
Greece	1.44	9.09	7.70	9.56	1.39	9	7	9
Hong Kong	17.94	6.31	5.31	5.98	17.00	6	5	6
Hungary	1.04	8.31	6.23	9.46	1.00	8	6	9
I. R. Iran	6.94	14.35	12.94	19.02	6.67	14	12	19
India	81.51	27.15	27.22	37.84	81.00	27	27	38
Indonesia	14.80	16.43	15.26	22.14	14.50	16	15	22

Table 15-4 (continued)

DMU	Convex DEA				Integer DEA			
	γ^*	Target			γ^*	Target		
		NG	NS	NB		NG	NS	NB
Ireland	4.11	4.11	0.94	2.34	4.00	4	2	3
Israel	6.07	6.07	2.80	5.00	6.00	6	4	6
Italy	1.92	20.06	19.36	22.72	1.90	20	19	22
Jamaica	1.05	2.11	1.23	1.86	1.00	2	1	2
Japan	1.61	25.69	25.56	35.11	1.56	25	25	35
Kazakhstan	3.57	7.94	6.24	9.33	3.33	7	6	10
Kenya	2.01	4.30	3.65	3.62	1.83	4	3	4
Korea	1.90	19.27	18.50	20.99	1.87	19	18	20
Latvia	1.55	2.46	1.28	2.82	1.25	2	1	2
Lithuania	2.16	3.30	2.03	3.83	2.00	3	2	3
Mexico	23.56	23.67	23.34	30.67	23.00	23	23	30
Mongolia	1.00	0.00	0.00	1.00	1.00	0	0	1
Morocco	4.74	9.47	7.53	11.71	4.50	9	7	11
Netherlands	2.21	14.00	13.00	13.18	2.16	14	12	13
New Zealand	1.31	3.92	2.62	3.87	1.00	3	3	4
Nigeria	33.00	9.65	7.72	11.97	30.50	9	7	11
Norway	1.00	5.00	0.00	1.00	1.00	5	0	1
Paraguay	7.68	2.62	2.44	1.00	6.00	2	2	1
Poland	4.83	14.49	13.12	17.36	4.67	14	13	17
Portugal	11.63	8.63	7.18	9.28	11.00	8	7	9
Romania	1.26	10.08	8.20	12.61	1.25	10	8	12
Russia	1.00	27.00	27.00	38.00	1.00	27	27	38
Serbia/Montenegro	7.16	5.06	4.19	4.81	7.00	5	4	4
Slovakia	2.13	4.73	3.32	5.54	2.00	4	4	4
Slovenia	2.77	2.27	1.19	2.39	2.40	2	1	2
South Africa	7.96	13.75	12.28	18.01	7.60	13	12	18
Spain	3.25	18.72	17.89	19.78	3.12	18	17	19
Sweden	2.10	8.41	4.59	5.38	2.00	8	7	7
Switzerland	4.98	6.74	5.74	6.39	4.67	6	6	6
Syrian Arab Rep.	42.02	6.29	5.06	6.74	41.00	6	5	7
Thailand	4.84	14.51	13.13	19.27	4.67	14	13	19
Trinidad/Tobago	10.15	1.74	0.66	1.89	7.00	1	1	1
Turkey	5.65	17.31	16.23	23.24	5.56	17	16	23
U. Arab Emirates	3.90	3.90	1.83	3.61	3.00	3	3	4
Ukraine	1.09	9.84	7.94	12.27	1.00	9	7	12
United States	1.00	35.00	39.00	29.00	1.00	35	39	29
Uzbekistan	1.91	3.83	3.30	2.87	1.50	3	3	4
Venezuela	39.30	11.33	9.59	14.10	38.00	11	9	14
Zimbabwe	2.34	2.62	2.44	0.98	1.86	2	2	1

Note that the targets of the convex DEA approach are generally fractional and that the optimal integer output targets given by the proposed model are not always a rounding up or down of the fractional output targets. Thus, for example, the integer DEA target for Belgium is (9, 8, 8) which is not the result of just rounding up or down its convex DEA target (9.44, 6.04, 6.90). Analogously, the integer target for Israel (6, 4, 6) differs significantly from the convex target (6.07, 2.80, 5.00).

Even when the integer values correspond to rounding up or down the fractional DEA targets there is no rule to ascertain which fractional values should be rounded up and which ones rounded down. The convex target for Colombia (11.61, 9.90, 14.91) is optimally rounded to (11, 9, 14) while for Dominican Republic (5.73, 4.67, 5.86) is optimally rounded to (5, 5, 6).

Summarizing, for this application, the consideration of the integer nature of the outputs leads to more valid targets, i.e. guaranteed to lie within the integer PPS and to be non-dominated by any other feasible operation point, as well as more reliable efficiency scores (from which a more effective ranking of DMU can be derived).

4. OTHER INTEGER DEA MODELS

The above section has illustrated how integrality constraints can be added to a radial DEA model in order to handle integer inputs and outputs. It has been also seen that this is compatible with the existence of weight restrictions as well as with the non-discretional character of the inputs or outputs.

Radial models are not the only DEA models that can be thus adapted. Actually, since the efficiency scores in radial DEA models do not capture all efficiency gains and since in the case of integer inputs and outputs the remaining (i.e. after the radial phase) slacks may be relatively important, other DEA formulations may be more appropriate. In this section we will expand the integer DEA approach presenting other DEA models that can also be adapted to handle integer inputs and outputs.

Perhaps the easiest integer DEA model that can be formulated is the integer additive model, which is mathematically related to Phase II of the integer radial VRS model, i.e.

Integer additive DEA model

$$\text{Max} \quad \sum_i s_i^- + \sum_k s_k^+$$

s.t.

$$\sum_j \lambda_j x_{ij} = x_i \quad \forall i \notin I'$$

$$\sum_j \lambda_j x_{ij} \leq x_i \quad \forall i \in I'$$

$$x_i = x_{iJ} - s_i^- \quad \forall i$$

$$\sum_j \lambda_j y_{kj} = y_k \quad \forall k \notin O' \qquad (9)$$

$$\sum_j \lambda_j y_{kj} \geq y_k \quad \forall k \in O'$$

$$y_k = y_{kJ} + s_k^+ \quad \forall k$$

$$\sum_j \lambda_j = 1$$

$$\lambda_j \geq 0 \ \forall j \quad s_i^-, x_i \geq 0 \ \forall i \quad s_k^+, y_k \geq 0 \ \forall k$$

$$x_i \text{ integer } \forall i \in I' \quad y_k \text{ integer } \forall k \in O'$$

One problem with additive DEA models is that the interpretation of the objective function is meaningful only when the inputs and output slacks are in the same units, which only occurs occasionally (e.g. Lozano and Salmerón 2005).

A preferable alternative may be the input-oriented non-radial Russell DEA model, which differs from the BCC-I model in that instead of using a single contraction factor for all the inputs it uses a different one for each input dimension. The Russell input-oriented measure of technical efficiency is the average of these factors along the different input dimensions. Integer constraints can be easily added to this type of models as follows.

Let
$\quad \theta_i \qquad \qquad$ reduction factor for input i

Input-oriented integer non-radial Russell VRS DEA model

$$\text{Min} \quad \frac{1}{m}\sum_i \theta_i - \varepsilon \cdot \sum_k s_k^+$$

s.t.

$$\sum_j \lambda_j x_{ij} = x_i \quad \forall i \notin I'$$

$$\sum_j \lambda_j x_{ij} \leq x_i \quad \forall i \in I'$$

$$x_i = \theta_i x_{iJ} \quad \forall i$$

$$\theta_i \leq 1 \quad \forall i$$

$$\sum_j \lambda_j y_{kj} = y_k \quad \forall k \notin O' \tag{10}$$

$$\sum_j \lambda_j y_{kj} \geq y_k \quad \forall k \in O'$$

$$y_k = y_{kJ} + s_k^+ \quad \forall k$$

$$\sum_j \lambda_j = 1$$

$$\lambda_j \geq 0 \ \forall j \quad x_i \geq 0 \ \forall i \quad s_k^+, y_k \geq 0 \ \forall k \quad \theta_i \text{ free}$$

$$x_i \text{ integer } \forall i \in I' \quad y_k \text{ integer } \forall k \in O'$$

The output-oriented non-radial Russell DEA model can analogously be adapted for integer data. In this model, instead of a single radial expansion scores applied to all the output dimensions, a different expansion factor for each output dimension is defined and an efficiency score is computed based on the average of these expansion factors.

Let

γ_k expansion factor for output k

Output-oriented integer non-radial Russell VRS DEA model

$$\text{Max} \quad \frac{1}{p}\sum_k \gamma_k + \varepsilon \cdot \sum_i s_i^-$$

s.t.

$$\sum_j \lambda_j x_{ij} = x_i \quad \forall i \notin I'$$

$$\sum_j \lambda_j x_{ij} \leq x_i \quad \forall i \in I'$$

$$x_i = x_{iJ} - s_i^- \quad \forall i$$

$$\sum_j \lambda_j y_{kj} = y_k \quad \forall k \notin O' \qquad (11)$$

$$\sum_j \lambda_j y_{kj} \geq y_k \quad \forall k \in O'$$

$$y_k = \gamma_k \cdot y_{kJ} \quad \forall k$$

$$\gamma_k \geq 1 \quad \forall k$$

$$\sum_j \lambda_j = 1$$

$$\lambda_j \geq 0 \ \forall j \quad s_i^-, x_i \geq 0 \ \forall i \quad y_k \geq 0 \ \forall k \quad \gamma_k \text{ free}$$

$$x_i \text{ integer } \forall i \in I' \quad y_k \text{ integer } \forall k \in O'$$

Unfortunately, the oriented non-radial Russell DEA models formulated above also suffer from the problem of the incommensurability of slacks, although less than radial models and much less than the additive model. To completely avoid the problem the following non-oriented integer non-radial DEA model can be used.

Let
$\quad \alpha_i \qquad\qquad$ relative reduction of input i
$\quad \beta_k \qquad\qquad$ relative increase of output k

Non-oriented integer non-radial VRS DEA model

$$\text{Max} \quad \frac{1}{m+p}\left(\sum_i \alpha_i + \sum_k \beta_k\right)$$

s.t.

$$\sum_j \lambda_j x_{ij} = x_i \quad \forall i \notin I'$$

$$\sum_j \lambda_j x_{ij} \leq x_i \quad \forall i \in I'$$

$$x_i = x_{iJ} \cdot (1-\alpha_i) \quad \forall i$$

$$\sum_j \lambda_j y_{kj} = y_k \quad \forall k \notin O' \qquad (12)$$

$$\sum_j \lambda_j y_{kj} \geq y_k \quad \forall k \in O'$$

$$y_k = y_{kJ} \cdot (1+\beta_k) \quad \forall k$$

$$\sum_j \lambda_j = 1$$

$$\lambda_j \geq 0 \quad \forall j \quad \alpha_i, x_i \geq 0 \quad \forall i \quad \beta_k, y_k \geq 0 \quad \forall k$$

$$x_i \text{ integer } \forall i \in I' \quad y_k \text{ integer } \forall k \in O'$$

This model is equivalent to the following formulation, which can be labelled the integer MIP (Measure of Inefficiency Proportions, Charnes et al 1985) model

Integer MIP VRS DEA model

$$\text{Max} \quad \sum_i \frac{s_i^-}{x_{iJ}} + \sum_k \frac{s_k^+}{y_{kJ}}$$

s.t.

$$\sum_j \lambda_j x_{ij} = x_i \quad \forall i \notin I'$$

$$\sum_j \lambda_j x_{ij} \leq x_i \quad \forall i \in I'$$

$$x_i = x_{iJ} - s_i^- \quad \forall i$$

$$\sum_j \lambda_j y_{kj} = y_k \quad \forall k \notin O' \quad \quad (13)$$

$$\sum_j \lambda_j y_{kj} \geq y_k \quad \forall k \in O'$$

$$y_k = y_{kJ} + s_k^+ \quad \forall k$$

$$\sum_j \lambda_j = 1$$

$$\lambda_j \geq 0 \ \forall j \quad s_i^-, x_i \geq 0 \ \forall i \quad s_k^+, y_k \geq 0 \ \forall k$$

$$x_i \text{ integer } \forall i \in I' \quad y_k \text{ integer } \forall k \in O'$$

Models (12) and (13) have the added advantage of being units-invariant. Units-invariance is a very desirable, from a practical point of view, feature of DEA models since the dependency of targets and efficiency gains on the units of measurement of the data does not contribute to increase the credibility of DEA in the eyes of the final user of the results. Even if the final user remains unaware of the possible lack of units-invariance in the results, that does not solve the issue. Note, however, that we are more interested in the units-invariance in relation to the continuous inputs and outputs than in relation to the integer ones since for the integer data with which we are concerned here there are usually no alternative units of measurement, i.e. they take small values (e.g. number of machines) that are absolute numbers.

Another units-invariant model that can be considered here is the RAM (Range Adjusted Measure of Inefficiency, Cooper et al 1999) model.

Let
$R_i = \max_j x_{ij} - \min_j x_{ij}$ range of input i values
$R'_k = \max_j y_{kj} - \min_j y_{kj}$ range of output k values

Integer RAM DEA model

$$\text{Max} \quad \sum_i \frac{s_i^-}{R_i} + \sum_k \frac{s_k^+}{R'_k}$$

s.t.

$$\sum_j \lambda_j x_{ij} = x_i \quad \forall i \notin I'$$

$$\sum_j \lambda_j x_{ij} \leq x_i \quad \forall i \in I'$$

$$x_i = x_{iJ} - s_i^- \quad \forall i$$

$$\sum_j \lambda_j y_{kj} = y_k \quad \forall k \notin O' \tag{14}$$

$$\sum_j \lambda_j y_{kj} \geq y_k \quad \forall k \in O'$$

$$y_k = y_{kJ} + s_k^+ \quad \forall k$$

$$\sum_j \lambda_j = 1$$

$$\lambda_j \geq 0 \ \forall j \quad s_i^-, x_i \geq 0 \ \forall i \quad s_k^+, y_k \geq 0 \ \forall k$$

$$x_i \text{ integer } \forall i \in I' \quad y_k \text{ integer } \forall k \in O'$$

As it can be seen, treatment of integer inputs and outputs in DEA consists basically in the addition of appropriate integrality constraints. This is easy to do in all LP DEA models transforming them in MILP but without significant increase in computational burden. Thus, other DEA models such as the normalized weighted additive model (Lovell and Pastor 1995) and the preference structure models of Thanassoulis and Dyson (1992) or Zhu (1996) can be similarly adapted to handle integer inputs and outputs.

5. CONCLUSIONS

In this chapter the necessity and convenience of taking into account the integer nature of certain inputs and outputs of DEA applications has been discussed and an effective approach for adapting common DEA models have been presented and illustrated. The proposed approach consists in introducing variables in the model to explicitly define the target inputs and outputs and imposing integrality constraints on these as required.

Note however, that not all DEA models can be adapted in this simple way. Thus, Non Linear Programming (NLP) DEA models such as the Russell Graph model (Färe et al 1985) or the MEP (Measure of Efficiency Proportions, Cooper et al 1999, Green et al 1997) model cannot be so easily adapted since the resulting model would be a difficult Mixed Integer NLP. Neither can other DEA models such as the enhanced Russell Graph (Pastor et al 1999) a.k.a. SBM (Slacks-Based Measure of Efficiency, Tone 2001) that are NLP but can be linearized.

REFERENCES

1. Banker, R.D., A. Charnes, W.W. Cooper (1984) "Some models for estimating technical and scale inefficiencies in data envelopment analysis", Management Science, 30, 1078-1092.
2. Charnes, A., W.W. Cooper, B. Golany, L. Seiford, L. and J. Stutz (1985) "Foundations of Data Envelopment Analysis and Pareto-Koopmans Empirical Production Functions", *Journal of Econometrics*, 30, 91-107.
3. Cooper, W.W., K.S. Park, J.T. Pastor (1999), "RAM: A Range Adjusted Measure of Inefficiency for Use with Additive Models, and Relations to Other Models and Measures in DEA", Journal of Productivity Analysis, 11, 5-42.
4. Dulá, J.H., F.J. López (2002) "Data Envelopment Analysis (DEA) in massive data sets," in *Handbook of Massive Data Sets*, J. Abello, P.M. Pardalos, M.G.C. Resende (eds.), Kluwer Academic Publishers, 419-437.
5. Färe, R., S. Grosskopf, C.A.K. Lovell (1985) *The Measurement of Efficiency of Production*, Kluwer-Nijhoff Publishing, Dordrecht
6. Green, R.H., W. Cook, J. Doyle (1997) "A note on the additive data envelopment analysis model", Journal of the Operational Research Society, 48- 446-448.
7. Lovell, C.A.K., J.T. Pastor (1995) "Units invariant and translation invariant DEA models", Operations Research Letters, 18, 147-151.
8. Lozano, S., J.L. Salmerón (2005) "Data Envelopment Analysis of OR/MS Journals", Scientometrics, 64, 2, 133-150.
9. Lozano, S., G. Villa, F. Guerrero, P. Cortés (2002), "Measuring the Performance of Nations At the Summer Olympics Using Data Envelopment Analysis", Journal of the Operational Research Society, 53, 5, 501-511.

10. Lozano, S., G. Villa, B. Adenso-Díaz (2004) "Centralised Target Setting for Regional Recycling Operations Using DEA", Omega, 32, 2, 101-110.
11. Lozano, S., G. Villa (2005) "Centralized DEA models with the possibility of downsizing", Journal of the Operational Research Society, 56, 4, 357-364.
12. Lozano, S., G. Villa (2006) "Data envelopment analysis of integer-valued inputs and outputs", Computers and Operations Research, 33, 10, 3004-3014.
13. Pastor, J.T., J.L. Ruiz, I. Sirvent (1999) "An enhanced DEA Russell graph efficiency measure", European Journal of Operational Research, 115, 596-607.
14. Thanassoulis, E., R.G. Dyson, (1992) "Estimating preferred target input-output levels using data envelopment analysis", European Journal of Operational Research, 56, 80-97.
15. Tone, K. (2001) "A slacks-based measure of efficiency in data envelopment analysis", European Journal of Operational Research, 130, 498-509.
16. Tulkens H. (1993) "On FDH Analysis: Some Methodological Issues and Applications to Retail Banking, Courts and Urban Transit", Journal of Productivity Analysis, 4, 183-210.
17. Zhu, J. (1996) "Data Envelopment Analysis with Preference Structure", Journal of the Operational Research Society, 47, 136-150.

Chapter 16

DATA ENVELOPMENT ANALYSIS WITH MISSING DATA
A Reliable Solution Method

Chiang Kao[1] and Shiang-Tai Liu[2]
[1]*Department of Industrial and Information Management, National Cheng Kung University, Tainan 701, Taiwan, Republic of China, ckao@mail.ncku.edu.tw*
[2]*Graduate School of Business and Management, Vanung University, Chung-Li, Tao-Yuan 320, Taiwan, Republic of China, stliu@vnu.edu.tw*

Abstract: In data envelopment analysis (DEA), the input and output data from all of the decision making units (DMUs) to be compared are required. If, for any reason, some data are missing, then the associated DMU must be eliminated to make the approach applicable. This study proposes a fuzzy set approach to deal with missing values. The value of a DMU in an input (or output) which is missing is represented by a triangular fuzzy number constructed from the values of other DMUs in that input (or output). A fuzzy DEA model is then used to calculate the efficiencies, which are usually also fuzzy numbers. We use a problem with complete data to investigate the effect of this approach when 1%, 2%, and 5% of the values are missing. While the conventional DMU-deletion method will overestimate the efficiencies of the remaining DMUs, the fuzzy set approach produces results which are very close to those calculated from complete data. The average error in estimating the true efficiency is less than 0.3%. Most importantly, the fuzzy set approach is able to calculate the efficiencies of all DMUs, including those with some values missing.

Key words: Data envelopment analysis, efficiency, missing data, fuzzy set

1. INTRODUCTION

Since the pioneering work of Charnes et al. (1978), data envelopment analysis (DEA) has been widely studied from both the theoretical and practical points of view. Different models have been developed to measure the efficiency of a group of decision making units (DMUs) which utilize the same inputs to produce the same outputs under different conditions. Applications for different types of organizations have also been reported (Seiford 1996, 1997, Cooper et al. 2000).

The basic idea of DEA is to allow each decision making unit to use different virtual multipliers, the most favorable, in calculating its relative efficiency expressed as the ratio of aggregated output to aggregated input. In selecting the multipliers, it is required that the ratio of aggregated output to aggregated input calculated from the multipliers selected by the DMU concerned should not exceed 1.0 for all DMUs. Let Xij, j=1,..., s and Yik, k=1,..., t denote the jth input and kth output, respectively, of DMU i, i=1,..., n. Banker et al. (1984) develop the following mathematical program to calculate the efficiency of DMU r:

$$E_r = \max. \sum_{k=1}^{t} u_k Y_{rk} / (v_0 + \sum_{j=1}^{s} v_j X_{rj})$$

$$\text{s.t.} \sum_{k=1}^{t} u_k Y_{ik} / (v_0 + \sum_{j=1}^{s} v_j X_{ij}) \leq 1, \quad i = 1,...,n \quad (1)$$

$$u_k, v_j \geq \varepsilon > 0, \quad v_0 \text{ unrestricted in sign,}$$

where u_k and v_j are the multipliers associated with output k and input j, respectively, to be determined from (1) and ε is a small non-Archimedean number (Charnes et al. 1979, Charnes and Cooper 1984) imposed to avoid DMU r from assigning zero weight to unfavorable factors. E_r is the relative efficiency of DMU r, where $E_r=1$ indicates efficiency and $E_r<1$ inefficiency. Model (1) is a linear fractional program which can be solved by transforming to a linear program (Charnes and Cooper 1962) and utilizing any linear programming solver. The underlying assumption of this model is variable returns to scale. If v_0 is set to zero, then the model boils down to one under the assumption of constant returns to scale

As indicated by its name, data envelopment analysis is based on data. To calculate the relative efficiency, data from all DMUs are required. If any observation of a DMU in the group is missing, then this DMU must be deleted from the group in order to calculate the efficiency of all other DMUs. A consequence of this is overestimation of the efficiency of some DMUs, because the number of DMUs for comparison is decreased. As more DMUs are deleted due to lack of data, the resulting efficiencies will be biased high to a larger extent. Therefore, it is desirable to keep those DMUs with some observations missing in the group by making some amendments.

The problem of missing data has been widely discussed in statistical analysis (Allison 2002, Rubin 2004, Schafer 1997). There are formulas for estimating missing values in randomized blocks, Latin squares, etc. (Snedecor and Cochran 1967). In general applications, a probability distribution such as triangular or beta is usually assumed for the data which are missing (Law and Kelton 1991). The studies of Simar and Wilson (1998, 2000) use this idea, although they are not developed for dealing with missing data. To the knowledge of the authors, the only journal article which proposes a methodology for handling missing data in DEA is the one by Kao and Liu (2000b). Instead of assuming a probability distribution for missing values, a membership function of fuzzy set theory is assumed. The fuzzy set approach allows the derivation of the membership function of the efficiency from the membership functions of the input and output data (Kao and Liu 2000a). With the membership functions of the efficiencies, generalized means can be calculated and rankings of the DMUs can be subsequently made. This approach has been successfully applied by Kao and Liu (2000b) to measure the efficiency of 24 university libraries in Taiwan, where 3 out of 144 observations were missing. Since the data in that study were really missing, whether the calculated efficiencies are correct or not cannot be verified. In this study we select a problem with complete data from the DEA literature, randomly delete some data to make it incomplete, and modify the approach of Kao and Liu (2000b) to calculate the efficiency. The results from the incomplete data sets are compared with those from the complete data set to investigate the reliability of the fuzzy set approach.

In the following, the fuzzy set approach of Kao and Liu (2000b) for handling missing data is first briefly reviewed. Secondly, different deletion rates are applied to the complete data set of a problem to calculate the efficiency from the incomplete data sets by using a modified approach of Kao and Liu (2000b). The efficiencies and ranks calculated from incomplete data are then compared with those calculated from the complete data. Finally, conclusion is made based on the comparison.

2. THE FUZZY SET APPROACH

When there are data in a group of DMUs which, for any reason, are missing, their values must be estimated so that the DEA approach can be applied. The simplest way is to find the most likely value to represent the missing data. However, how to find a representative value is a problem. In statistical analysis it is usually to assume that the missing data follows a probability distribution. Unfortunately, this approach involves sophisticated mathematical derivation and cumbersome simulation analysis, and none of the existing studies in DEA has successfully accomplished this task. Alternatively, Kao and Liu (2000b) tackle this problem by assuming the missing data to be fuzzy numbers and then applying the fuzzy DEA approach of Kao and Liu (2000a).

$$\tilde{E}_r = \max. \sum_{k=1}^{t} u_k \tilde{Y}_{rk} / (v_0 + \sum_{j=1}^{s} v_j \tilde{X}_{rj})$$

$$\text{s.t.} \sum_{k=1}^{t} u_k \tilde{Y}_{ik} / (v_0 + \sum_{j=1}^{s} v_j \tilde{X}_{ij}) \leq 1, \quad i = 1,...,n \quad (2)$$

$$u_k, v_j \geq \varepsilon > 0, \quad v_0 \text{ unrestricted in sign.}$$

Since the missing data $\tilde{X}_{ij}$ and $\tilde{Y}_{ik}$ are fuzzy numbers, the efficiency $\tilde{E}_r$ calculated from them must be fuzzy as well. The characteristic of a fuzzy number is described by its membership function. Let $\mu_{\tilde{X}_{ij}}$ and $\mu_{\tilde{Y}_{ik}}$ denote the membership functions of $\tilde{X}_{ij}$ and $\tilde{Y}_{ik}$, respectively. According to Zadeh's extension principle (Yager 1986, Zadeh 1978), the membership function for $\tilde{E}_r$ can be expressed as:

$$\mu_{\tilde{E}_r}(z) = \sup_{x,y} \min. \{\mu_{\tilde{X}_{ij}}(x_{ij}), \mu_{\tilde{Y}_{ik}}(y_{ik}), \forall i,j,k \mid z = E_r(x,y)\}, \quad (3)$$

where $E_r(x,y)$ is the efficiency of DMU r calculated from Model (1).

The membership function can be viewed either vertically or horizontally. Vertically, $\mu_{\tilde{E}_r}(z)$ shows the possibility of occurrence for the value z, while horizontally, at a level α in the range of 0 and 1, it shows the range of values whose possibility of occurrence is greater than or equal to α. In fuzzy set terminology it is called α-level set, or α-cut, which is expressed as:

$$(X_{ij})_\alpha = [(X_{ij})_\alpha^L, (X_{ij})_\alpha^U] \quad (4a)$$
$$= [\min._{x_{ij}} \{x_{ij} \in S(\tilde{X}_{ij}) \mid \mu_{\tilde{X}_{ij}}(x_{ij}) \geq \alpha\},$$
$$\max._{x_{ij}} \{x_{ij} \in S(\tilde{X}_{ij}) \mid \mu_{\tilde{X}_{ij}}(x_{ij}) \geq \alpha\}]$$

$$(Y_{ik})_\alpha = [(Y_{ik})_\alpha^L, (Y_{ik})_\alpha^U] \quad (4b)$$
$$= [\min._{y_{ik}} \{y_{ik} \in S(\tilde{Y}_{ik}) \mid \mu_{\tilde{Y}_{ik}}(y_{ik}) \geq \alpha\},$$
$$\max._{y_{ik}} \{y_{ik} \in S(\tilde{Y}_{ik}) \mid \mu_{\tilde{Y}_{ik}}(y_{ik}) \geq \alpha\}]$$

where $S(\tilde{X}_{ij})$ and $S(\tilde{Y}_{ik})$ are the supports of $\tilde{X}_{ij}$ and $\tilde{Y}_{ik}$, respectively. The membership function $\mu_{\tilde{E}_r}$ can be constructed by deriving its α-cut at different α values. According to (3), $\mu_{\tilde{E}_r}(z)$ is the minimum of $\mu_{\tilde{X}_{ij}}(x_{ij})$

and $\mu_{\widetilde{Y}_{ik}}(y_{ik})$, $\forall i, j, k$. We need $\mu_{\widetilde{X}_{ij}}(x_{ij}) \geq \alpha$, $\mu_{\widetilde{Y}_{ik}}(y_{ik}) \geq \alpha$, and at least one $\mu_{\widetilde{X}_{ij}}(x_{ij})$ or $\mu_{\widetilde{Y}_{ik}}(y_{ik})$ equal to α, $\forall i, j, k$, such that $z = E_r(x, y)$ to satisfy $\mu_{\widetilde{E}_r}(z) = \alpha$. Furthermore, all α-cuts form a nested structure with respect to α (Zimmermann 1996); viz., $[(X_{ij})^L_{\alpha_1}, (X_{ij})^U_{\alpha_1}] \subseteq [(X_{ij})^L_{\alpha_2}, (X_{ij})^U_{\alpha_2}]$ and $[(Y_{ik})^L_{\alpha_1}, (Y_{ik})^U_{\alpha_1}] \subseteq [(Y_{ik})^L_{\alpha_2}, (Y_{ik})^U_{\alpha_2}]$, for $0 < \alpha_2 < \alpha_1 \leq 1$. To find the membership function $\mu_{\widetilde{E}_r}$, it suffices to find the lower and upper bounds of the α-cut of $\widetilde{E}_r$, which, based on (3), can be solved as:

$$(E_r)^L_\alpha = \min_{\substack{(X_{ij})^L_\alpha \leq x_{ij} \leq (X_{ij})^U_\alpha \\ (Y_{ik})^L_\alpha \leq y_{ik} \leq (Y_{ik})^U_\alpha \\ \forall i,j,k}} \begin{cases} E_r = \max. \sum_{k=1}^t u_k y_{rk} / (v_0 + \sum_{j=1}^s v_j x_{rj}) \\ \text{s.t.} \sum_{k=1}^t u_k y_{ik} / (v_0 + \sum_{j=1}^s v_j x_{ij}) \leq 1, \ i = 1, \ldots, n, \\ u_k, v_j \geq \varepsilon > 0, v_0 \text{ unrestricted in sign.} \end{cases} \quad (5a)$$

$$(E_r)^U_\alpha = \max_{\substack{(X_{ij})^L_\alpha \leq x_{ij} \leq (X_{ij})^U_\alpha \\ (Y_{ik})^L_\alpha \leq y_{ik} \leq (Y_{ik})^U_\alpha \\ \forall i,j,k}} \begin{cases} E_r = \max. \sum_{k=1}^t u_k y_{rk} / (v_0 + \sum_{j=1}^s v_j x_{rj}) \\ \text{s.t.} \sum_{k=1}^t u_k y_{ik} / (v_0 + \sum_{j=1}^s v_j x_{ij}) \leq 1, \ i = 1, \ldots, n, \\ u_k, v_j \geq \varepsilon > 0, v_0 \text{ unrestricted in sign.} \end{cases} \quad (5b)$$

In measuring the relative efficiency of DMU r, its smallest value is derived by setting the output level of this DMU and the input levels of all other DMUs to their lowest possible values and setting the input level of this DMU and the output levels of all other DMUs to their highest possible values.Conversely, to find the highest relative efficiency of a DMU, one will set the output level of this DMU and the input levels of all other DMUs to their highest possible values and set the input level of this DMU and the output levels of all other DMUs to their lowest possible values. Therefore, the two-level mathematical model (5) can be simplified to the following conventional one level model (Kao 2006)[1]:

---

[1] See also Chapter 3.

$$(E_r)_\alpha^L = \max. \sum_{k=1}^{t} u_k (Y_{rk})_\alpha^L / (v_0 + \sum_{j=1}^{s} v_j (X_{rj})_\alpha^U)$$

$$\text{s.t.} \sum_{k=1}^{t} u_k (Y_{rk})_\alpha^L / (v_0 + \sum_{j=1}^{s} v_j (X_{rj})_\alpha^U) \leq 1 \qquad (6a)$$

$$\sum_{k=1}^{t} u_k (Y_{ik})_\alpha^U / (v_0 + \sum_{j=1}^{s} v_j (X_{ij})_\alpha^L) \leq 1, \quad i=1,\ldots,n, i \neq r$$

$$u_k, v_j \geq \varepsilon > 0, \quad v_0 \text{ unrestricted in sign.}$$

$$(E_r)_\alpha^U = \max. \sum_{k=1}^{t} u_k (Y_{rk})_\alpha^U / (v_0 + \sum_{j=1}^{s} v_j (X_{rj})_\alpha^L)$$

$$\text{s.t.} \sum_{k=1}^{t} u_k (Y_{rk})_\alpha^U / (v_0 + \sum_{j=1}^{s} v_j (X_{rj})_\alpha^L) \leq 1 \qquad (6b)$$

$$\sum_{k=1}^{t} u_k (Y_{ik})_\alpha^L / (v_0 + \sum_{j=1}^{s} v_j (X_{ij})_\alpha^U) \leq 1, \quad i=1,\ldots,n, i \neq r$$

$$u_k, v_j \geq \varepsilon > 0, \quad v_0 \text{ unrestricted in sign.}$$

The α-cut of $\widetilde{E}_r$ is obtained from (6) as $(E_r)_\alpha = [(E_r)_\alpha^L, (E_r)_\alpha^U]$ and the membership function of $\widetilde{E}_r$ is constructed from $(E_r)_\alpha$ at different α values.

The most difficult part of this approach is the determination of the membership function. Kao and Liu (2000b) use a triangular function to represent the missing value, where the three vertices of the function are represented by the smallest, largest, and median observations of the factor corresponding to the missing value. As a result, different DMUs with missing values occurring for the same factor have the same membership function, which, intuitively, is not appropriate. In this study, we change the left, right, and top vertices of the triangular function to the values corresponding to the smallest, largest, and median ranks which have appeared in other inputs (or outputs) for the DMU with missing values. Consider a case of four inputs (outputs will be treated the same way), and the value of the fourth input of a DMU is missing. Suppose the ranks for the values of this DMU in the first three inputs are 3, 7, and 8. Then the values corresponding to ranks 3, 8, and 7 of the fourth input are used as the left, right, and top vertices, respectively, of the membership function. If, excluding the inputs corresponding to the missing values, there are two inputs left, then the ranks of the values in these two inputs will be used for determining the left and right vertices, and the average rank for determining the top vertex. If there is only one input left, then its rank will be used for determining all three vertices. In this case, the fuzzy number degenerates to a deterministic value. Finally, if there is no input left, then this DMU should be deleted from the group for comparison. Alternatively, one can use the ranks in outputs for substitution.

Notably, when some data are missing, not only will the DMUs with values missing have fuzzy efficiencies, but also other DMUs with complete data. It is also possible for DMUs with some values missing to have deterministic efficiencies. In this case the efficiency must be 1. All these are dependent on the nature of the frontier facets, whether they are deterministic or fuzzy.

3. A CASE ANALYSIS

Kao and Liu (2000b) applied the fuzzy set approach to measure the efficiency of 24 university libraries in Taiwan where three values were missing. Although the efficiencies have been calculated, there is no way to verify whether the results are correct or not, and thus is no way to prove whether this approach can produce correct results. For this reason, this study uses a case with complete data and creates the condition of incomplete data by deleting a number of observations to investigate how reliable the fuzzy set approach is.

Table 16-1. Input and output data of the forest reorganization problem

Factor	1	2	3	4	5	6	7
DMU	Input 1	Input 2	Input 3	Input 4	Output 1	Output 2	Output 3
1	67.55	82.83	44.37	60.85	26.04	85.00	23.95
2	85.78	123.98	55.13	<u>108.46</u>	43.51	173.93	6.45
3	80.33	104.65	53.30	79.06	27.28	<u>132.49</u>	42.67
4	205.92	183.49	144.16	59.66	14.09	196.29	16.15
5	51.28	117.51	32.07	84.50	16.20	144.99	0.00
6	82.09	104.94	46.51	127.28	44.87	108.53	0.00
7	123.02	82.44	87.35	98.80	43.33	125.84	404.69
8	<u>71.77</u>	88.16	69.19	123.14	44.83	74.54	6.14
9	<u>61.95</u>	99.77	33.00	86.37	45.43	79.60	1252.62
10	25.83	105.80	9.51	227.20	19.40	120.09	0.00
11	27.87	107.60	14.00	146.43	25.47	131.79	0.00
12	72.60	132.73	44.67	173.48	5.55	135.65	24.13
13	84.83	104.28	159.12	171.11	11.53	<u>110.22</u>	49.09
14	202.21	187.74	149.39	93.65	44.97	184.77	0.00
15	66.65	<u>104.18</u>	257.09	13.65	139.74	115.96	0.00
16	51.62	11.23	49.22	33.52	40.49	14.89	3166.71
17	36.05	193.32	59.52	8.23	46.88	190.77	822.92

The case to be investigated is the forest reorganization problem by Kao and Yang (1992). That problem is considered by Seiford (1996) as one of the most novel applications in DEA. Its data set is composed of 17 DMUs, where each DMU uses four inputs to produce three outputs, with a total of

119 observations. Table 16-1 shows the data set of this problem. The missing rates considered are 1%, 2%, and 5%. For a total of 119 observations, these rates correspond to 1, 2, and 6 observations. The observations to be deleted are selected randomly. In order to make a consistent analysis, the observations deleted in the case with the smaller missing rate are also deleted in the case with the larger missing rate. This type of experimental design provides a more accurate picture of the effect of the missing rate on efficiency estimation.

The six observations selected randomly, as underlined in Table 16-1, are located at (2,4), (3,6), (8,1), (9,1), (13,6), and (15,2). For the case of 1% missing rate, we assume the observation at (2,4), i.e., the fourth input of the second DMU, is missing. The values of DMU 2 in inputs 1, 2, and 3 are ranked tenth, thirteenth, and fourteenth, respectively, in the corresponding inputs. Therefore, the values corresponding to ranks 10, 13, and 14 in the fourth input, which occur at DMUs 7, 6, and 11, respectively, are chosen as the left, top, and right vertices of the triangular function, and the function constructed is (98.80, 127.28, 146.43). By applying Model (6), one obtains the α-cut of the fuzzy efficiency $\widetilde{E}_r$ at a specific α value, $[(E_r)_\alpha^L, (E_r)_\alpha^U]$.

The average of the central values of all α-cuts serves as an estimate of the deterministic efficiency of DMU r. This average value can also be used for ranking. The experience of Chen and Klein (1997) is that three or four cuts are sufficient to obtain a good estimation. In this study we calculate the α-cut at five α values: 0, 0.25, 0.5, 0.75, and 1.

When the observation at (2,4) is missing, with its value replaced by the fuzzy number (98.80, 127.28, 146.43), there are five DMUs, viz., 1, 3, 6, 8, and 12, whose efficiencies become fuzzy. In Table 16-2, the first pair of columns, with the heading "0% missing," shows the efficiencies and their corresponding ranks calculated from complete data. There are ten efficient DMUs and their ranks are not specified. The second pair of columns shows the estimated efficiencies and their corresponding ranks when the value at (2,4) is assumed missing. The total absolute difference between the original efficiency and the estimated efficiency of the 17 DMUs, as shown at the bottom of Table 16-2, is 0.0251. Its average is 0.0015. The absolute difference of each DMU divided by the original efficiency is the error in estimating the true efficiency. As shown at the bottom of Table 16-2, the average error for the 17 DMUs is only 0.1597%, indicating an accurate estimation. In addition to efficiency scores, their ranks are also very close to the original ones. There are only two DMUs whose ranks are different from the original case, with a total difference of 2 and an average difference of 0.1176.

For the case of 2% missing rate, we assume the observation at (3,6), in addition to that at (2,4), is missing. Following the same procedure, we obtain the triangular function (120.09, 131.79, 135.65) for the missing value at

(3,6). This triangular function is constructed from the values of DMUs 10, 11, and 12, corresponding to ranks 8, 10, and 12, respectively, in the second output. When the values at (2,4) and (3,6) are assumed fuzzy, there are six DMUs, viz., 1, 3, 6, 8, 12, and 13, whose efficiencies are fuzzy numbers. Note that although DMU 2 has a fuzzy observation at the fourth input, it has a deterministic efficiency of 1. The estimates of the efficiencies and their corresponding ranks are shown in the third pair of columns of Table 16-2. Interestingly, the results are better than the case of 1% missing rate. The total absolute difference between the original efficiencies and the estimated efficiencies of the 17 DMUs is only 0.0088, with an average difference as small as 0.0005. From the absolute difference of each DMU, we obtain 0.0598% as the average error in estimating the true efficiency. The rankings are exactly the same as that of the original case. The reason is quite straightforward. As more values are missing, the resulting efficiency of each DMU is fuzzier. However, this does not imply that the generalized mean of the fuzzy efficiency is farther away from the true efficiency.

Table 16-2. Results of the fuzzy set approach

	0% missing		1% missing		2% missing		5% missing	
DMU	Eff.	Rank	Eff.	Rank	Eff.	Rank	Eff.	Rank
1	0.8283	14	0.8340	14	0.8340	14	0.8359	14
2	1.0000		1.0000		1.0000		1.0000	
3	0.9581	12	0.9773	11	0.9609	12	0.9718	11
4	1.0000		1.0000		1.0000		1.0000	
5	1.0000		1.0000		1.0000		1.0000	
6	0.9572	13	0.9573	13	0.9573	13	0.9582	13
7	1.0000		1.0000		1.0000		1.0000	
8	0.7941	16	0.7941	16	0.7941	16	0.7993	16
9	1.0000		1.0000		1.0000		1.0000	
10	1.0000		1.0000		1.0000		1.0000	
11	1.0000		1.0000		1.0000		1.0000	
12	0.8172	15	0.8173	15	0.8173	15	0.8173	15
13	0.7441	17	0.7442	17	0.7442	17	0.7610	17
14	0.9644	11	0.9644	12	0.9644	11	0.9644	12
15	1.0000		1.0000		1.0000		1.0000	
16	1.0000		1.0000		1.0000		1.0000	
17	1.0000		1.0000		1.0000		1.0000	
Total absolute difference			0.0251	2	0.0088	0	0.0445	2
Average difference			0.0015	0.1176	0.0005	0	0.0026	0.1176
Average error			0.1597%		0.0598%		0.2786%	

Finally, for the case of 5% missing rate, we assume the observations at (8,1), (9,1), (13,6), and (15,2), in addition to those at (2,4) and (3,6), are missing. The triangular fuzzy numbers associated with these four values are

(51.28, 82.09, 82.09), (51.28, 51.62, 67.55), (74.54, 120.09, 144.99), and (82.44, 104.28, 193.32). Following the same procedure for missing rates of 1% and 2%, we obtain the efficiency estimates and their corresponding ranks as shown in the last pair of columns of Table 16-2. This time there are eleven DMUs whose efficiencies are fuzzy numbers. However, the total absolute difference between the original efficiencies and the estimates is not much, a value of 0.0445. The average difference for each DMU is only 0.0026 and the average error in estimating the true efficiency is 0.2786%. Only two DMUs have ranks that are different from those calculated from the original data.

In sum, the errors in estimating the true efficiency for three missing rates are 0.1597%, 0.0598%, and 0.2786%. Despite the mild increasing trend, they indicate a very accurate estimation.

4. A COMPARISON

Conventionally, when some values are missing, the corresponding DMUs are deleted from the group for comparison. In the case of the 1% missing rate, where the value at (2,4) is assumed missing, DMU 2 will be deleted, and the remaining 16 DMUs will apply Model (1) to calculate their efficiencies. Similarly, in the case of the 2% missing rate, DMUs 2 and 3 will be deleted. Finally, for the case of the 5% missing rate, DMUs 2, 3, 8, 9, 13, and 15 will be deleted, leaving 11 DMUs to calculate their efficiencies.

The first part of Table 16-3 shows the original efficiencies of the 17 DMUs and the relative ranks of the DMUs of three cases: 1%, 2%, and 5% missing rates. The entries with a dash indicate that the corresponding DMUs have been deleted from the group. Empty entries correspond to efficient DMUs, so there is no need to show their ranks. The second part of Table 16-3 shows the relative efficiencies of 16 DMUs and their corresponding ranks when DMU 2 is deleted. As expected, when a DMU is deleted, the new efficiencies of the remaining 16 DMUs are greater than or equal to their original counterparts. The total difference and average difference are 0.2063 and 0.0129, respectively, where the average is approximately nine times larger than that of the fuzzy set approach. Recall that the average from the fuzzy set approach is 0.0015. The error in estimation measured from the absolute difference of each DMU is 1.5388%, which is approximately ten times larger. Meanwhile, the total difference and average difference in ranks are 6 and 0.375, respectively. Compared with the average difference of the fuzzy set approach, the ratio is 3.19.

For the case of 2% missing rate, where DMUs 2 and 3 are deleted, the results in Table 16-3 show that the efficiencies are the same as the case of the 1% missing rate. This is because the additional DMU being deleted, viz.,

DMU 3, is an inefficient one. Due to ignorance of the inaccurate estimation of DMU 3 in the 1% case, the total difference in this case is smaller than that of the 1% case. The total and average differences are 0.1723 and 0.0115, respectively, where the average is twenty times larger than that of the fuzzy set approach. Regarding the average error in estimating the true efficiency, it is 1.4048%, a value which is also twenty times larger than that of the fuzzy set approach.

Table 16-3. Results of the DMU-deletion approach

DMU	0% Eff.	Relative rank 1%	2%	5%	1% missing Eff.	Rank	2% missing Eff.	Rank	5% missing Eff.	Rank
1	0.8283	13	12	10	0.8442	14	0.8442	13	0.8442	11
2	1.0000	-	-	-	-	-	-	-	-	-
3	0.9581	11	-	-	0.9921	10	-	-	-	-
4	1.0000				1.0000		1.0000		1.0000	
5	1.0000				1.0000		1.0000		1.0000	
6	0.9572	12	11	9	0.9856	11	0.9856	10	1.0000	8
7	1.0000				1.0000		1.0000		1.0000	
8	0.7941	15	14	-	0.8073	15	0.8073	14	-	-
9	1.0000				1.0000		1.0000		-	-
10	1.0000				1.0000		1.0000		1.0000	
11	1.0000				1.0000		1.0000		1.0000	
12	0.8172	14	13	11	0.8728	13	0.8728	12	0.8728	10
13	0.7441	16	15	-	0.7956	16	0.7956	15	-	-
14	0.9644	10	10	8	0.9721	12	0.9721	11	0.9728	9
15	1.0000			-	1.0000		1.0000		-	-
16	1.0000				1.0000		1.0000		1.0000	
17	1.0000				1.0000		1.0000		1.0000	
Total absolute difference					0.2063	6	0.1723	4	0.1227	4
Average difference					0.0129	0.3750	0.0115	0.2667	0.0112	0.3636
Average error					1.5388%		1.4048%		1.2787%	

For the last case of the 5% missing rate, there are 11 DMUs left after deleting 6 of them. As expected, the efficiency of every DMU is greater than or equal to those of the original case. The total difference from the 11 DMUs is 0.1227, with an average of 0.0112. This value is approximately four times larger than the value of 0.0026, calculated from the fuzzy set approach. The average error in estimating the true efficiency is 1.2787%, which is approximately 4.6 times larger than that of the fuzzy set approach. The total and average differences in ranks are 4 and 0.3636, respectively. The average difference in ranks is also greater than that of the fuzzy set approach: 0.3636 versus 0.1176, a ratio of 3.09.

The results from three missing rates show that the fuzzy set approach is superior to the conventional DMU-deletion approach, because it not only produces better estimates of the efficiencies but also preserves the available information of all DMUs. It is also worthwhile to note that as the missing rate increases the error in estimating the true efficiency for the DMU-deletion approach shows a decreasing trend, indicating that this approach may perform better for higher missing rates.

5. CONCLUSION

The basis of DEA is data and missing data thus hinders the calculation of efficiency. In practice, the DMUs with values missing are usually deleted to make the DEA approach applicable. However, this has two consequences, one is the loss of information contained in those deleted DMUs and the other is an overestimation of the efficiencies of the remaining DMUs. In this study we represent the missing values by fuzzy numbers and apply a fuzzy DEA approach to calculate the efficiencies of all DMUs. Since each fuzzy number is constructed from the available input/output information of the DMU in concern, it appropriately represents the value which is missing.

To investigate whether the proposed approach can produce satisfactory estimates for true efficiencies, we select a typical problem with complete data and delete different numbers of input/output values to result in cases of incomplete data. The errors in estimating true efficiencies for three missing rates are less than 0.3% for the fuzzy set approach and approximately 1.5% for the DMU-deletion approach, and thus the former obviously outperforms the latter. Most importantly, the efficiencies of those DMUs with some values missing can also be calculated.

The conclusions of this study are based on one problem with three missing rates. While the proposed approach is intuitively attractive and the results are promising, a comprehensive comparison which takes into account different problem sizes, missing rates, and other problem characteristics is necessary in order to draw conclusions applicable to all situations.

REFERENCES

1. Allison, P.D. (2002), *Missing Data*. LA: Sage Publications.
2. Banker, R.D. A. Charnes, and W.W. Cooper (1984), Some models for estimating technical and scale inefficiencies in data envelopment analysis, *Management Science* 30, 1078-1092.
3. Charnes, A. and W.W. Cooper (1962), Programming with linear fractional functionals, *Naval Research Logistics Quarterly* 9, 181-186.

4. Charnes, A. and W.W. Cooper (1984), The non-Archimedean CCR ratio for efficiency analysis: A rejoinder to Boyd and Färe, *European J. Operational Research* 15, 333-334.
5. Charnes, A., W.W. Cooper, and E. Rhodes (1978), Measuring the efficiency of decision making units, *European J. Operational Research* 2, 429-444.
6. Charnes, A., W.W. Cooper, and E. Rhodes (1979), Short communication: Measuring the efficiency of decision making units, *European J. Operational Research* 3, 339.
7. Chen, C.B. and C.M. Klein (1997), A simple approach to ranking a group of aggregated fuzzy utilities, *IEEE Trans. Systems, Man and Cybernetics Part B* 27, 26-35.
8. Cooper, W.W., L.M. Seiford and K. Tone (2000), *Data Envelopment Analysis: A Comprehensive Text with Models, Applications, References and DEA-Solver Software.* Boston: Kluwer Academic Publishers.
9. Kao, C. (2006), Interval efficiency measures in data envelopment analysis with imprecise data, *European J. Operational Research* 174, 1087-1099.
10. Kao, C. and S.T. Liu (2000a), Fuzzy efficiency measures in data envelopment analysis, *Fuzzy Sets and Systems* 113, 427-437.
11. Kao, C. and S.T. Liu (2000b), Data envelopment analysis with missing data: an application to university libraries in Taiwan, *J. Operational Research Society* 51, 897-905.
12. Kao, C. and Y.C. Yang (1992), Reorganization of forest districts via efficiency measurement. *European J. Operational Research* 58, 356-362.
13. Law, A.M. and W.D. Kelton (1991), *Simulation Modeling & Analysis*, NY: McGraw-Hill.
14. Rubin, D.B. (2004), *Multiple Imputation for Nonresponse in Survey.* NY: Wiley Interscience.
15. Schafer, J.L. (1997), *Analysis of Incomplete Multivariate Data.* London: Chapman & Hall.
16. Seiford, L.M. (1996), Data envelopment analysis: the evolution of the state of the art (1978-1995). *J. Productivity Analysis* 7, 99-137.
17. Seiford, L.M. (1997), A bibliography for data envelopment analysis (1978-1996). *Annals of Operations Research* 73, 393-438.
18. Simar, L. and P. Wilson (1998), Sensitivity of efficiency scores: How to bootstrap in nonparametric frontier models. *Management Science* 44, 49-61.
19. Simar, L. and P. Wilson (2000), A general methodology for bootstrapping in nonparametric frontier models. *J. Applied Statistics* 27, 779-802.

20. Snedecor, G.W. and W.G. Cochran (1967), *Statistical Methods*, sixth edition, Ames, Iowa: The Iowa State University Press.
21. Yager, R.R. (1986), A characterization of the extension principle, *Fuzzy Sets and Systems* 18, 205-217.
22. Zadeh, L.A. (1978), Fuzzy sets as a basis for a theory of possibility, *Fuzzy Sets and Systems* 1, 3-28.
23. Zimmermann, H.J. (1996), *Fuzzy Set Theory and Its Applications*, 3rd edition, Boston: Kluwer-Nijhoff.

Chapter 17

PREPARING YOUR DATA FOR DEA

Joe Sarkis
Graduate School of Management, Clark University, 950 Main Street, Worcester, MA, 01610-1477, jsarkis@clarku.edu

Abstract: DEA and its appropriate applications are heavily dependent on the data set that is used as an input to the productivity model. As we now know there are numerous models based on DEA. However, there are certain characteristics of data that may not be acceptable for the execution of DEA models. In this chapter we shall look at some data requirements and characteristics that may ease the execution of the models and the interpretation of results. The lessons and ideas presented here are based on a number of experiences and considerations for DEA. We shall not get into the appropriate selection and development of models, such as what is used for input or output data, but focus more on the type of data and the numerical characteristics of this data.

Key words: Data Envelopment Analysis (DEA), Homogeneity, Negative, Discretionary

1. SELECTION OF INPUTS AND OUTPUTS AND NUMBER OF DMUS

Selection of inputs and outputs and number of DMUs is one of the core difficulties in developing a productivity model and in preparation of the data. In this brief review, we will not focus on the managerial reasoning for selection of input and output factors, but more on the computational and data aspects of this selection process.

Typically, the choice and the number of inputs and outputs, and the DMUs determine how good of a discrimination exists between efficient and inefficient units. There are two conflicting considerations when evaluating the size of the data set. One consideration is to include as many DMUs as possible because with a larger population there is a greater probability of capturing high performance units that would determine the efficient frontier and improve discriminatory power. The other conflicting consideration with a large data set is that the homogeneity of the data set may decrease, meaning that some exogenous impacts of no interest to the analyst or beyond control of the manager may affect the results (Golany and Roll 1989; Haas and Murphy, 2003). Also, the computational requirements would tend to increase with larger data sets. Yet, there are some rules of thumb on the number of inputs and outputs to select and their relation to the number of DMUs.

Homogeneity

There are methods to look into homogeneity based on pre-processing analysis of the statistical distribution of data sets and removing "outliers" or clustering analysis, and post-processing analysis such as multi-tiered DEA (Barr, et al., 1994) and returns-to-scale analysis to determine if homogeneity of data sets is lacking. However, multi-tiered approaches require large numbers of DMU to do this.

Another set of three strategies to adjust for non-homogeneity were proposed by Haas and Murphy (2003). The first, a multi-stage approach by Sexton et al. (1994) is one technique. In the first stage they perform DEA using raw data producing a set of efficiency scores for all DMUs. In the second stage they run a stepwise multiple regression on that set of efficiency scores using a set of site (exogenous) characteristics that are expected to account for differences in efficiency. In the third stage they adjust DMU outputs to account for the differences in site characteristics and perform a second DEA to produce a new set of efficiency scores based on the adjusted data. The adjusted output levels used in the second DEA are derived by multiplying the level of each output by the ratio of the DMU's unadjusted efficiency score to its expected efficiency score. The magnitude of error method (actual minus forecast) is the second technique suggested by Haas and Murphy (2003). In this approach the adjustment is to estimate inputs and outputs using regression analysis. The factors causing non-homogeneity are treated as independent variable(s). DEA is then applied using the differences between actual and forecast inputs and outputs, rather than the original inputs and outputs. The ratio of actual to forecast method (actual/forecast) is the third technique suggested by Haas and Murphy (2003). This time instead of using differences, they use the ratio of actual input to forecast input and actual outputs to forecast outputs executing DEA. But simulation experiments showed that none of these three strategies performed well. At this time alternative approaches for addressing non-homogeneity are limited.

To further determine homogeneity and heterogeneity of data sets, using clustering analysis may be appropriate. Clustering techniques such as using Ward's method maximizes within-group homogeneity and between-group heterogeneity and forms cluster and dendograms to help identify homogeneous groups. A good example of this preliminary data analysis approach is found in Cinca and Molinero (2004).

Size of Data Set

Clearly, there are advantages to having larger data sets to complete a DEA analysis, but there are minimal requirements as well. Boussofiane et al. (1991) stipulate that to get good discriminatory power out of the CCR and BCC models the lower bound on the number of DMUs should be the multiple of the number of inputs and the number of outputs. This reasoning is derived from the issue that there is flexibility in the selection of weights to assign to input and output values in determining the efficiency of each DMU. That is, in attempting to be efficient a DMU can assign all of its weight to a single input or output. The DMU that has one particular ratio of an output to an input as highest will assign all its weight to those specific inputs and outputs to appear efficient. The number of such possible inputs is the product of the number of inputs and the number of outputs. For example, if there are 3 inputs and 4 outputs the minimum total number of DMUs should be 12 for some discriminatory power to exist in the model.

Golany and Roll (1989) establish a rule of thumb that the number of units should be at least twice the number of inputs and outputs considered. Bowlin (1998) and Friedman and Sinuany-Stern (1998) mention the need to have three times the number of DMUs as there are input and output variables or that the total number of input and output variables should be less than one third of the number of DMUs in the analysis: $(m + s) < n/3$. Dyson et al. (2001) recommend a total of two times the product of the number of input and output variables. For example with a 3 input, 4 output model Golany and Roll recommend using 14 DMUs, while Bowlin/Friedman and Sinuany recommend at least 21 DMUs, and Dyson et al. recommend 24. In any circumstance, these numbers should probably be used as minimums for the basic productivity models.

These rules of thumb attempt to make sure that the basic productivity models are more discriminatory. If the analyst still finds that the discriminatory power is lost due to the fewer number of DMUs, they can either reduce the number of input and output factors, or the analyst can turn to a different productivity model that has more discriminatory power. DEA-based productivity models that can help discriminate among DMUs more effectively regardless of the size of the data set include models introduced or developed by Andersen and Petersen (1993), Rousseau and Semple (1995), and Doyle and Green (1994).

2. REDUCING DATA SETS FOR INPUT/OUTPUT FACTORS THAT ARE CORRELATED

With extra large data sets, some analysts may wish to reduce the size by eliminating the correlated input or output factors. To show what will happen in this situation, a simple example of illustrative data is presented in Table 17-1. In this example, we have 20 DMUs, 3 inputs and 2 outputs. The first input is perfectly correlated with the second input. The second input is calculated by adding 2 to the first input for each DMU. The outputs are randomly generated numbers.

Table 17-1. Efficiency Scores from Models with Correlated Inputs

DMU	Input 1	Input 2	Input 3	Output 1	Output 2	Efficiency Score (3 Inputs, 2 Outputs)	Efficiency Score (2 Inputs, 2 Outputs)
1	10	12	7	34	7	1.000	1.000
2	24	26	5	6	1	0.169	0.169
3	23	25	3	24	10	1.000	1.000
4	12	14	4	2	5	0.581	0.576
5	11	13	5	29	8	0.954	0.948
6	12	14	2	7	2	0.467	0.460
7	44	46	5	39	4	0.975	0.975
8	12	14	7	6	4	0.400	0.400
9	33	35	5	2	10	0.664	0.664
10	22	24	4	7	6	0.541	0.541
11	35	37	7	1	4	0.220	0.220
12	21	23	5	0	6	0.493	0.493
13	22	24	6	10	6	0.437	0.437
14	24	26	8	20	9	0.534	0.534
15	12	14	9	5	4	0.400	0.400
16	33	35	2	7	6	0.900	0.900
17	22	24	9	2	3	0.175	0.175
18	12	14	5	32	10	1.000	1.000
19	42	44	7	9	7	0.352	0.352
20	12	14	4	5	3	0.349	0.346

The basic CCR model is executed for the 3-input case (where the first two inputs correlate) and a 2 input case where Input 2 is removed from the analysis. The efficiency scores are in the last two columns. Notice that in this case, the efficiency scores are also almost perfectly correlated.[1] The only differences that do occur are in DMUs 4, 5, 6, and 20 (for three decimal places). This can save some time in data acquisition, storage, and calculation, but the big caveat is that even when a perfectly correlated factor is included it may provide a slightly different answer. What happens to the results may depend on the level of correlation that is acceptable and whether the exact efficiency scores are important.

Jenkins and Anderson (2003) describe a systematic statistical method, using variances and covariances, for deciding which of the original correlated variables can be omitted with least loss of information, and which should be retained. The method uses conditional variance as a measure of information contained in each variable and delete the variable that loses the least information. They also found that even omitting variables that are highly correlated and which contain little additional information can have a major influence on the computed efficiency measures.

Reducing Number of Input and Output Factors – Principal Component Analysis

Given this initial caveat that correlated data may still provide some information and that removing data can cause information loss, the minimization of information loss has been a concern among researchers. Yet, researchers are investigating how to complete this data reduction with minimal information loss. Principal component analysis (PCA) has been applied to data reduction for empirical survey type data. PCA is a multivariate statistical tool for data reduction. It is designed with the goal of reducing a large set of variables to a few factors that may represent a fewer underlying factors for which the observed data are partial surrogates. The aim is to form new factors that are linear combinations of the original factors (items), and explain best the deviation of each observed datum from that variable's mean value.

A number of researchers have considered using PCA with DEA. Ueda and Hoshiai (1997; Adler and Golany, 2001) considered using a few PCA factors of the original data to obtain a more discerning ranking of DMUs. Zhu (1998) in his approach compared DEA efficiency measures with an efficiency measure made up as the ratio of a reduced set of PCA factors of output variables divided by a reduced set of PCA factors for input variables. Adler and Golany (2002) introduce three separate PCA–DEA formulations which utilize the results of PCA to develop objective, assurance region type constraints on the DEA weights. So PCA can be used at various junctures of

---

[1] A simple method to determine correlation is by evaluating the correlation of the data in a statistical package or even on a spreadsheet with correlation functions.

DEA analysis to more effectively provide discriminatory power of DEA, as well, without losing additional information.

Yet, a shortcoming of PCA is that it is dependent on the actual values of the data. Any changes in the data values will result in different factors being computed. Even though PCA is robust where small perturbations in the data will cause only a small perturbation in the calculated values of all the factors for a specific DEA run, it may not be too sensitive. On the other hand, if the study is replicated later with the same DMUs, or repeated with another set of comparable DMUs, the factors, and the ensuing DEA may not be strictly comparable, because the composition of the factors may have changed.

3. IMBALANCE IN DATA MAGNITUDES

One of the best ways of making sure there is not much imbalance in the data sets is to have them at the same or similar magnitude. A way of making sure the data is of the same or similar magnitude across and within data sets is to *mean normalize* the data. The process to mean normalize is taken in two simple steps. First step is to find the mean of the data set for each input and output. The second step is to divide each input or output by the mean for that specific factor.

Table 17-2. Raw Data Set for the Mean Normalization Example

DMU	Input 1	Input 2	Output 1	Output 2
1	1733896	97	1147	0.82
2	2433965	68	2325	0.45
3	30546	50	1998	0.23
4	1052151	42	542	0.34
5	4233031	15	1590	0.67
6	3652401	50	1203	0.39
7	1288406	65	1786	1.18
8	4489741	43	1639	1.28
9	4800884	90	2487	0.77
10	536165	19	340	0.57
Column Mean	2425119	53.9	1505.7	0.67

For example, let us look at a small set of 10 random data points (10 DMUs) with 2 inputs and 2 outputs as shown in Table 17-2. The magnitudes range from 10^0 to 10^6; in many cases this situation may be more extreme.[2]

---

[2] For example, if total sales of a major company (usually in billions of dollars) was to be compared to the risk associated with that company (usually a "Beta" score of approximately 1).

Assume that Table 17-2 is a raw data set for the problem at hand and a simple output-oriented constant returns to scale (CCR) model is to be applied.

One of the difficulties faced by some mathematical programming software that may be used to execute the DEA models (for example LINDO [3]) is that there may be "scaling" issues. This imbalance could cause problems in the execution of the software and round-off error problems may occur. Mean normalization will allow you to address this problem.

In Table 17-2, the column means are given in the final row. The mean is determined by the simple mean equation (1) that sums up the value of each DMU's input or output in that column and then divides the summation by the number of DMUs.

$$\overline{V}_i = \frac{\sum_{n=1}^{N} V_{ni}}{N} \qquad (1)$$

where $\overline{V}_i$ is the mean value for column i (an input or output), N is the number of DMUs and V_{ni} is the value of DMU n for a given input or output i. In Table 17-2, $\overline{V}_1 = 2425119$, $\overline{V}_2 = 53.9$, $\overline{V}_3 = 1505.7$, $\overline{V}_4 = 0.67$.

Table 17-3. Mean Normalized Data Set for the Mean Normalization Example

DMU	Input 1	Input 2	Output 1	Output 2	CCR
1	0.7150	1.7996	0.7618	1.2239	0.432
2	1.0036	1.2616	1.5441	0.6716	0.709
3	0.0126	0.9276	1.3270	0.3433	1.000
4	0.4339	0.7792	0.3600	0.5075	0.419
5	1.7455	0.2783	1.0560	1.0000	1.000
6	1.5061	0.9276	0.7990	0.5821	0.448
7	0.5313	1.2059	1.1862	1.7612	0.952
8	1.8513	0.7978	1.0885	1.9104	0.865
9	1.9796	1.6698	1.6517	1.1493	0.570
10	0.2211	0.3525	0.2258	0.8507	1.000
Column Mean	1	1	1	1	

The second step is to divide all of the values of a given column by this final row of mean values. The general equation for each cell of Table 17-3 is:

$$VNorm_{ni} = \frac{V_{ni}}{\overline{V}_i} \qquad (2)$$

---

[3] http://www.lindo.com/

where VNorm_{ni} is the normalized value for the value associated with DMU n and input or output in column i. Table 17-3 shows the mean normalized data set to four decimal places. As we can see in the last row of Table 17-3, the mean for each column of a mean normalized data set is 1. The last column of Table 17-3 contains the respective efficiency score for each DMU using the basic CCR model. These efficiency scores are the exact same efficiency score results as for the data set in Table 17-2.

4. NEGATIVE NUMBERS AND ZERO VALUES[4]

Many times the data set will have negative numbers. Basic DEA models are not capable of completing an analysis with negative numbers and all numbers must be non-negative and preferably strictly positive (no zero values). This has been defined as the "positivity" requirement of DEA. Charnes et al. (1991) provide a model to relax this requirement, but we shall not focus on it here.

One of the more common methods for eliminating the problems of non-positive values in DEA (for ratio models) has been through the addition of a sufficiently large positive constant to the values of the input or output that has the non-positive number. And in some cases it has been advised (Bowlin 1998) to make the negative numbers or zero values a smaller number in magnitude than the other numbers in the data set. This might overcome some of the difficulties of this limitation, yet the results may still change depending on the scale (adjusting constant value) used by the models. This problem has been defined as "translation variance". It has been found that some ratio-based DEA models are translation invariant, for example, the BCC model (Ali and Seiford 1990), but even they are limited as to what values can be scaled or translated.

Before addressing some of the solutions, let us investigate a scenario for insurance policy providers who decided to invest in a number of Information Technology (IT) systems. This example can show us a number of difficulties in the original preparation of data. Assume that there are 8 new IT systems in place in various departments for a given insurance company. The only data given to an analyst is labor cost savings, operational cost savings, decreases in policy mistakes, and decrease in turnaround time for a policy. The analyst has decided that the cost savings will be used to determine operational and labor costs and serve as inputs to the productivity model, while outputs will be the improvement in policy accuracy and time to process policies. In some cases there are improved results (e.g. processing mistakes decrease with

---

[4] Chapter 4, in this book, provides some more detail on dealing with negative data.

positive values) and in some cases there are worse performing results (e.g. processing mistakes increase and have negative values). Thus, the data, as it is currently presented needs to be "adjusted" (see Table 17-4).

Table 17-4. Unadjusted Data Set for Information Technology Systems

DMU	INPUTS		OUTPUTS	
	Labor Cost Savings (000's)	Operational Cost Savings (000's)	Processing Mistake Decrease (number of policies)	Decrease in Turnaround Time (minutes per policy)
IT1	240	128	2	0
IT2	-325	1006	0	59
IT3	-1400	1017	27	26
IT4	-363	500	0	75
IT5	-55	256	77	66
IT6	4450	-246	12	-14
IT7	6520	79	-3	-14
IT8	3900	341	20	23

One of the first steps in this adjustment process is to eliminate any zero or negative values in the data set. Another issue is to set up the data such that inputs have the characteristic of smaller values being better (less resource needed for a given output), and outputs with a characteristic of having larger values being better. In this situation we have two adjustments that need to be completed before execution of the model and such that the results make managerial sense.

To show the variation in our results and the problem of translation variance in DEA (at least the basic CCR model), we shall illustrate two examples of arbitrarily selecting translation constants and approaches where no negative numbers exist and the inputs are better when smaller in value and outputs are better when larger in value. As inputs, we will use labor and operational costs. To convert the "cost savings" in the original data into labor cost and operational cost estimates, we will arbitrarily select a large number, say, 10,000 (in 000's), and subtract each number in the first two columns from this value.

We arrive at the data set in the second and third columns in Table 17-5 defined now as Labor Costs and Operational Costs for the inputs. Notice that now we have all positive values for the inputs since it is better for smaller costs than for larger costs. For example IT1 does better on labor costs (less cost value) than IT2, but IT2 does better on operational costs. This is a similar situation as in previous Table 17-4 where IT1 was better in labor cost savings than IT2 (with higher cost savings values), and IT2 was better in operational cost savings.

Table 17-5. Scaled Data Set 1

DMU	INPUTS		OUTPUTS		
	Labor Costs	Operational Costs	Processing Mistake Decrease (number of policies)	Decrease in Turnaround Time (minutes per policy)	CCR Technical Efficiency Score
IT1	9760	9872	12	20	0.232
IT2	10325	8994	10	79	0.881
IT3	11400	8983	37	46	0.564
IT4	10363	9500	10	95	1.000
IT5	10055	9744	87	86	1.000
IT6	5550	10246	22	6	0.458
IT7	3480	9921	7	6	0.232
IT8	6100	9659	30	43	0.805

For the outputs we looked at increasing the decreases in mistakes and increasing the decreases in turnaround time, so we arbitrarily added a value of 20 to each. This translation eliminated any zero or negative values while still retaining the characteristic that larger values are better. Executing the CCR model shows IT4 and IT5 as efficient units followed by IT2 as the highest scoring inefficient unit.

Table 17-6. Scaled Data Set 2

DMU	INPUTS		OUTPUTS		
	Labor Costs	Operational Costs	Processing Mistake Decrease (number of policies)	Decrease in Turnaround Time (minutes per policy)	CCR Technical Efficiency Score
IT1	6760	1372	6	15	0.184
IT2	7325	494	4	74	1.000
IT3	8400	483	31	41	1.000
IT4	7363	1000	4	90	1.000
IT5	7055	1244	81	81	1.000
IT6	2550	1746	16	1	0.546
IT7	480	1421	1	1	0.181
IT8	3100	1159	24	38	1.000

Let us now change the translation constants separately for each input and output factor. In this next example, the values are still arbitrarily chosen, but vary for each input and output. We selected numbers that were about 500 above the maximum values for each of the output values. That is, for the labor cost savings column of Table 17-4 we subtracted all the values from 7000, and for the operational cost savings we subtracted all the values from 1500. For the inputs we added enough units to make them positive. So we added values of 4 and 15 to each value of the processing mistake decrease and turnaround decrease times respectively, to arrive at the values that appear in the Outputs columns of Table 17-6.

This time, when we execute the basic models we see that there are more efficient units that exist and that some increased in efficiency (e.g. IT2) while others decreased (e.g. IT1). This is the major difficulty encountered when there is scale translation of results. Now the issue is how to address this situation.

A number of steps can be taken to make sure that data scaling or translation errors do not play a role in the final results of this analysis. First, instead of using arbitrary numbers, every attempt should be made to acquire and apply actual values for the initial baseline[5]. This may mean access to additional raw data and data from previous periods, thus the analyst will have to go searching.

The original data set in Table 17-4 is based on a real world problem, where the actual baselines were known for each input and output factor. The values within each column were subtracted from their respective baseline. These results provided appropriate characteristics for the inputs, but the outputs still needed to have larger values as more preferable. These columns were scaled by taking the inverse of their actual values, thus maintaining the integrity of the relationships of the data. As we see in Table 17-7 these new values contain the necessary characteristics for completing the analysis. There is also a caveat when taking the inverse of data as a translation. This translation may also cause a variation in the efficiency scores (Lewis and Sexton 2000)[6]. Thus, managerial subjectivity and preference come into play again, depending on how the managers wish to view and interpret the input and output values.

---

[5] A baseline is defined as the initial value from which a translation adjustment will be made.

[6] An output that has larger numbers as less preferable have been termed as "undesirable outputs" (Yaisawarng and Klein 1994) and "anti-isotonic" (Sheel, 2001), and taking the inverse of these values has been recommended. An example of this type of output is pollution. Another way to achieve a similar result of adjusting for undesirable outputs is to make this output value an input value but this does not guarantee the same efficiency score as taking the inverse of the output and leaving it in the numerator. More detail on the 'bad' and undesirable outputs is discussed in Chapter 6 of this book.

Table 17-7. Scaled Data Set 3

DMU	INPUTS		OUTPUTS		
	Labor Costs	Operational Costs	Processing Mistake Decrease (number of policies)	Decrease in Turnaround Time (minutes per policy)	CCR Technical Efficiency Score
IT1	7260	4372	3.597	0.7143	0.833
IT2	7825	3494	3.571	0.7457	1.000
IT3	8900	3483	3.953	0.7278	1.000
IT4	7863	4000	3.571	0.7547	0.915
IT5	7555	4244	4.926	0.7496	1.000
IT6	3050	4746	3.731	0.7072	0.887
IT7	980	4421	3.534	0.7072	1.000
IT8	3600	4159	3.846	0.7262	1.000

Another way to address data scaling or translation error concerns is to explicitly include before and after performance. That is, instead of subtracting the data from period to period, or adjusting with inverses, the purest method may be to use the previous period's performance as an input and this period's performance as an output for those measures where larger values are better, and the opposite for those measures where smaller values are better. This will require the additional input and output factors to be included (double the number) and may hurt the discriminatory power of some productivity models if not enough DMUs exist.

Yet another remedy for making sure that translation variance does not impact your solution is to use a different type of DEA model. For example, Ali and Seiford (1990) and Pastor (1996) have shown that a displacement does not alter the efficient frontier for certain DEA formulations (specifically, the additive model for both inputs and outputs and the BCC model for outputs) and thus these approaches are translation invariant. Thus, *absolute value constants can be added to any input and output in the additive model, and any output in the BCC model in order to solve the issue of negative or zero-valued problems.* Similar to what we showed in our example, the same adjustment must be made to the input or output values for all DMUs included in the data set in order not to alter the efficiency frontier.

Bowlin (1998) mentions the substitution of a very small positive value for the negative value if the variable is an output. He suggests this approach because the characteristic of DEA models is that they try to put each DMU in the best light possible and therefore, emphasize (weight highest) those outputs on which the DMU performs best. Thus, Bowlin argues, an output variable with a very small value would not be expected to contribute to a high efficiency score which would also be true of a negative value. Thus, this type of translation would generally not inappropriately affect the efficiency score. Of course this will mean that this value must not be larger than any other output value in the data set.

5. MISSING DATA

When evaluating DMUs and attempting to get the necessary inputs and outputs, the analyst will come across situations where data is missing. The usual alternative is to eliminate any DMUs that lack data for any input or output. Clearly, when eliminating DMUs from consideration there is an issue of whether the true efficient frontier is developed. The remedies for missing data are still quite limited. In all the situations presented by researchers and practitioners of DEA, it is still a relatively subjective approach in filling a gap from missing data.

One approach, from a managerial perspective, is to get a "best estimate" from managers as to what they believe would be the value for the missing data point. This is a simple way to accomplish this task, but is very subjective. To attempt at getting around the issue of asking for one value, sometimes it may be beneficial to ask for the most optimistic, pessimistic and likely values and run the models based on those values to see how sensitive the solution would be to these subjective rankings. An expected value may be determined in this situation. In project management when estimating the completion time of a project, a beta distribution (see later) is used to arrive at an expected value when optimistic, most likely, and pessimistic times are used. One method to arrive at an expected value is to apply the subjective values from a manager into an expected value calculation based on this probability distribution.

For example, assume that a data point is missing for the amount of time it took to process an insurance policy in an insurance underwriting department. A group of managers may be asked to define the optimistic estimate of time of completion, a most likely time, and a pessimistic time. The beta distribution expected value calculation is defined as:

$$V_e = \frac{V_o + 4V_m + V_p}{6} \tag{3}$$

where V_e is the estimated value, V_o is the optimistic value, V_m is the most like value and V_p is the pessimistic value.

Clearly, there are a number of assumptions here, first, that the beta distribution truly represents the data structure, and the managers are capable of giving estimates. An alternative approach is to look at the data across DMUs and to determine what structure does exist and use estimation models based on those structures. In this situation, the optimistic value would be the best value among the given input or output factor, the pessimistic would be the worst value and the average will be the most likely value. Yet, in this situation you cannot ignore the values of the other inputs and outputs. Another issue that arises is that the missing data point may be the best point for that DMU which may severely impact its efficiency score. In this case, variations in data and sensitivity analysis will be required.

Recently, others have proposed alternative approaches, based on more complex techniques, to determine missing values. One such technique is by Kao and Liu (2000), which relies on a Fuzzy Mathematics approach to arrive at estimates of missing value data. This technique requires a significant effort to arrive at possible values that may be used and is still dependent on a series of value variations to determine the relative sensitivity of the solutions[7]. The application of imprecise data approaches may also be useful in this situation, Chapters 2 and 3 in the book cover many of the issues related to imprecise data and qualitative data.

REFERENCES

1. Adler N and Golany B (2001), Evaluation of Deregulated Airline Networks using Data Envelopment Analysis Combined with Principal Component Analysis with an Application to Western Europe. *European Journal of Operational Research* **132**: 18–31.
2. Adler, N., and Golany, B., (2002), Including Principal Component Weights to Improve Discrimination in Data Envelopment Analysis, *Journal of the Operational Research Society* **53**, 985–991.
3. Ali, A.I. and Seiford, L.M. (1990) Translation Invariance in Data Envelopment Analysis. *Operations Research Letters* **9**, 403–405.
4. Andersen, P., and Petersen, N.C. (1993) A Procedure for Ranking Efficient Units in Data Envelopment Analysis. *Management Science* **39**, 1261-1264.
5. Barr, R., Durchholz, M. and Seiford, L. (1994) Peeling the DEA Onion: Layering and Rank-Ordering DMUS using tiered DEA. Southern Methodist University Technical Report, 1994/2000.
6. Boussofiane, A., Dyson, R.G., and Thanassoulis, E. (1991) Applied Data Envelopment Analysis. *European Journal of Operational Research* **52**, 1-15.

---

[7] See Chapter 16.

7. Bowlin, W.F. (1998) Measuring Performance: An Introduction to Data Envelopment Analysis (DEA). *Journal of Cost Analysis* **7**, 3-27.
8. Charnes, A., Cooper, W.W., and Thrall, R.M. (1991) A Structure for Characterizing and Classifying Efficiency and Inefficiency in Data Envelopment Analysis. *Journal of Productivity Analysis* **2**, 197–237.
9. Cinca, C.S., and Molinero C.M., (2004), "Selecting DEA specifications and ranking units via PCA," *Journal of the Operational Research Society*, **55**, 5,521-528
10. Doyle, J., and Green, R. (1994) Efficiency and Cross-efficiency in DEA: Derivations, Meanings and Uses. *Journal of the Operational Research Society* **45**, 567-578.
11. Dyson, R.G., Allen, R., Camanho, A.S., Podinovski, V.V., Sarrico, C.S., and Shale, E.A., (2001) Pitfalls and Protocols in DEA, *European Journal of Operational Research*, **132**, 245-259.
12. Golany, B. and Roll, Y. (1989) An Application Procedure for DEA. *Omega* **17**, 237-250.
13. Haas D.A., and Murphy F.H., (2003), "Compensating for non-homogeneity in decision-making units in data envelopment analysis," *European Journal of Operational Research*, **44**, 3, 530-544.
14. Jenkins L. and Anderson M. (2003), "A multivariate statistical approach to reducing the number of variables in data envelopment analysis," *European Journal of Operational Research*, **147**, 1, 51-61.
15. Kao, C. and Liu, S.T. (2000) Data Envelopment Analysis with Missing Data: An Application to University Libraries in Taiwan. *Journal of the Operational Research Society* **51**, 897-905.
16. Lewis, H.F., and Sexton, T.R. (2000) Data Envelopment Analysis with Reverse Inputs. Working Paper, State University of New York at Stony Brook.
17. Pastor, J.T. (1996) Translation Invariance in DEA: A Generalization. *Annals of Operations Research* **66**, 93-102.
18. Rousseau, J.J., and Semple, J.H. (1995) Radii of Classification Preservation in Data Envelopment Analysis: A Case Study of 'Program Follow-Through. *Journal of the Operational Research Society*, **46**, 943-957.
19. Scheel, H. (2001), Undesirable outputs in efficiency valuations, *European Journal Of Operational Research* **132**, 400-410.
20. Sexton, T.R., Sleeper, S., Taggart Jr., R.E., 1994. Improving pupil transportation in North Carolina. *Interfaces* **24** 87–103.
21. Ueda T and Hoshiai Y (1997). Application of principal component analysis for parsimonious summarization of DEA inputs and/or outputs. *Journal of the Operational Research Society of Japan* **40**: 466–478.

22. Yaisawarng, S., and. Klein, J.D. (1994) The Effects of Sulfur-Dioxide Controls on Productivity Change in the United-States Electric-Power Industry. *Review of Economics and Statistics* **76,** 447-460.
23. Zhu, J., (1998) Data Envelopment Analysis vs. Principal Components Analysis: An Illustrative Study of Economic Performance of Chinese Cities. *European Journal of Operational Research,* **111**, 50-61.

An original version of this chapter can be found in Chapter 12, pp.115-124, in Avkiran, N.K. 2006. *Productivity Analysis in the Service Sector with Data Envelopment Analysis*. NK Avkiran, Brisbane. It is reproduced here with the express permission of NK Avkiran.

ABOUT THE AUTHORS

Nicole Adler is an Assistant Professor at the Hebrew University of Jerusalem's School of Business Administration. She read for a B.Sc. in accounting and economics at the University of Ulster, an M.Sc. in operations research at the London School of Economics and then for a Ph.D. at Tel Aviv University, completing her thesis in 1999, under the guidance of Professors Boaz Golany and Gabriel Handler. Her major research interests include data envelopment analysis and applied game theory, in particular with reference to the transportation field. Her work has been published in various journals, including *Transportation Science*, *Transportation Research A*, *B* and *E* and the *European Journal of Operational Research*. She is a member of INFORMS and the Air Transport Research Society and serves on the editorial board of *Transportation Research B*.

Yiwen Bian is an assistant professor of Operations Management, the Sydney Institute of Language & Commerce, Shanghai University. He received a M.A. Ph.D. in engineering from Beijing University of Aeronautics and Astronautics, and University of Science and Technology of China, respectively. His researches focus on performance evaluation and decision making. Dr. Bian's recent research has appeared in *Omega, International Journal of Management Science* and *International Journal of Management and Decision Making*.

About the Authors

Yao Chen is Assistant Professor at College of Management, University of Massachusetts at Lowell. She received her Ph.D. of Business Administration from the University of Massachusetts at Amherst. Her current research interests include efficiency and productivity issues of information systems, information technology's impact on operations performance, and methodology development of Data Envelopment Analysis. Her work has appeared in such journals as *European Journal of Operational Research, International Journal of Production Economics, Information Technology & Management Journal, Annals of Operations research, OMEGA, Socio-Economic Planning Science,* and others.

Wade Cook is the Gordon Charlton Shaw Professor of Management Science in the Schulich School of Business, York University, Toronto, Canada, where he serves as Department Head of Management Science and as Associate Dean of Research. He holds a doctorate in Mathematics and Operations Research. He has published several books and more than 100 articles in a wide range of academic and professional journals, including *Management Science, Operations Research, Journal of the Operational Society, European Journal of Operational Research, IIE Transactions*, etc.. His areas of specialty include Data Envelopment Analysis, and Multi Criteria Decision Modeling. He is a former Editor of the *Journal of Productivity Analysis*, and of *INFOR*, and is currently an Associate Editor of *Operations Research*. Professor Cook has consulted widely with various companies and government agencies.

José H. Dulá is an Associate Professor of Production and Operations Management at Virginia Commonwealth University. His Ph.D. is in Industrial and Operations Engineering from the University of Michigan with a Master's from Stanford University. He applies his interests in polyhedral set theory, computational geometry, and linear programming to understanding DEA geometry and large scale applications. Examples of journals where his publications appear include Operations Research, Mathematical Programming, INFORMS Journal on Computing, and European Journal of Operational Research.

About the Authors

Rolf Färe is Professor of Agricultural and Resource Economics, and Economics at Oregon State University. Prior to coming to OSU in 1998, He was a Professor of Economics at Southern Illinois University at Carbondale. Ronald W. Shephard (known for Shephard's Lemma) is the person that made it possible for me to come to the United States from Sweden in the 1970s.

Boaz Golany is a Professor at the Industrial Engineering and Management Faculty and the holder of Samuel Gorney Chair in Engineering in the Technion - Israel Institute of Technology where he serves as the Dean since 2006. He has a B.Sc. in Industrial Engineering and Management from the Technion (1982), and a Ph.D. from the Business School of the University of Texas at Austin (1985). He was awarded the Naor Prize of the Israeli Operations Research Society in 1982 and the Yigal Alon Fellowship from the Israeli Education Ministry in 1986. In 1991 he was a recipient of the Technion Academic Excellence Award and in 1994 he became a Senior Research Fellow of the IC^2 Institute at the University of Texas at Austin. He has served as an Area Editor and member of the editorial Board for the *Journal of Productivity Analysis, IIE Transactions* and *Omega*. Dr. Golany has published over 80 papers in academic journals and books. His publications are in the areas of Industrial Engineering, Operations Research and Management Science. He has been active in various professional societies including The Institute of Management Science (in the US) and the Israeli Operations Research Society (where he served as the Treasurer in 1989-1992 and as the vice-President 1996-2002).

Shawna Grosskopf is a Professor at the Department of Economics, Oregon State University. She obtained here Ph.D. from Syracuse University. Her research Fields include Public Economics, Performance Measurement, Labor Economics. Professor Grosskopf teaches courses in public economics, microeconomics, and performance measurement. Her research includes theoretical work in performance measurement with a broad array of empirical applications including public sector performance, education, health and environmental productivity. She has published in American Economic Review, Journal of Econometrics, and Journal of Productivity Analysis among others.

Zhongsheng Hua is Professor of Operations Management, and Chairman of Department of Information Management and Decision Science, University of Science and Technology of China (USTC), Hefei, Anhui Province, People's Republic of China. He holds an M.A. and Ph.D., both in engineering from USTC. Professor Hua's research focuses on forecasting, decision analysis, production and operations management, and supply chain management. His research has appeared in *Computers & Operations Research, Information Science, Journal of the Operational Research Society, Omega (The International Journal of Management Science), International Journal of Production Research, International Journal of Production Economics, Robotics and CIM, European Journal of Operational Research, Expert Systems With Application, Fuzzy Sets and Systems, Socio-Economic Planning Sciences* etc. He has also published two books and two book chapters.

Chiang Kao is a professor in the Department of Industrial and Information Management, National Cheng Kung University, Taiwan, Republic of China. He received his BS degree from National Taiwan University, Taiwan, and holds an MS and a Ph.D. degree both earned from Oregon State University, USA. His research interest is in the applications of operations research, with more than 100 articles published in journals such as *INFORMS Journal on Computing, Interfaces, European Journal of Operational Research, Journal of the Operational Research Society, Annals of Operations Research, Computers and Operations Research, OMEGA, IEEE Trans. on Systems, Man and Cybernetics, International Journal of Production Research, Fuzzy Sets and Systems, Forest Science, Journal of Banking and Finance, Engineering Optimization, International Journal of Systems Science, Computers in Industry,* etc. He has also published a book titled *Management Performance Evaluation: Data Envelopment Analysis* in Chinese. He taught at the Computer Science Department, Southwest Texas State University, USA, for one year, was a Visiting Scholar at Purdue University, USA, and a Guest Professor at Aachen University of Technology, Germany.

About the Authors

Monique Le Moing is a Computer Scientist at INRA (National Institute for Agricultural Research) and is highly skilled in statistics and informatics. She is taking part in the empirical implementation of Isabelle Piot-Lepetit's research developments. This work has led her to generalise a set of DEA programmes written by Isabelle Piot-Lepetit in SAS language, and to transcribe them into a free software package. The original feature of the DEAS software package is that it enables work to be carried out within a context of joint production in the presence of undesirable outputs and environmental regulations restricting the disposal or polluting waste. The DEAS software package will be available for download from the end of 2006 (http://www.rennes.inra.fr/deas/).

Liang Liang received his MSc degree in Computer Science at Hefei Polytechnic University in 1987 and PhD degree in Systems Engineering from Southeast University of China in 1991. Dr. Liang Liang is Professor in Management School, University of Science and Technology of China. His main research interests include Decision Science, Data Envelopment Analysis and supply chain theory and practice. He has published 60 papers in Chinese journals and 20 papers in English journals. Professor Liang also serves on the editor boards of International Journal of Mass Customization and Asian- Pacific Management Reviews.

Shiang-Tai Liu is currently a professor at the Graduate School of Business and Management, Vanung University, Taiwan. He received his B.S., M.S, and Ph.D. degrees from National Cheng Kung University, Taiwan. His main research interests are in the area of mathematical programming under uncertainty and performance evaluation. He has published articles in *European Journal of Operational Research*, *Fuzzy Sets and Systems*, *IEEE Transactions on Systems, Man, and Cybernetics-Part B*, *International Journal of Production Research*, *Journal of Banking and Finance*, *Internal Journal of Approximate Reasoning*, *OMEGA*, *Cybernetics and Systems*, etc.

Sebastián Lozano is Professor of Quantitative Methods in Management at the Department of Industrial Management, University of Seville. He holds a Ph.D. from the University of Seville and a Master of Engineering in Industrial Management from the Katholieke Universiteit Leuven. His research interests include DEA, metaheuristics and soft computing and their applications in operations management. He has published in the Journal of the Operational Research Society, European Journal of Operational Research, OMEGA, Journal of Productivity Analysis, International Journal of Production Research, Computers and Industrial Engineering, Computers and Operations Research and OR Spectrum, among others.

Hiroshi Morita is Professor of Information Science and Technology at Osaka University. Prior to joining Osaka University, he was Associate Professor at Kobe University, Assistant Professor at Osaka City University and Osaka Prefectural University. His research interests include issues of performance evaluation under uncertainty, system analysis, production scheduling, quality management and data envelopment analysis. His works have appeared in Omega, JSME International Journal, International Journal of Information Technology and Decision Making, International Transactions in Operational Research.

Jesús T. Pastor is Professor of Statistics and Operations Research at the Universidad Miguel Hernandez of Elche, Spain. He earned an MBA and a Ph.D. in mathematical sciences from Valencia University, Spain. Prof. Pastor research interests are found in operations research, and more specifically in location science, banking and efficiency analysis. He has served on the editorial review or advisory board of more than 20 journals. He has authored or coauthored 9 books in various fields of mathematics, while his research has appeared in such journals as *Operations Research, OMEGA, Journal of the Operational Research Society, Operations Research Letters, European Journal of Operations Research, Location Science, Environment and Planning, Economics Letters, Operations Research Letters, Top, International Journal of Information*

Technology and Decision Making, Annals of Operations Research, Journal of Productivity Analysis and European Finance Review.

Isabelle Piot-Lepetit is a Doctor of Economic Sciences (University of Bordeaux I, France) and holds an Accreditation to Supervise Research diploma awarded by the University of Paris I Panthéon-Sorbonne in 2005. Isabelle Piot-Lepetit works as an Economic Research Scientist at INRA (National Institute for Agricultural Research). Her research fields relate to the incorporation of environmental constraints and the introduction of dynamics into the micro-economic representation of behaviour. Among other things, she shows the value of the non-parametric analysis framework of the DEA (Data Envelopment Analysis) approach in representing measures of environmental policies in the European Union. Her research enables the impacts of regulations on the performance of agricultural producers and on changes in production structures to be quantified.

Adi Raveh studied mathematics and statistics (B.Sc.) and statistics for an M.A. and Ph.D. at the Hebrew University. Since 1982 he is affiliated with the Hebrew University jointly at the Business School and the Statistics Department. He has been a visiting professor at Stanford University and Baruch College New York. He has been an invited scholar at Columbia University and Research Fellow (ASA/NSF) at the Census Bureau in Washington D.C. His main fields of interest are in general data analysis and in particular multivariate analysis using non-metric concepts, time-series analysis and forecasting

John Ruggiero is a J. Robert Berry Endowed Faculty Fellow and Associate Professor of Economics at the University of Dayton. He received his B.A. in economics and mathematics from the State University of New York at Cortland and his M.S. and Ph.D. in economics from Syracuse University. Dr. Ruggiero's current research interests are in theoretical and applied performance evaluation. He has published dozens of articles in leading and high quality journals in Operations Research and Economics. Currently, Dr. Ruggiero is an Associate Editor of the Asia-Pacific Journal of Operational Research and serves a reviewer for many journals.

José L. Ruiz is Associate Professor of Statistics and Operations Research at the Universidad Miguel Hernández of Elche, Spain. He earned an MBA in Mathematical Sciences from the Universidad of Granada and a Ph.D. from the Universidad Miguel Hernández. He develops his research at the Universitary Institute for Research "Centro de Investigación Operativa" (CIO) of the Universidad Miguel Hernández. His main research interest is in the analysis of efficiency and productivity. He has published his research in journals such as *Operations Research, European Journal of Operational Research, Fuzzy Sets and Systems* or *International Journal of Information Technology and Decision Making*.

Joseph Sarkis is a Professor of Operations and Environmental Management within the Graduate School of Management at Clark University. He earned his Ph.D. from the University of Buffalo. He has published over 200 articles as peer reviewed journal articles, published proceedings, and book chapters. He has applied Data Envelopment Analysis to a number of areas including business and the environment, logistics, supply chain, information systems, technology and operations management topics. He recently published an edited book titled "Greening the Supply Chain." He is a member of a number of editorial boards for a wide range of internationally recognized journals. Currently he is the editor of Management Research News, published by Emerald Publishers. He is also a member of a number of professional organizations.

About the Authors

Gabriel Villa is Associate Professor at the Department of Industrial Management, University of Seville. He holds a Ph.D. from the University of Seville. His primary research interest is in Data Envelopment Analysis. He has published in the Journal of the Operational Research Society, Journal of Productivity Analysis, OMEGA, C&OR and Lecture Notes in Computer Science, among others.

Gerald Whittaker received a B.S. in chemistry and an M.S. in resource economics, both from Oregon State University. He received a J.D. from Lewis and Clark Law School, Portland, Oregon. Whittaker has worked in economic consulting, and as an economist with the Economic Research Service, USDA. He is currently a research hydrologist with the Agricultural Research Service, USDA.

Feng Yang is a lecturer at Management school, University of science and technology of China (USTC) (currently a visiting scholar at the Schulich school of Business, York University, Canada). He received a B.Sc. in finance from Department of statistics and finance at USTC and a Ph.D. from Management School at USTC. His major research interests include data envelopment analysis, supply chain management, and others. His work has been published in various journals, including *Annals of Operations Research, Computers and Operations Research*.

Ekaterina Yazhemsky, born in 1976 in Ekaterinburg Russia, received her B.A. in statistics in 1999 and M.B.A. in finance and operations management in 2003, both at the Hebrew University of Jerusalem. Her research experience began in 1999 as an economist in the Research Department of The Central Bank of Israel. Since 2003, Ekaterina has worked as a research assistant at the Hebrew University of Jerusalem's School of Business Administration. Presently, Ekaterina is a Ph.D. candidate in operations management and operations research under the guidance of Dr. Nicole Adler and Prof. Jonathan Kornbluth. Her doctoral training is focused on extending the applicability of data envelopment analysis, specifically in the development of discrimination-improving models and in the presentation of the results of DEA graphically.

Joe Zhu is Associated Professor of Operations, Department of Management at Worcester Polytechnic Institute, Worcester, MA. His research interests include issues of performance evaluation and benchmarking, supply chain design and efficiency, and Data Envelopment Analysis. He has published over 50 articles in journals such as *Management Science, Operations Research, IIE Transactions, Annals of Operations Research, Journal of Operational Research Society, European Journal of Operational Research, Information Technology and Management Journal, Computer and Operations Research, OMEGA, Socio-Economic Planning Sciences, Journal of Productivity Analysis, INFOR, Journal of Alternative Investment* and others. He is the author of *Quantitative Models for Evaluating Business Operations: Data Envelopment Analysis with Spreadsheets* (Kluwer Academic Publishers, 2003). He developed the *DEAFrontier* software which is a DEA add-in for Microsoft Excel. Professor Zhu has also co-authored three books on modeling performance measurement, improving service operations, and evaluating hedge funds. He is a co-editor of the *Handbook on Data Envelopment Analysis* (Springer, 2004). He is an Associate Editor of OMEGA and The Asia-Pacific Journal of Operational Research. He is also a member of Computers & Operations Research Editorial Board. For more information about his research, please visit www.deafrontier.com.

Index

A

assurance region, 139
attractiveness, 241, 245, 257

B

benchmarking, 11, 151, 193, 208, 241
best practice, 189
bounded, 35, 38, 44
buyer, 10, 189, 191, 192, 193, 194, 197, 200, 201, 202, 204, 206, 207

C

classification invariance, 8, 63, 72, 103
context-dependent, 1, 10, 241, 242, 243, 245, 246, 247, 250, 251, 253, 255, 258, 259

D

data mining, 1
DEA, 1, 9, 11, 13, 33, 61, 62, 82, 83, 84, 120, 121, 139, 140, 151, 153, 154, 155, 171, 185, 186, 189, 190, 228, 229, 243, 258, 259, 261, 270, 288, 289, 291, 292, 302, 303, 319
DEA efficiency, 5
directional distance function, 117, 123

directional distance model, 74
discretionary, 1, 8, 9, 67, 70, 85, 86, 87, 88, 89, 90, 92, 95, 96, 97, 98, 99, 100, 103, 105, 113, 114, 115, 116, 278
disposability, 8, 9, 103, 105, 106, 107, 108, 110, 111, 112, 117, 119, 123, 125, 127, 128, 129, 136, 156, 159, 223
disposable, 8, 103, 104, 105, 106, 127, 129, 131, 224, 225
dynamic production, 209

E

efficiency, 1, 5, 8, 9, 10, 11, 13, 17, 24, 28, 29, 31, 32, 33, 34, 35, 36, 41, 42, 44, 45, 46, 49, 50, 51, 52, 53, 54, 57, 58, 59, 61, 62, 63, 65, 66, 67, 68, 69, 70, 72, 73, 74, 75, 76, 77, 79, 80, 81, 82, 83, 85, 86, 87, 88, 89, 90, 91, 92, 95, 96, 97, 99, 100, 103, 104, 105, 106, 111, 113, 116, 117, 120, 121, 125, 129, 133, 134, 135, 136, 137, 138, 140, 150, 151, 152, 153, 155, 156, 157, 158, 159, 161, 165, 166, 169, 170, 172, 173, 177, 181, 182, 184, 185, 186, 187, 189, 190, 191, 192, 194, 195, 196, 197, 199, 200, 202, 204, 206, 207, 208, 217, 227, 228, 229, 241, 242, 243, 244, 250, 251, 252, 253, 254, 255, 258, 259,

261, 262, 266, 267, 269, 270, 276,
277, 278, 279, 281, 282, 283, 286,
289, 291, 292, 293, 294, 295, 297,
298, 299, 300, 301, 302, 303, 306,
307, 309, 312, 315, 316, 318, 319
efficient unit, 65, 66, 67, 69, 72, 73,
79, 81, 144, 150, 175, 185, 258, 276,
314, 315
envelopment, 1, 3, 4, 6, 8, 13, 35, 36,
37, 39, 42, 57, 59, 61, 62, 65, 66, 67,
68, 69, 103, 121, 139, 140, 154, 155,
158, 159, 160, 167, 168, 171, 184,
186, 187, 189, 241, 259, 289, 291, 303
environmental regulation, 123
evaluate, 10, 15, 16, 27, 57, 64, 65, 72,
83, 86, 92, 96, 103, 105, 113, 125,
155, 194, 195, 197, 218, 219, 241,
243, 250
exact data, 8, 28, 35, 36, 37, 42, 43,
44, 45, 46, 49, 50, 51, 53, 54, 56,
57, 58, 59
extended facets, 65
extreme efficient unit, 78

F

flexible, 1, 261, 262
frontier, 4, 5, 6, 8, 9, 10, 25, 26, 29, 32,
57, 63, 65, 66, 68, 73, 74, 75, 76, 77,
78, 80, 81, 82, 85, 86, 89, 90, 97, 99,
104, 129, 134, 135, 140, 144, 155,
156, 160, 161, 170, 172, 180, 184,
185, 241, 242, 243, 244, 245, 250,
251, 253, 254, 274, 276, 297, 303,
304, 306, 316, 317

H

homogeneity, 305, 306
hospital, 11, 261, 262
hyperbolic output efficiency measure, 8,
103, 104, 105, 106

I

imprecise, 8, 14, 27, 28, 34, 35, 36, 37,
38, 40, 52, 57, 59, 61, 120, 303, 318
integer, 11, 29, 30, 264, 265, 266, 271,
272, 273, 274, 275, 276, 277, 278,
279, 280, 281, 283, 284, 285, 286,
287, 289
intermediate Products, 209
interval, 1, 35, 38, 70, 303
interval scale variables, 64, 70, 71, 80

L

Likert scale, 14, 16, 29, 30
linear programming, 3, 17, 19, 20, 35, 39,
65, 67, 103, 107, 129, 165, 191, 203,
207, 221, 241, 243, 246, 247, 248,
249, 252, 266, 292

M

manure management, 123
multi-dimensional scaling, 171
multiplier, 3, 8, 18, 35, 36, 37, 39, 40, 41,
42, 59, 61, 65, 66, 72, 73, 78, 83, 140,
158, 159, 160, 165, 195, 197
municipal, 9, 139, 140

N

negative, 1, 2, 8, 63, 64, 66, 67, 69, 70,
72, 74, 76, 79, 81, 82, 83, 109, 110,
124, 126, 137, 163, 174, 176, 213,
226, 312, 313, 314, 316
network, 1, 132, 207, 209, 211, 217,
219, 228
non-parametric, 9, 155, 227, 229

O

ordinal, 1, 15, 35, 44, 46, 47

P

performance, 1, 5, 8, 9, 10, 13, 14, 16, 36, 65, 85, 92, 99, 104, 106, 110, 111, 113, 116, 120, 121, 123, 125, 129, 133, 134, 136, 137, 138, 154, 167, 168, 186, 187, 189, 190, 191, 192, 194, 196, 205, 206, 207, 208, 241, 242, 243, 245, 246, 247, 248, 250, 251, 253, 256, 258, 259, 261, 262, 270, 272, 278, 306, 316
principal component analysis, 9, 139, 140, 146, 153, 154, 175, 182, 187, 320

Q

qualitative, 2, 7, 13, 14, 15, 16, 34, 61, 318

R

R&D projects, 15, 16
ranking, 14, 16, 27, 29, 54, 68, 99, 117, 119, 146, 187, 258, 281, 298, 303, 309, 319
ratio scale variables, 65, 70
ratios, 10, 70, 140, 156, 158, 171, 173, 175, 176, 177, 180, 181, 185, 187
response time, 3, 4, 5

S

seller, 10, 189, 191, 192, 193, 194, 196, 197, 199, 200, 201, 202, 204, 206, 207
slack, 5, 6, 10, 22, 44, 46, 59, 67, 80, 87, 89, 156, 177, 241, 243, 244, 250, 251, 252, 253, 254, 255, 256, 257, 258, 259
structure, 1, 2, 191, 270, 317
supply chain, 1, 3, 10, 189, 190, 191, 192, 193, 194, 196, 200, 201, 202, 204, 205, 206, 207, 208

T

target setting, 8, 63, 75, 76
technology adoption, 209
translation invariance, 8, 63, 64, 66, 67, 72, 82, 109, 116

U

undesirable, 1, 103, 121, 319
units invariance, 8, 63

V

value judgment, 241

W

weighted additive model, 65, 66, 72, 76, 144, 287

Printed by Printforce, the Netherlands